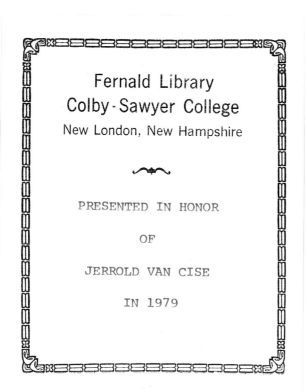

Aromatic Amino Acids
in the Brain

*The Ciba Foundation for the promotion of international cooperation in
medical and chemical research is a scientific and educational charity established by
CIBA Limited – now CIBA-GEIGY Limited – of Basle. The Foundation operates
independently in London under English trust law.*

*Ciba Foundation Symposia are published in collaboration with
Associated Scientific Publishers (Elsevier Scientific Publishing Company, Excerpta Medica,
North-Holland Publishing Company) in Amsterdam.*

Associated Scientific Publishers, P.O. Box 211, Amsterdam

Aromatic Amino Acids in the Brain

Ciba Foundation Symposium 22 (new series)

Symposium on aromatic amino acids in the brain

1974

Elsevier · Excerpta Medica · North-Holland

Associated Scientific Publishers · Amsterdam · London · New York

QP
356.3
S9

ISBN Excerpta Medica 90 219 4023 x
ISBN American Elsevier 0-444-15019-6

Library of Congress Catalog Card Number 73-91643

Published in 1974 by Associated Scientific Publishers, P.O. Box 211, Amsterdam, and American Elsevier, 52 Vanderbilt Avenue, New York, N.Y. 10017.

Suggested series entry for library catalogues: Ciba Foundation Symposia.
Suggested publisher's entry for library catalogues: Associated Scientific Publishers.

Ciba Foundation Symposium 22 (new series)

76019

Printed in The Netherlands by Mouton & Co, The Hague

Contents

Participants

Symposium on Aromatic Amino Acids in the Brain held at the Ciba Foundation on 15–17th May, 1973

R. J. WURTMAN (*Chairman*) Laboratory of Neuroendocrine Regulation, Department of Nutrition and Food Science, Massachusetts Institute of Technology, Cambridge, Massachusetts 02139, USA

*J. AXELROD Laboratory of Clinical Science, Mental Health Intramural Research Program, National Institute of Mental Health, 9000 Rockville Pike, Bethesda, Maryland 20014, USA

S. H. BARONDES Department of Psychiatry, School of Medicine, University of California San Diego, PO Box 109, La Jolla, California 92037, USA

M. BULAT Department of Pharmacology, Chicago Medical School, 2020 West Ogden Avenue, Chicago, Illinois 60612, USA

A. CARLSSON Department of Pharmacology, University of Göteborg, Fack, S-400 33 Göteborg 33, Sweden

G. CURZON Department of Neurochemistry, Institute of Neurology, The National Hospital, Queen Square, London W1C 3BG

D. ECCLESTON MRC Brain Metabolism Unit, University of Edinburgh, 1 George Square, Edinburgh EH8 9JZ

D. FELIX Institut für Hirnforschung der Universität Zürich, August-Forel-Strasse 1, CH-8008 Zürich, Switzerland

J. FERNSTROM Department of Nutrition and Food Science, Massachusetts Institute of Technology, Cambridge, Massachusetts 02139, USA

E. M. GÁL Department of Psychiatry, State Psychopathic Hospital, The University of Iowa, 500 Newton Road, Iowa City, Iowa 52240, USA

G. L. GESSA Istituto di Farmacologia, Università di Cagliari, Via Porcell 4, 09100 Cagliari, Sardinia

* Unable to attend.

J. GLOWINSKI   Collège de France, Laboratoire de Biologie Moléculaire, Groupe de Neuropharmacologie Biochimique, 11 place Marcelin-Berthelot, 75 Paris 5e, France

D. G. GRAHAME-SMITH   MRC Clinical Pharmacology Unit, University Department of Clinical Pharmacology, Radcliffe Infirmary, Woodstock Road, Oxford OX2 6HE

S. KAUFMAN   Laboratory of Neurochemistry, Mental Health Intramural Research Program, National Institute of Mental Health, 9000 Rockville Pike, Bethesda, Maryland 20014, USA

A. LAJTHA   New York State Research Institute for Neurochemistry and Drug Addiction, Ward's Island, New York 10035, USA

L. MAÎTRE   Biological Research Laboratories, Pharmaceuticals Division, CIBA-GEIGY Limited, CH-4002 Basel, Switzerland

P. MANDEL   Centre de Neurochimie, Centre National de la Recherche Scientifique, 11 rue Humann, 67085 Strasbourg Cedex, France

A. T. B. MOIR   Scottish Home and Health Department, St Andrew's House, Edinburgh EH1 3DE

H. N. MUNRO   Department of Nutrition and Food Science, Massachusetts Institute of Technology, Cambridge, Massachusetts 02139, USA

S. S. OJA   Department of Biomedicine, University of Tampere, Teiskontie 37, SF-33520 Tampere 52, Finland

S. ROBERTS   Department of Biological Chemistry, University of California, School of Medicine, The Center for the Health Sciences, Los Angeles, California 90024, USA

M. SANDLER   Bernhard Baron Memorial Research Laboratories, Department of Chemical Pathology, Queen Charlotte's Maternity Hospital, Goldhawk Road, London W6 0XG

D. F. SHARMAN   ARC Institute of Animal Physiology, Babraham, Cambridge CB2 4AT

T. L. SOURKES   Departments of Psychiatry and Biochemistry, McGill University, 1033 Pine Avenue West, Montreal 112, Quebec, Canada

N. WEINER   Department of Pharmacology, University of Colorado Medical Center, 4200 East Ninth Avenue, Denver, Colorado 80220, USA

Editors: G. E. W. WOLSTENHOLME and DAVID W. FITZSIMONS

Editors' note

As far as possible, we have followed the Recommendations of the IUPAC–IUB Commission on Biochemical Nomenclature and the conventions of the *Biochemical Journal* (see Instructions to Authors). In view of the variety of names in common use, we append a list of the trivial and systematic names of some compounds mentioned in the book.

dopa:	3-(3,4-dihydroxyphenyl)alanine (I)
dopamine:	2-(3,4-dihydroxyphenyl)ethylamine, i.e. 4-(2-aminoethyl)benzene-1,2-diol
homogentisic acid:	2,5-dihydroxyphenylacetic acid
homovanillic acid	4-hydroxy-3-methoxyphenylacetic acid
6-hydroxydopa:	3-(2,4,5-trihydroxyphenyl)alanine
5-hydroxyindoleacetic acid (5HIAA):	5-hydroxy-3-indolylacetic acid (II)
p-hydroxymandelic acid:	hydroxy(4-hydroxyphenyl)acetic acid
5-hydroxytryptamine (5HT, serotonin);	2-(5-hydroxy-3-indolyl)ethylamine, i.e. 3-(2-aminoethyl)-5-indolol
5-hydroxytryptophol:	2-(5-hydroxy-3-indolyl)ethanol
α-ketoglutarate:	2-oxoglutarate
melatonin:	*N*-acetyl-2-(5-methoxy-3-indolyl)ethylamine
α-methyldopa:	3-(3,4-dihydroxyphenyl)-2-methylalanine
MK 486	L-3-(3,4-dihydroxyphenyl)-2-hydrazino-2-methylpropionic acid
noradrenaline (norepinephrine):	2-amino-1-(3,4-dihydroxyphenyl)ethanol, i.e. 4-(2-amino-1-hydroxyethyl)benzene-1,2-diol
NSD 1015	3-hydroxybenzylhydrazine, i.e. 3-hydrazinomethylphenol
NSD 1055	4-bromo-3-hydroxybenzyloxyamine, i.e. 5-aminooxy-3-bromo-phenol
octopamine:	2-amino-1-(4-hydroxyphenyl)ethanol
phenylethanolamine:	2-amino-1-phenylethanol
pterin:	2-amino-4-pteridinol
tryptamine:	2-(3-indolyl)ethylamine
tyramine:	4-(2-aminoethyl)phenol
m-tyramine:	3-(2-aminoethyl)phenol

(I) $CH_2 \cdot CH(NH_2) \cdot COOH$ (II)

Chairman's introduction

R. J. WURTMAN

Laboratory of Neuroendocrine Regulation, Department of Nutrition and Food Science, Massachusetts Institute of Technology, Cambridge, Massachusetts

It is no secret to this community that aromatic amino acids have a special significance in the functions of the brain. By 'aromatic amino acids' we are forced to restrict our attention to three such compounds: phenylalanine, tyrosine and tryptophan. (It is not that histidine is not important, but rather that the time at our disposal is not infinite.) These compounds are extremely important in normal brain function, in the pathophysiology of various disease states and in the responses of the brain to various drugs. Like the other amino acids present in dietary protein, they circulate in the blood and are taken up into brain, where they charge transfer RNAs and are subsequently incorporated into peptides and proteins. Furthermore, these amino acids can be hydroxylated within specific neurons, generating other amino acids which in turn are decarboxylated to yield biogenic monoamines that function as neurotransmitters. One can make an impressive list of brain diseases in which they participate, starting with phenylketonuria, the prototypic inborn error of metabolism. Other brain diseases have, in recent years, been shown to respond favourably to treatment with hydroxylated amino acids not normally found in the circulation—L-dopa and L-5-hydroxytryptophan. The lack of adequate amounts of dietary protein (Shoemaker & Wurtman 1971) or the chronic consumption of proteins like corn that contain unfavourable proportions of these amino acids (Fernstrom & Wurtman 1971) can interfere with normal brain function and behaviour; these disturbances coincide with changes in the concentrations of the monoamine neurotransmitters in the brain and urine (Hoeldtke & Wurtman 1973).

Considering this basic recognition of the importance to brain of the aromatic amino acids, some of us thought that it might be useful to try to bring together representatives of the three communities of scientists who study these compounds. The first such community comprises those who are concerned with the factors that control the concentrations of these aromatic amino acids in the brain, their flux between brain and extracellular fluid, and their catabolism in brain. Amino acid transport into brain appears to be mediated by group-specific transport systems that differ markedly from, for example, the insulin-sensitive mechanisms that operate in skeletal muscle. As we shall see, phenylalanine, tyrosine and tryptophan can apparently be decarboxylated by brain enzymes to yield the corresponding simple amines; they can also be transaminated, and the indole ring of tryptophan can be opened by an oxygenase, originally described from small intestine, that also cleaves D- and L-5-hydroxytryptophan, 5-hydroxytryptamine and melatonin (N-acetyl-5-methoxytryptamine) (Hirata & Hayaishi 1972).

The second group focuses on the use of these amino acids as precursors of the monoamine neutrotransmitters dopamine, noradrenaline and 5-hydroxytryptamine. Its concerns include the relationships between the neuronal concentrations of tyrosine or tryptophan and the rates at which these amino acids are hydroxylated to form dopa and 5-hydroxytryptophan respectively. Present information suggests the operation of two different mechanisms controlling the hydroxylations of tyrosine and tryptophan: the mechanism for tyrosine is thought to depend not on precursor availability but on the activity of a rate-limiting enzyme, tyrosine hydroxylase. As will be discussed, tyrosine hydroxylase activity may be controlled by end-product (i.e. catecholamine) inhibition and seems to increase or decrease when the physiological activity of 'catecholaminergic' neurons changes. In contrast, it appears that the major factor controlling the rate at which brain neurons synthesize 5-hydroxytryptamine is neuronal concentration of tryptophan; hence, physiological actions such as eating which change this concentration thereby modify brain levels of 5-hydroxyindoles. In part of this symposium we shall consider the validity of these generalizations and the possible biological consequences of these two, very different, control systems.

The third community consists of those who are concerned with the varieties of proteins that the brain synthesizes, the loci of protein synthesis within neurons and other cells, and the extent to which the amino acid charging of brain tRNA and other factors control the rates at which specific proteins and proteins as a group are made. There is abundant evidence that the availability of amino acids, especially tryptophan, is of major importance in determining the overall rate of protein synthesis in another mammalian organ, the liver (Munro 1970). The extent to which hepatic messenger RNA can perform its function— the extent to which it aggregates with pairs of ribosomes, forming polysomes—is largely determined by the availability of amino acids. Tryptophan seems normally to be the limiting amino acid, probably because it is the least-abundant amino acid in most foods and in the protein reservoirs in our own tissues. One wonders whether brain protein synthesis might also be limited by tryptophan availability—especially within neurons that also use this scarce amino acid for the biosynthesis of 5-hydroxytryptamine.

These communities have tended to overlap in the pursuit of several specific problems that will be discussed here; for example, all three groups have used drugs like *p*-chlorophenylalanine, α-methyltryptophan and, more recently, L-dopa which affect the concentrations of amino acids in brain and also the rates of synthesis of both the monoamine neurotransmitters and proteins.

An emergent generalization is that the concentrations of precursors in mammalian cells may be crucial in controlling the rates of many reactions. Those of us who came into mammalology with a background of work on *E.coli* had learnt to believe that the quantities of enzyme proteins present in the cell, or the allosteric state of these proteins, controlled the cell's biochemical activity; thus genetics were the key to an understanding of physiology. Now we see more and more evidence that the availability of substrate—a nutritional factor—can limit enzymic reactions, at least in mammalian cells. An undercurrent throughout this symposium will be 'brain nutrition'—the control of brain concentrations of three nutrients (phenylalanine, tyrosine and tryptophan) which it cannot provide for itself, and the extent to which these concentrations control the synthesis of compounds essential for normal brain function.

References

FERNSTROM, J. D. & WURTMAN, R. J. (1971) Effect of chronic corn consumption on serotonin content of rat brain. *Nat. New Biol.* **234**, 62–64

HIRATA, F. & HAYAISHI, O. (1972) New degradative routes of 5-hydroxytryptophan and serotonin by intestinal tryptophan 2,3-dioxygenase. *Biochem. Biophys. Res. Commun.* **47**, 1112–1119

HOELDTKE, R. D. & WURTMAN, R. J. (1973) The excretion of catecholamines and catecholamine metabolites in kwashiorkor. *Am. J. Clin. Nutr.* **26**, 205–210

MUNRO, H. N. (1970) A general survey of mechanisms regulating protein metabolism in mammals in *Mammalian Protein Metabolism*, vol. 4, pp. 3–130, Academic Press, New York

SHOEMAKER, W. J. & WURTMAN, R. J. (1971) Perinatal undernutrition: accumulation of catecholamines in rat brain. *Science (Wash. D.C.)* **171**, 1017–1019

Control of plasma amino acid concentrations

H. N. MUNRO

Department of Nutrition and Food Science, Massachusetts Institute of Technology, Cambridge, Massachusetts

Abstract   The daily flux of amino acids in the body is extensive. About 300 g of protein is synthesized each day in an adult man. This requires the uptake and release of 150 g of essential amino acids, yet the minimum dietary requirement for essential amino acids is only 6 g while the customary diet contains 45 g. This indicates considerable and efficient recycling of the essential amino acids released by protein breakdown. Since plasma contains a total of 0.2 g of essential amino acids, recycling of amino acids between tissues will cause rapid turnover of the free amino acids in plasma.

Not all amino acid molecules released by the turnover of body proteins are transferred to the blood plasma. Considerable amounts can be reused for protein synthesis within cells; consequently, the equation of amino acid uptake by (or release from) tissues with the total flux of amino acids within the tissue is not valid. This difficulty can be overcome if an amino acid released by protein breakdown is not reused for synthesis. For example, some histidine molecules in actin and myosin of skeletal muscle are methylated after protein synthesis. The resulting 3-methylhistidine cannot be reutilized and is quantitatively excreted in the urine; thus it is a measure of muscle protein turnover. Recognition of non-reusable products of protein degradation in other tissues would similarly be most useful.

After a meal containing protein, the liver monitors the access of the incoming amino acids to the systemic blood. For most essential amino acids, hepatic degradation is regulated in relation to adequacy of intake and rises sharply when intake exceeds requirements. In spite of this protection, the amount of a given essential amino acid in the systemic plasma increases progressively when its intake exceeds requirements. The dietary intake of an amino acid at which its plasma concentration starts to rise has been used to estimate the demand for that amino acid. Studies on young and old rats receiving various intakes of tryptophan show that the response of the amount of tryptophan in plasma to different dietary intakes of tryptophan varies with the age of the animal and the time of plasma sampling after meals. When the tryptophan intake exceeds the required amount, the activity of tryptophan oxygenase in liver displays a diurnal rhythm. In older animals, carbohydrate in the meal causes transfer of tryptophan from plasma to muscle, a phenomenon common to other amino acids.

The aim of this paper is to survey the factors regulating the plasma concentrations of amino acids, specifically, the aromatic amino acids. I shall begin by attempting to quantify roughly the daily flux of amino acids in the body of an adult man and in some individual tissues. This will demonstrate how extensive is the reutilization of amino acids released by protein turnover within the body; this factor has caused problems in the interpretation of some metabolic studies. I shall briefly describe the use of amino acids which are not reutilized to circumvent this and shall then consider how amino acid metabolism adjusts to changes in protein intake so that plasma concentrations are maintained within tolerable limits. This regulation will be illustrated by reference to the use of various intakes of dietary tryptophan.

DAILY FLUX OF AMINO ACIDS IN THE BODY

The daily flux of amino acids in various compartments of the body of an adult man weighing 70 kg is depicted in Fig. 1. The quantities shown are derived from a variety of sources described elsewhere (Munro 1972a) and should be regarded as first approximations in an attempt to arrive at some conception of the magnitude of amino acid turnover in the body. Balance experiments show that the

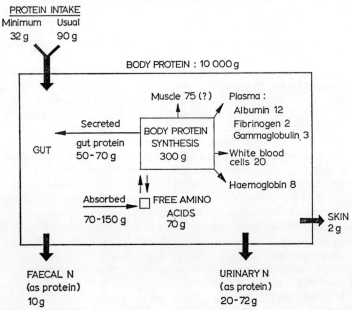

FIG. 1. Diagram of daily amino acid flux in the body of a 70 kg man (see Munro 1972a).

average adult maintains nitrogen equilibrium on an intake of 32 g of high quality protein (over 40 g is required by people whose requirements exceed the average by two standard deviations). In contrast, the customary protein intake in Western countries is much higher; on average it is over 90 g daily. This intake joins about 70 g of protein secreted into the gastrointestinal tract as digestive juice, and especially as shed mucosal cells (Fauconneau & Michel 1970), so that the total daily load for absorption is commonly as high as 150 g. The pools of free amino acids in the tissues which receive this load contain at least 70 g and exchange readily and extensively with body protein. Experiments with ^{15}N indicate that about 300 g of protein are synthesized daily in the body of an adult man. Fig. 1 shows estimates of the rates of synthesis of the proteins in some specific tissues. Some 200 g of body protein made daily are thus accounted for. The data for plasma proteins, haemoglobin and white cells are based on reasonably reliable estimates of daily turnover rates. The amount of protein secreted into the gastrointestinal tract is less reliable, because shed mucosal cells comprise most of the daily output of endogenous gut protein This protein loss is difficult to quantitate with adequate precision (Fauconneau & Michel 1970). The diagram also bears a figure for muscle protein turnover. This figure is based on the output of amino acids from muscle into the blood by fasting adults which, if continued for 24 h, would represent breakdown of 75 g of muscle (Cahill 1972). While this rate of loss must be compensated after meals by a net uptake of amino acids by muscle in order to maintain equilibrium within the tissue, we shall see later that muscle protein turnover is not necessarily represented by the rate of amino acid output during fasting and may even be double the value shown in Fig. 1.

Some components in daily amino acid flux are expressed in Table 1 as quantities of essential and non-essential amino acids, including tryptophan, phenylalanine and tyrosine (Table 2). The *average* minimum amount of dietary protein required in order to maintain nitrogen equilibrium in the adult (32 g daily) need only contain 6 g of essential amino acids (Munro 1972b). In contrast, of the 90 g of protein in the customary Western diet about half is present as essential amino acids. These 45 g join the essential amino acids entering the gut as endogenous proteins. Consequently, about 75 g of essential amino acids are absorbed daily. Compare this with the total amounts of essential and non-essential amino acids in the plasma, namely 0.2 and 0.5 g, respectively. So it is to be expected that free amino acids will be rapidly transported out of the blood into the tissues. The tissues contain some 70 g of free amino acids, but four non-essential amino acids (glycine, glutamic acid, glutamine and alanine) represent 80% of this, whereas only about 10 g of essential amino acids are present. Nevertheless, the daily turnover of 300 g of body protein requires incorporation

TABLE 1

Intake, tissue content and turnover of essential and non-essential amino acids for a 70 kg adult

Amino acid source	Amino acids/70 kg body wt		
	Total (g)	Essential (g)	Non-essential (g)
Daily diet			
Minimum amino acid needs[a]	32	6	26
Western diet[b]	90	45	45
Absorbed (with secreted gut protein)[b]	150	75	75
Free amino acid pools[c]			
Plasma	0.7	0.2	0.5
Tissues	70	10	60
Daily body protein turnover[d]	300	150	150

[a] From Munro (1972b).
[b] These data are taken from Fauconneau & Michel (1970). It is assumed that the mixed proteins of the diet and the intestinal secretions contain about 50% essential amino acids, which is a reasonable estimate.
[c] These are calculated for a 70 kg adult from data on free amino acid concentrations in the blood and tissues of the rat (Munro 1970). A few data on human tissues confirm their applicability.
[d] From San Pietro & Rittenberg (1953), assuming that 50% of the amino acids incorporated into and released from body protein are essential amino acids.

TABLE 2

Intake, plasma content and turnover of tryptophan, phenylalanine and tyrosine for a 70 kg man

	Amounts/70 kg man		
	Trp (g)	Phe (g)	Tyr (g)
Daily diet			
Minimum needs[a]	0.2	–1.0–	
Western diet[b]	1.0	4.2	3.1
Absorbed (with gut protein)	1.5	5.9	4.7
Plasma amino acid pool[c]	0.020	0.012	0.014
Daily protein turnover[d]	3.3	13.0	10.5

[a] Requirement for nitrogen equilibrium (Munro 1972b) calculated for a 70 kg man. The requirement for phenylalanine includes the amount needed for tyrosine formation.
[b] The average Western diet is taken to contain 90 g protein (Table 1), with 60 g derived from animal sources and 30 g from vegetable sources. Based on the amino acid content of foods (FAO Handbook 1970) and the proportions of these eaten, the approximate percentages of tryptophan, phenylalanine and tyrosine in food protein are respectively 1.2, 4.6 and 3.7 for animal foods and 1.0, 4.8 and 2.6 in vegetable foods. These percentages are not appreciably altered by variations in the major types of animal and vegetable protein sources used for the computations.
[c] From data for human plasma (Munro 1969), assuming plasma volume to be 2% of body weight.
[d] From amino acid content of carcass protein of mammals (Munro & Fleck 1969), and a daily protein turnover of 300 g (Table 1).

of 150 g of essential amino acids daily into tissue protein. Most of this must be derived from recycling of essential amino acids released by the tissues, since the diet customarily provides only about 45 g essential amino acids and nitrogen equilibrium can still be achieved when the total content of essential amino acids in the diet is as low as 6 g (Munro 1972b). This implies very efficient recycling of essential amino acids within the body.

A similar picture emerges when single essential amino acids such as trypto-phan and phenylalanine and the semi-essential amino acid tyrosine are con-sidered (Table 2). The minimum dietary requirement for the first two are 1g/day or less, whereas the daily turnover within the body is about 15–20 times the requirement. The amounts of these three amino acids in the plasma are small compared with this daily turnover, about 1/150 for tryptophan and 1/1000 for both phenylalanine and tyrosine. This implies that the proportion of tryptophan in the plasma is unusually large in view of its function in body protein synthesis. These metabolic parameters computed for humans can be amplified by analogous calculations for the rat. Table 3 shows that the ratio of tryptophan to phenylala-nine to tyrosine in the total free amino acid pool of the body and free in the tissues is about 1:4:4 whereas in plasma it is 1:1.2:1.4. If, instead of total plasma tryptophan, we compute the ratio of phenylalanine and tyrosine to non-albumin-bound tryptophan in plasma, the ratio becomes 1:4.8:5.8, as in the tissues. This implies that the bound tryptophan in plasma represents an excess (a reservoir) and that the tissues probably do not have a corresponding binding protein.

TABLE 3

Requirements, body protein content and free concentrations of tryptophan, phenylalanine and tyrosine in the rat

	Amount (μmol/100 g body or tissue wt.)			Ratios		
	Trp	Phe	Tyr	Trp	Phe	Tyr
Daily requirements[a]	55	–450–		1.0	–9.0–	
Body protein content[a]	980	5 800	3 550	1.0	6.0	3.6
Body free amino acid[a]	2	9	8	1.0	4.5	4.0
Tissue free amino acids[b]						
Liver	5.6	21	25	1.0	3.8	4.5
Muscle	3.7	15	20	1.0	4.1	5.4
Brain	2.3	5.6	9.5	1.0	2.4	4.1
Plasma	6.5	7.7	9.4	1.0	1.2	1.4
Plasma (non-bound)	(1.6)			(1.0)	(4.8)	(5.8)

[a] From Munro (1970). The daily requirements are those for rapid growth.
[b] From Williams et al. (1950), with the assumption that 25% of plasma tryptophan is not bound to albumin (Fernstrom et al., this volume, pp. 153–166).

AMINO ACID REUTILIZATION IN THE BODY

The preceding calculations constitute indirect evidence that the essential amino acids are efficiently reutilized. There is, however, much direct evidence of this phenomenon. If an amino acid such as [U-^{14}C]arginine is injected into rats and the rate of loss of labelled arginine from liver protein is measured, the apparent half-life of mixed liver protein is 4.5 days, whereas if [*guanidino*-^{14}C]arginine is used, the half-life falls to 3.3 days (Arias *et al.* 1969). This is because [*guanidino*-^{14}C]arginine participates in the arginine–ornithine cycle for urea synthesis after release from liver protein and in the process the labelled guanidino-carbon atom of free arginine is replaced by ^{12}C, thereby greatly reducing the reuse of the ^{14}C label in protein synthesis. Thus the intracellular pool is made up of amino acids entering the tissue from the plasma and also of amino acids released by protein turnover. It has been estimated that recycling in the rat liver can account for 50% of the free amino acid pool in the fed state and 90% after a short fast (Gan & Jeffay 1967).

The complexities of recycling on the interpretation of data obtained from analysis of free amino acid concentrations in plasma are illustrated by some recent studies of amino acid metabolism in skeletal muscle. By measuring arterio-venous differences in amino acid concentration in the forearms of human subjects, Cahill and his colleagues (Pozefsky *et al.* 1969; Marliss *et al.* 1971) were able to quantitate the amounts of individual amino acids added to or removed from the limb muscles as the blood passed through. After an overnight fast, there is a considerable net output of amino acids from the musculature equivalent to a daily loss of 75 g from the total muscle protein of the body (Cahill 1972). Much of this released amino nitrogen is made up of alanine and glutamine, owing to transamination of the bulk of the amino acids released from muscle protein. The alanine and glutamine are quantitatively absorbed by the liver where both amino acids are donors of $-NH_2$ for urea synthesis and where alanine also provides a source of carbon for gluconeogenesis (Fig. 2). When insulin is injected into fasting subjects, the release of amino nitrogen into the bloodstream is sharply reduced. Pozefsky *et al.* (1969) consider this to be due to a reduction in rate of *degradation* of muscle proteins. However, it could also happen if the well known stimulation of muscle protein *synthesis* by insulin results in more rapid removal of amino acids from the intracellular pool, so that less is available for release into the blood (Fig. 2). These hypotheses could be tested directly if an amino acid that is *not* reutilized for protein synthesis is released from muscle breakdown. We have identified such an amino acid— 3-methylhistidine (Young *et al.* 1972). This is present in both the actin and myosin of muscle and is produced by methylation of histidine after the peptide

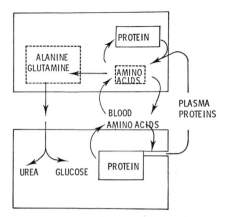

FIG. 2. Recycling of amino acids within skeletal muscle and between muscle and liver (mainly as alanine and glutamine).

chains of these proteins have been made (Fig. 3). Injected 3-methylhistidine is metabolized by the rat to *N*-acetyl-3-methylhistidine, and the free and conjugated 3-methylhistidine are then rapidly and quantitatively excreted in the urine. Thus, release of 3-methylhistidine from muscle into the blood and its excretion in urine provide an absolute measure of the rate of degradation of muscle protein, provided that other tissues are negligible sources and that the diet is free from sources of 3-methylhistidine such as meat. Recently, for example, we measured the urinary excretion of 3-methylhistidine by three obese patients on a 20-day fast (Young *et al.* 1973). Over this period, total body protein content fell only 10–15% whereas urinary output of 3-methylhistidine nitrogen declined by about 40%. Thus, the catabolism of muscle protein must

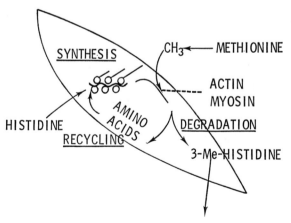

FIG. 3. The use of 3-methylhistidine in measuring rate of muscle protein breakdown without reutilization of amino acids.

be considerably reduced during fasting, through adaptation of protein break-down. Other non-reusable amino acids would be most useful in evaluating changes in amino acid metabolism. It should be noted that calculations based on 3-methylhistidine content of human muscle suggest that the amount excreted is derived from breakdown of 150 g of muscle protein daily, more than twice the figure calculated from arteriovenous differences by Pozefsky *et al.* (1969). This implies extensive recycling within the muscle cell of the amino acids released by turnover.

The interpretation of amino acid exchanges between tissues and blood has also been attempted for the human brain. Felig *et al.* (1973) have measured the arteriovenous difference in the brain for 15 amino acids including six essential amino acids in fasting adults and observed uptake of all amino acids by the brain. On the assumption of a blood flow of 1.2 1/min (Lewis *et al.* 1960), this means that about 30 g of amino acids are taken up every 24 h. It is difficult to interpret such data, since they imply either that an equivalent quantity of amino nitrogen is released from the brain at other times of day, notably after meals, or else that the uptake is balanced by a loss into blood not drained by the jugular vein, such as the spinal cord and its cerebrospinal fluid. In this connection, Aoki *et al.* (1972) have found that red cells participate extensively as donors of glutamate to muscle. In future, arteriovenous studies may need to include red cells as well as plasma before a balance sheet of uptake and release of amino acids can be drawn.

THE LIVER AND REGULATION OF FREE AMINO ACID CONCENTRATIONS

One reason for the limited requirement for essential amino acids is that the key catabolic enzymes for seven of them are restricted to the liver (Miller 1962). Consequently, essential amino acids liberated by protein breakdown within other tissues such as muscle will be reutilized with considerable efficiency if these amino acids do not leak into the blood in large amounts and reach the liver. It is, therefore, not surprising that the liver functions especially in monitoring the intake of essential amino acids and catabolizing the excess over requirements, whereas muscle liberates mainly non-essential amino acids for transport to the liver.

Although the recycling of amino acids within the body is extensive compared with requirements (Table 1), the mechanisms for detecting and dealing with amounts of dietary amino acids in excess of needs are nevertheless surprisingly precise. The liver provides the major site of regulation. Amino acids absorbed after a meal rich in protein can raise the amounts of free amino acids in the portal vein considerably during the absorptive period, whereas their concentrations in the systemic circulation change much less. Studies on dogs bearing cannulae in the blood supply to and from the liver show that this is due to the enormous capacity of the liver to remove amino acids from the portal blood and thus to regulate their flow into the systemic circulation (Elwyn 1970). Consequently, amino acid metabolism after meals causes diurnal rhythms in liver protein metabolism which affect both the rate of synthesis of liver proteins and the amounts of enzymes involved in the catabolism of amino acids (Fishman *et al.* 1969).

As a result of these adaptive responses by the liver to meals, the systemic circulation is protected against excessive changes in the amounts of free amino acids entering the body. There is reason to believe that, in the case of the essential amino acids, this catabolic response is finely adjusted to the amount entering the body in relation to requirements. For example, Brookes *et al.* (1972) added successively greater amounts of $[^{14}C]$lysine to the diet of growing rats; the amounts of the other dietary amino acids were kept constant. Consumption of amounts of total dietary $[^{14}C]$lysine below requirements led to a constant low release of $^{14}CO_2$. However, when lysine intake was raised above the requirement for optimal growth, the proportion of $[^{14}C]$lysine released as $^{14}CO_2$ rose steeply and further addition of lysine to the diet increased $^{14}CO_2$ production still more. Consequently, there was a sharp inflection in $^{14}CO_2$ output at the point where lysine intake was just sufficient to support maximal growth rate. This pattern of response took several days to develop, presumably because feedback of information from the body as a whole to the liver, where lysine is degraded, required this period to establish increased hepatic concentrations of free lysine.

RESPONSE OF PROTEIN METABOLISM TO DIFFERENT INTAKES OF TRYPTOPHAN

Although the liver is thus equipped to respond sensitively to intakes of amino acids in excess of requirements, some of the extra amino acid enters the systemic blood. Consequently, the sudden increase in hepatic oxidation of an essential amino acid at the point at which requirements are met usually coincides with an increase in the blood concentration of the amino acid. This may be the signal from the tissues required by the liver to turn on the catabolic process. Recently, we have examined several aspects of tryptophan metabolism in rats with different intakes using both young rats (54 g initial weight) with a high requirement and older rats (300–400 g) with a lower requirement for tryptophan (Young & Munro 1973). Groups of rats from each age group were given diets providing amino acids in place of protein, and the intake of tryptophan was varied from 0 to 0.33 % of the diet. The requirement for tryptophan in the young rat has

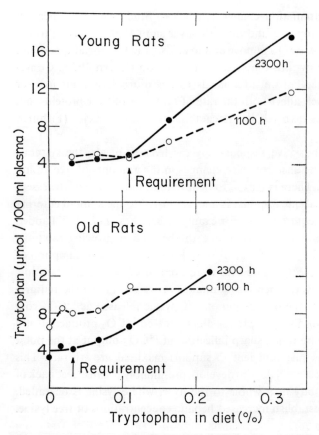

FIG. 4. Concentrations of tryptophan in the plasma of young and mature rats fed amino acid diets containing various amounts of tryptophan, other nutrients being adequate and constant. The arrows indicate requirements for tryptophan reported in the literature for weanling and adult rats (Munro & Young, unpublished data).

been taken as 0.11% (Rama Rao *et al.* 1959) and for mature rats as 0.03% (Smith & Johnson 1967) (see Fig. 4). Plasma tryptophan was measured twice a day (at 1100 and 2300) after nine days on these diets. In young rats, the plasma tryptophan concentration increased at both these times as soon as the requirement had been met (Fig. 4). In the mature animals, the increase was evident for blood samples at 2300, the rat's habitual time of feeding, whereas the results obtained at 1100 did not provide this clear evidence of a requirement-related increment. Thus, blood amino acid responses can be used more sensitively during the absorptive phase as monitors of requirements. It would be interesting to know whether changing the proportion of plasma tryptophan bound to albumin by altering the amount of non-essential fatty acid in the plasma (see

Fernstrom *et al.*, this volume, pp. 153–166) results in a change of the plasma tryptophan response curve, thus indicating an effect of albumin binding on the use of tryptophan.

From Fig. 4 it can be seen that the older rats had a much lower plasma tryptophan concentration at 2300 than at 1100 for all except the greatest intakes of tryptophan. A constituent of the food which the rat begins to eat just before 2300 must have caused this reduction. We know from other work that dietary carbohydrate decreases the amounts of free amino acids in the plasma owing to the insulin-dependent deposition of the amino acids in muscle. Therefore, we measured the concentrations of free tryptophan in muscle and showed that the reduction in plasma concentration did correlate with a rise in muscle concentration at 2300 (Table 4). This presumably means that the carbohydrate content of the meal caused transfer of tryptophan into muscle irrespective of the tryptophan content of the diet. If the diet is inadequate in tryptophan, this will presumably limit tryptophan supply to other tissues even more—a feature of carbohydrate action without adequate dietary amino acid supply which I have pointed out previously (Munro 1964). It is clear from Fig. 4 that, unlike the older rats, the young rats did not show a reduction in the amount of plasma tryptophan when they consumed meals containing little or no tryptophan. This correlates with Fernstrom's observation (personal communication) that carbohydrate administration affects plasma tryptophan concentrations differently in young and mature rats.

We also measured the activity of tryptophan oxygenase in the liver and observed that activity rose after meals (from observations at 2300) only in those animals that were receiving more than their dietary requirements of tryptophan (Table 4). This indicates that the enzyme is sensitive to the amount of dietary

TABLE 4

Effect of dietary tryptophan concentration on tryptophan content of plasma and muscle, and on activity of tryptophan oxygenase in older rats

% Trp in diet	Plasma Trp ($\mu mol/100$ ml)		Muscle Trp ($\mu mol/100$ g)		Tryptophan oxygenase (units of activity/g liver)	
	2300	1100	2300	1100	2300	1100
0.000	3.2	6.6	17.5	14.3	2.0	1.1
0.016	4.3	8.5	17.0	14.2	2.5	1.8
0.033	4.1	8.0	18.0	15.6	2.1	1.8
0.066	5.2	8.2	16.3	12.3	3.2	1.6
0.108	6.2	10.9	17.0	12.3	3.2	1.5
0.220	12.2	10.5	17.4	13.8	3.2	1.4

Data are from Young & Munro (1973).

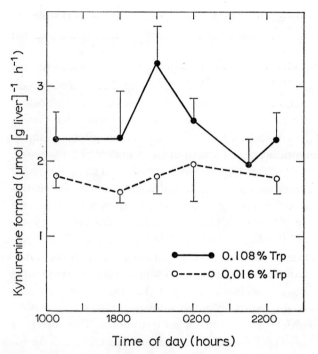

FIG. 5. Hepatic activity of tryptophan oxygenase, as shown by production of kynurenine, at different times for mature rats fed small or large amounts of tryptophan; other dietary factors are adequate and constant (Young & Munro 1973).

tryptophan with respect to requirements, and that it exercises this monitoring action at the time of absorption of the meal. We verified this conclusion by feeding rats large and small amounts of dietary tryptophan and observing diurnal changes in tryptophan oxygenase due to variations in food intake. Fig. 5 shows that the superoptimal intake induced a meal-related increase in enzyme activity that was absent when suboptimal amounts of tryptophan were fed.

ACKNOWLEDGEMENT

The original data reported here were supported by USPHS grant AM 15364.

References

AOKI, T. T., BRENNAN, M. F., MULLER, W. A., MOORE, F. D. & CAHILL, G. F. (1972) Effect of insulin on muscle glutamate uptake. Whole blood versus plasma glutamate analysis. *J. Clin. Invest.* **51**, 2889–2894

ARIAS, I. M., DOYLE, D. & SCHIMKE, R. T. (1969) Studies on the synthesis and degradation of proteins of the endoplasmic reticulum of rat liver. *J. Biol. Chem.* **244**, 3303–3315

BROOKES, I. M., OWEN, F. N. & GARRIGUS, U. S. (1972) Influence of amino acid level in the diet upon amino acid oxidation by the rat. *J. Nutr.* **102**, 27–34

CAHILL, G. F. (1972) Carbohydrates in *Symposium on Total Parenteral Nutrition* (Vanamee, P. & Shils, M. E., eds.), pp. 45–51, American Medical Association, Chicago

ELWYN, D. (1970) The role of the liver in regulation of amino acid and protein metabolism in *Mammalian Protein Metabolism*, vol. 4 (Munro, H. N., ed.), pp. 523–571, Academic Press, New York

FAUCONNEAU, G. & MICHEL, M. C. (1970) The role of the gastrointestinal tract in the regulation of protein metabolism in *Mammalian Protein Metabolism*, vol. 4 (Munro, H. N., ed.), pp. 481–522, Academic Press, New York

FELIG, P. & WAHREN, J. (1971) Amino acid metabolism in exercising man. *J. Clin. Invest.* **50**, 2703–2705

FELIG, P., WAHREN, J. & AHLBORG, G. (1973) Uptake of individual amino acids by the human brain. *Proc. Soc. Exp. Biol. Med.* **142**, 230–231

FISHMAN, B., WURTMAN, R. J. & MUNRO, H. N. (1969) Daily rhythms in hepatic polysome profiles and tyrosine transaminase: role of dietary protein. *Proc. Natl. Acad. Sci. U.S.A.* **64**, 677–682

GAN, J. C. & JEFFAY, H. (1967) Origin and metabolism of the intracellular amino acid pools in rat liver and muscle. *Biochim. Biophys. Acta.* **198**, 448–459

LEWIS, B. M., SOKOLOFF, L., WECHSLER, R. L., WENTZ, W. B. & KETY, S. S. (1960) A method for the continuous measurement of cerebral blood flow in man by means of radioactive krypton. *J. Clin. Invest.* **39**, 707–716

MARLISS, E., AOKI, T. T., POZEFSKY, T., MOST, A. & CAHILL, G. F. (1971) Muscle and splanchnic glutamine and glutamate metabolism in post absorptive and starved man. *J. Clin. Invest.* **50**, 814–817

MILLER, L. L. (1962) The role of the liver and the non-hepatic tissues in the regulation of free amino acids levels in the blood in *Amino Acid Pools* (Holden, J. T., ed.), pp. 708–721, Elsevier, Amsterdam

MUNRO, H. N. (1964) General aspects of the regulation of protein metabolism by diet and hormones in *Mammalian Protein Metabolism*, vol. 1 (Munro, H. N. & Allison, J. B., eds.), pp. 381–481, Academic Press, New York

MUNRO, H. N. (1969) in *Mammalian Protein Metabolism*, vol. 3 (Munro, H. N., ed.), pp. 113–182, Academic Press, New York

MUNRO, H. N. (1970) Free amino acid pools and their role in regulation in *Mammalian Protein Metabolism*, vol. 2 (Munro, H. N., ed.), pp. 286–299, Academic Press, New York

MUNRO, H. N. (1972a) Basic concepts in the use of amino acids and protein hydrolysates for parenteral nutrition in *Symposium on Total Parenteral Nutrition* (Vanamee, P. & Shils, M. E., eds.), pp. 7–35, American Medical Association, Chicago

MUNRO, H. N. (1972b) Amino acid requirements and metabolism and their relevance to parenteral nutrition in *Parenteral Nutrition* (Wilkinson, A., ed.), pp. 34–67, Churchill, Livingstone, Edinburgh & London

MUNRO, H. N. & FLECK, A. (1969) Analysis of tissues and body fluids for nitrogenous constituents in *Mammalian Protein Metabolism*, vol. 3 (Munro, H. N., ed.), pp. 423–525, Academic Press, New York

POZEFSKY, T., FELIG, P., TOBIN, J., SOELDNER, J. S. & CAHILL, G. F. (1969) Amino acid balance across the tissues of the forearm in post-absorptive man: effects of insulin at two dose levels. *J. Clin. Invest.* **48**, 2273–2280

RAMA RAO, P. B., METTA, V. C. & JOHNSON, B. C. (1959) The amino acid composition and the nutritive value of proteins. I. Essential amino acid requirements of the growing rat. *J. Nutr.* **69**, 387–391

SAN PIETRO, A. & RITTENBERG, D. (1953) A study of the rate of protein synthesis in humans.

II. Measurement of the metabolic pool and the rates of protein synthesis. *J. Biol. Chem.* **201**, 457–473

SMITH, E. B. & JOHNSON, B. C. (1967) Studies of amino acid requirements of adult rats. *Br. J. Nutr.* **21**, 17–27

WILLIAMS, J. N., SCHURR, P. E. & ELVEHJEM, C. A. (1950) The influence of chilling and exercise on free amino acid concentrations in rat tissues. *J. Biol. Chem.* **182**, 55–59

YOUNG, V. R., ALEXIS, S. D., BALIGA, B. S. & MUNRO, H. N. (1972) Metabolism of administered 3-methylhistidine: lack of muscle transfer ribonucleic acid charging and quantitative excretion as 3-methylhistidine and its N-acetyl derivative. *J. Biol. Chem.* **247**, 3592–3600

YOUNG, V. R., HAVERBERG, L. N. & MUNRO, H. N. (1973) Use of 3-methylhistidine excretion as an index of progressive reduction in muscle protein catabolism during starvation. *Metab. (Clin. Exp.)* **22**, 1429–1436

YOUNG, V. R. & MUNRO, H. N. (1973) Plasma and tissue tryptophan levels in relation to tryptophan requirements of weanling and adult rats. *J. Nutr.* **103**, 1756–1763

Discussion

Fernstrom: Does Fig. 5 show the amount of kynurenine produced *in vitro*?

Munro: Yes. Those results give us an index of the total amount of active enzyme because it was assayed in the presence of cofactors. It does not indicate the amount of substrate passage in the whole animal. Obviously, substrate concentrations are important in relation to enzymes as well as the total amount of enzymes present.

Fernstrom: When we fed rats diets containing carbohydrate, or carbohydrate plus protein, we found that whereas tyrosine aminotransferase is rapidly activated by protein consumption tryptophan oxygenase is not. Nonetheless, we noticed that the oxygenase activity reaches a peak around the time of day that the animal eats, so that while the enzyme may not be activated by tryptophan it seems to be active at an appropriate time of day, that is, when large amounts of tryptophan are entering the portal vein from the gut.

Munro: This was not apparent when we studied animals receiving low intakes of tryptophan. The adaptation of lysine oxidation to excessive intake indicates that equilibration takes about ten days (Brookes *et al.* 1972). So, the information demanding these responses must be somehow relayed to the liver. The minimum requirement in the adult human of 6 g total essential amino acids (cf. p. 7 and Table 1) is able to prime the 150 g cycle each day. The evidence regarding the sensitivity of enzymes degrading amino acids to appropriate loads will depend on the animal preparation used.

Grahame-Smith: For a long time I have tried to understand the correlation between activity of regulating enzymes (tryptophan oxygenase, tyrosine aminotransferase) with the activity of the enzymes found in liver homogenates *in vitro*.

Various discrepancies appear between the metabolism of the amino acids *in vivo* and the activity of the enzymes in the liver tissue. What is known about the correlation?

Munro: It is often not satisfactory. For example, corticosteroids can increase the amount of tryptophan oxygenase but the amount of $^{14}CO_2$ released from [^{14}C]tryptophan *in vivo* does not increase proportionally with the changes in the enzyme activity *in vitro* (Kim & Miller 1969).

Curzon: Kim & Miller used [*methylene*-^{14}C]tryptophan. Joseph (1973), in my laboratory, found that the amount of $^{14}CO_2$ exhaled quadrupled after injecting a tracer dose of [2-^{14}C]tryptophan if 5 mg/kg cortisol had been injected 150 min previously. This increase was proportionately at least as great as the increase in kynurenine synthesis from tryptophan by liver tissue *in vitro*.

Munro: It seems that Kim & Miller did not obtain an increase in output of $^{14}CO_2$, because the load they gave was inappropriate.

Wurtman: We have examined the daily rhythms in plasma tryptophan concentration and hepatic activity of tryptophan oxygenase under control conditions and after various dietary manipulations (Ross *et al.* 1973). We have found virtually no evidence that a physiological increase in tryptophan oxygenase activity is associated with accelerated catabolism of its circulating substrate. Indeed, the enzyme and the plasma amino acid tend most often to rise and fall together. This is the complete opposite of what one would expect if enzyme activity, assayed *in vitro*, was proportional to the *in vivo* degradation of tryptophan. I suspect that changes in the activity of tryptophan oxygenase (or tyrosine aminotransferase) induced by drugs or hormones reveal little about the physiological actions of the enzyme.

Grahame-Smith: But Dr Munro said that, under physiological conditions, the amounts of the amino acids were regulated by the activity of these hepatic enzymes.

Munro: Fig. 4 indicates that, above basal dietary requirements, the amount of tryptophan in the plasma rises. Also, the activity of liver tryptophan oxygenase increases at the same time (Fig. 5). The cause of both is not the higher plasma concentration but probably the larger intake through the portal vein, which we did not sample. So we are observing (i) the adaptation of the enzyme after the meal and (ii) the increased systemic plasma concentration indicating imperfect control of the amount that gets through.

Fernstrom: The oxygenase might be there to protect the systemic circulation from a large dose of tryptophan as it comes from the gut through the liver; your data seem to indicate that.

Wurtman: Although we can relate the concentrations of aromatic amino acids in plasma to those in the brain (as Dr Fernstrom will discuss later), there doesn't appear to be any correlation between cerebral tryptophan (or tyrosine) concentration and the activity of hepatic tryptophan oxygenase (or tyrosine aminotransferase).

Munro: The sharp inflection in plasma concentrations in humans and other species responds to tryptophan loading in precisely the way seen in the rat (e.g. Young *et al.* 1971).

Wurtman: Dr Munro, how much amino acid from dietary protein passed through the liver unchanged (see for example Elwyn 1970)?

Munro: In Elwyn's studies, the pattern of plasma amino acids was much altered by passage through the liver. As we know, seven of the essential amino acids are degraded in liver exclusively and the remaining three (the branched chain amino acids) in muscle, so the pattern that emerged from the large load Elwyn administered to his dogs showed an excess of branched chain amino acids and a much smaller increment in the other essential amino acids.

Mandel: Amino acids are probably reused to a much greater extent in the brain than elsewhere, since after two months on a protein-free diet consisting of carbohydrate, lipid, minerals and vitamins, the concentration of the essential amino acids (except tyrosine, phenylalanine and tryptophan) in rat brains is constant. The concentration of the three aromatic amino acids drops, probably because they are used for transmitter synthesis and then deaminated by monoamine oxidase. In contrast, the concentrations of the amino acids in muscles or the liver drop tremendously after two or three weeks on such a diet.

Lajtha: In brain there are big differences between the immature and the mature organ, that is between the period of active mitosis and the time when no cells divide. Let us consider only the mature brain. In the adult brain, unlike muscle, the rate of neither protein breakdown nor protein synthesis changes with the decreasing availability of amino acids, nor does the flux of free amino acids between plasma and brain alter. In spite of this, the brain maintains its protein content during protein starvation much better than does muscle. The major mechanism responsible for this seems to be amino acid transport, which is very active in the brain and is capable of maintaining reasonably constant quantities in the free amino acid pool in the brain. In this way, amino acids liberated from muscle protein can be used for protein synthesis in the brain. Brain proteins are still metabolized at a high rate, and free amino acids in the brain maintain their flux with plasma during protein deficiency; the brain can keep its homeostasis through use of the amino acids from the rest of the organism.

Wurtman: Can the uptake of amino acids by the brain be related to the plasma amino acid pattern after two months of starvation?

Mandel: The amount of amino acids in the plasma drops tremendously.

Wurtman: Is the drop the same for other amino acids as it is for tryptophan, phenylalanine and tyrosine?

Mandel: In plasma, yes.

Wurtman: Are there any reliable arteriovenous differences in plasma amino acids across the brain that might provide information about whether the net brain uptake of amino acids, which are precursors for neurotransmitters, is different from that of other amino acids?

Munro: Felig has catheterized the appropriate artery and vein in man to measure such differences and has found a reproducible difference in the fasting subject which did not vary in fed or fasting subjects (Felig *et al.* 1973).

Glowinski: Both tryptophan and tyrosine are involved in the synthesis of neurotransmitters but it would be hazardous to generalize that changes in the cerebral concentrations of these amino acids always reflect effects on the synthesis of 5-hydroxytryptamine or of catecholamines. The quantities of these amino acids in aminergic neurons are small compared with those found in tissues. For instance, we have failed to detect significant changes in tyrosine concentrations in the rat striatum and in the rat cerebral cortex after selective degeneration of the nigro-striatal (Agid *et al.* 1973) and of the dorsal ascending noradrenergic pathways (A. M. Thierry, unpublished findings, 1973), respectively.

Mandel: That is why a protein-free diet for several weeks is necessary to decrease the amount of the amino acid precursors of transmitters and of the transmitters themselves.

Wurtman: Your data, Dr Munro, show that the amount of amino acid per g body weight which the rat has to consume to maintain the concentrations of amino acids in the tissues is five times that in humans. What inferences for human dosage can we draw from treatment of rats with L-dopa or 5-hydroxytryptophan?

Munro: Probably that the capacity of the rat to remove L-dopa is related to that of humans by a factor of five.

Gál: Some time ago, we compared starvation with tryptophan deficiency in the rat brain (Gál & Drewes 1961, 1962). There seems to be some difference between the balance attained in starvation, when, as Dr Lajtha mentioned, the brain is receiving amino acids from the periphery, and the removal of an amino acid which was not supplied in the deficient diet. This amino acid (i.e. tryptophan) must be mobilized from reserves to ensure the continued synthesis of protein in sensitive areas of the organism. This becomes evident from the difference in the cerebral concentration of 5-hydroxytryptamine between starved and control rats which is much more pronounced between tryptophan-deficient and control animals. The deficient animals have appreciably lower concentrations of 5-hydroxytryptamine even though activity of tryptophan 5-hydroxylase is not affected.

Wurtman: The difference between the effect of protein starvation and the deletion of a single amino acid from the diet on brain 5-hydroxytryptamine might be related to the competition for brain uptake between other neutral amino acids and tryptophan. In starvation, the competitors of tryptophan are also removed.

Fernstrom: Both Pozefsky *et al.* (1969) and Wool (1965) found no change in tryptophan uptake by muscle after treatment with insulin.

Munro: But protein synthesis is stimulated by insulin. The uptake of some amino acids is masked by the rate at which the subsequent protein synthesis removes them. The picture of transport is complicated by this factor.

Mandel: How is the pool of aminoacyl-tRNAs related to protein synthesis? Aminoacyl-tRNAs are more directly involved in protein biosynthesis than the pool of the amino acids.

Munro: Technically, it is difficult to measure the charging of individual tRNA species. Some years ago, Allen *et al.* (1969) administered a mixture of amino acids lacking tryptophan or other essential amino acids to rats. The charging of tRNA was diminished most in the absence of tryptophan and much less (or not at all) by deficiency of other essential amino acids. This confirmed our finding (Pronczuk *et al.* 1968) that administration of the same mixtures to rats affected hepatic protein synthesis and polysome patterns chiefly in tryptophan deficiency and not when other amino acids were lacking. Since then, we have administered diets that were imbalanced with regard to other essential amino acids (Pronczuk *et al.* 1970). In these conditions where the liver amino acid pool differs from the normal pool produced by overnight fasting, it has been possible to produce changes in the liver polysome patterns by giving the missing amino acid, such as threonine or isoleucine. We concluded that whichever amino acid is present in sufficiently low concentration is a limiting factor in charging tRNA, so that the particular aminoacyl-tRNA becomes the rate-limiting factor in protein synthesis in the liver.

Lajtha: We can calculate the rate of turnover in the free amino acid pool, that is, the flux of essential amino acids to and from protein in the brain. For many amino acids, we obtained a half-life of 10 min or less. In other words, protein synthesis uses all the leucine in the free amino acid pool every 20 min. The similarly calculated half-life of tRNA in most cases is less than one minute, which means a very high rate of turnover.

Munro: It isn't usually realized how low the tRNA content of the cell is. For every ribosome in rat liver, there are about eight tRNA molecules; allowing for the 64 codons this would provide about 0.1 tRNA molecule of each type for one ribosome. This implies structural features of protein synthesis as yet unknown to us, for simple physical principles should not allow protein synthesis except on rare occasions. So tRNA must be reused at great speed. Furthermore, it must be restrained from straying from the site of protein synthesis.

Wurtman: Dr Barondes will discuss the charging of the tRNAs (see later, pp. 265–275). It has been claimed that a much higher concentration of tryptophan is required to charge its specific tRNA in the brain than is so for tyrosine or phenylalanine. An important question is whether the anatomical separation of neuronal tryptophan into perikaryal and nerve-terminal pools determines the pattern of its use. Are there loci within the neuron at which the tRNA charging enzyme and tryptophan hydroxylase compete for a particular tryptophan molecule? Is tryptophan incorporated into protein only in the perikaryon and 5-hydroxylated only in the terminals, or do both biotransformations occur at some sites? These questions will come up again later in our discussions (see for example pp. 42, 170 and Barondes pp. 265–275).

Mandel: Allowing for the overall pool of tRNA in the calculation of protein synthesis might be unwise since, as we have shown for instance in the silkworm gland, the tRNA pool adapts to protein synthesis. In the silkworm gland (which produces fibroin), the four tRNAs used mainly for fibroin synthesis make up about 90% of the total tRNA. So in addition to reutilization, there is probably also adaptation towards these specific tRNAs which are needed for protein synthesis.

Roberts: We should be cautious in assuming that tRNA availability is the limiting factor in protein synthesis or in polysome aggregation in these experiments where one amino acid may be limiting or others may be present in excessive quantities. As yet, no satisfactory data are available to prove the charging of tRNA is rate-limiting. Other explanations can be proposed on the basis of alterations in the properties of messenger RNA, including its availability, ease or degree of attachment to ribosomes and turnover. Later, I shall discuss the effect of amino acid depletion in the brain upon polyribosome function (see pp. 299–318). A somewhat unexpected factor affecting this function may be the degree of ribonuclease activity in polyribosome preparations under different conditions of amino acid balance or imbalance.

Munro: The only condition under which tRNA has been shown to affect the rate of protein chain formation has been the artificial one of reticulocyte synthesis of globin *in vitro*, when it is possible to break up the globin into successive peptides. The successive peptides are labelled as the chain extends. By limiting threonine, for example, in the serum in which the red cells are suspended, both the overall process and specific phases of chain elongation may be slowed down (Hunt *et al.* 1969).

References

AGID, Y., JAVOY, F. & GLOWINSKI, J. (1973) Hyperactivity of the remaining dopaminergic neurons after partial destruction in the nigro-striatal dopaminergic system in the rat. *Nat. New Biol.* **245**, 150–151

ALLEN, R. E., RAINES, P. L. & REGEN, D. M. (1969) Regulatory significance of transfer RNA charging levels. *Biochim. Biophys. Acta* **190**, 323–336

BARONDES, S. H. (1974) Do tryptophan concentrations limit protein synthesis at specific sites in the brain? *This Volume*, pp. 265–275

BROOKES, I. M., OWENS, F. N. & GARRIGUS, U. S. (1972) Influence of amino acid level in the diet upon acid oxidation by the rat. *J. Nutr.* **102**, 27–35

ELWYN, D. (1970) The role of the liver in regulation of amino acid and protein metabolism in *Mammalian Protein Metabolism*, vol. 4 (Munro, H. N., ed.), pp. 523–571, Academic Press, New York

FELIG, P., WAHREN, J. & AHLBORG, G. (1973) Uptake of individual amino acids by the human brain. *Proc. Soc. Exp. Biol. Med.* **142**, 230–231

GÁL, E. M. & DREWES, P. A. (1961) *Proc. Soc. Exp. Biol. Med.* **106**, 295–297

GÁL, E. M. & DREWES, P. A. (1962) *Proc. Soc. Exp. Biol. Med.* **110**, 368–371

HUNT, R. T., HUNTER, A. R. & MUNRO, A. J. (1969) The control of haemoglobin synthesis: factors controlling the output of α and β chains. *Proc. Nutr. Soc.* **28**, 248–254

JOSEPH, H. M. (1973) Ph. D. Thesis, University of London

KIM, J. H. & MILLER, L. L. (1969) The functional significance of changes in activity of the enzymes, tryptophan pyrrolase and tyrosine transaminase, after induction in intact rats and in the isolated, perfused rat liver. *J. Biol. Chem.* **244**, 1410–1416

POZEFSKY, T., FELIG, P., TOBIN, J. D., SOELDNER, J. S. & CAHILL, G. F. (1969) Amino acid balance across tissues of the forearm in post-absorptive man: effect of insulin at two dose levels. *J. Clin. Invest.* **48**, 2273–2282

PRONCZUK, A. W., BALIGA, B. S., TRIANT, J. W. & MUNRO, H. N. (1968) Comparison of the effect of amino acid supply on hepatic polysome profiles *in vivo* and *in vitro*. *Biochim. Biophys. Acta* **157**, 204–206

PRONCZUK, A. W., ROGERS, Q. R. & MUNRO, H. N. (1970) Liver polysome patterns of rats fed amino acid imbalanced diets. *J. Nutr.* **100**, 1249–1258

ROBERTS, S. (1974) Effects of amino acid imbalance on amino acid utilization, protein synthesis and polyribosome function in cerebral cortex, *This Volume*, pp. 299–318

ROSS, D. S., FERNSTROM, J. D. & WURTMAN, R. J. (1973) The role of dietary protein in generating daily rhythms in rat liver tryptophan pyrrolase and tyrosine transaminase. *Metabolism* **22**, 1175–1184

WOOL, I. G. (1965) Relation of effects of insulin on amino acid transport and on protein synthesis. *Fed. Proc.* **24**, 1060–1070

YOUNG, V. R., HUSSEIN, M. A., MURRAY, E. & SCRIMSHAW, N. S. (1971) Plasma tryptophan response curve and its relation to tryptophan requirements in young adult men. *J. Nutr.* **101**, 45–59

Amino acid transport in the brain *in vivo* and *in vitro*

A. LAJTHA

New York State Research Institute for Neurochemistry and Drug Addiction, New York

Abstract   The access of amino acids to the living brain is restricted by the cerebral barrier system. This restriction is particularly strong for the non-essential amino acids that are present in high concentrations in the brain. These cerebral barriers are weaker in the immature brain where they are present but not fully developed; the main characteristics do not change during development.

Amino acids may be divided into several overlapping transport classes. The members of these classes participate to various degrees in the uptake, efflux and exchange of amino acids. Efflux is an active process with different properties from those of uptake. The various transport classes are heterogeneously distributed among the structural elements of the nervous system, enabling separate control of metabolite composition in each structure. Each structural compartment possesses a corresponding transport compartment which varies in its sensitivity to the influences of physiological and pharmacological factors. The specificity of control of metabolites by transport mechanisms is further increased by differences in sensitivity to ionic and energetic changes. Sodium ions are required for transport; ATP does not seem to be the primary energy source.

Specific transport mechanisms have different functions, for instance the high affinity transport of neurotransmitter amino acids near the synaptic region as is observed in isolated systems. These isolated systems, unlike the living brain, have a great capacity for uptake of the neurotransmitter amino acids; the differences are not due to changes in ions, energy or water spaces. Transport mechanisms with high activity in the brain serve as controls for the distribution of metabolites.

The free amino acid pool of the brain has a unique composition for this organ. It is characterized by the presence of compounds that are found mainly in brain, such as 4-aminobutyric acid, and is rich in glutamate and related compounds (for example, glutamine, aspartate, acetylaspartate and glutathione). Although the constitution of the pool differs from species to species, in various brain areas and developmental stages, the main features are common to all brains with only minor variations between species (Himwich & Agrawal 1969; Levi *et al.* 1967) including man (Perry *et al.* 1971); whole brain is similar in this respect to those brain fractions that have been analysed. In most cases, composition is under strong homeostatic control; the cerebral barriers prevent major changes. In contrast to the situation in the central nervous system, the amino acid pool in the peripheral nervous system varies greatly between species and between nerves (Marks *et al.* 1970; Csanyi *et al.* 1973). For example, some nerves of the invertebrates contain the highest concentrations in the whole organism of a few compounds (such as alanine, aspartate, taurine, glycine and cystathionine). Access of amino acids to the brain is restricted by the barrier system of the organ. Although there is no net change, the free amino acid pools of plasma and brain exchange rapidly. Thus, the homeostatic control of cerebral composition is based on a dynamic process rather than the impermeability of a membrane, at least as far as metabolites are concerned. Our interest has focused on the way in which transport processes determine the concentration and distribution of amino acids in the brain. Although other processes (such as amino acid metabolism) are undoubtedly important, indications are that transport is one of the main determinants of tissue concentrations and of the distribution of amino acids in the various cellular and subcellular structures. Knowledge of the controls of cerebral metabolites will further our understanding of the functional significance of metabolite distribution in normal as well as pathological brain metabolism. In aminoacidurias, mental retardation accompanies alterations in cerebral composition, but any direct connection between metabolite concentrations and brain function or dysfunction still eludes us, and only few data suggest a role for metabolite distribution in determining rates and the pathways of metabolism. However, the findings that a deficit of amino acids in the diet during development can lead to permanent deficit of brain cells and brain protein (Zamenhof *et al.* 1968) and that changes in the amounts of amino acid in the brain influence incorporation, metabolism and transport of amino acids (Roberts 1968) stress the importance of the concentrations of metabolites.

I shall briefly discuss the heterogeneity of the processes for amino acid transport and some factors that influence them in the brain.

With the sensitive analytical techniques available, the distribution of amino acids in brain regions has been measured. We still know little about the reasons for the abundance and heterogeneous distribution of particular compounds. Neurotransmitter amino acids, especially in the regions where they act, are particularly abundant; distribution studies yielded the first clues in amino acid–transmitter research. Even this, however, is an enigma, since the amino acids are active far below their physiological concentrations. Studies of the mechanisms controlling distribution, as discussed here, and of the effects of altered cerebral amino acid levels are beginning to yield information on the different roles amino acids play in various brain structures.

FLUX OF AMINO ACIDS IN THE BRAIN

The rapid appearance of the label in the brain upon injection of tracer doses of labelled amino acids into plasma indicates a continuous and fast flow (exchange) between the free amino acids in circulation in the plasma and the brain. When, for example, a bolus containing a tracer amount of a labelled amino acid was injected into the right common carotid artery of a rat 15 s before the animal was killed, the exchange of many amino acids was as much as 30–50% of the label that had passed, perhaps in a single passage, through the brain (Oldendorf 1971). The results of infusion of tracer doses in the rat indicate that the major portion of each amino acid is in this dynamic state (Seta *et al.* 1973), although under physiological conditions the presence of amino acids belonging to the same transport class in the plasma may result in lower rates of exchange compared to those measured on injection of a bolus containing only one amino acid. I have estimated that the half-life of many cerebral amino acids is less than 30 min (Lajtha 1964). The flux between the free and protein-bound amino acids within the brain is also equally rapid: the half-life of 90% or more of brain proteins in the mouse appears to be less than 18 days (Lajtha & Toth 1966) and that of a significant portion is less than five days (Seta *et al.* 1973). If protein turns over at this rate and to this extent, the amount of amino acid that is incorporated and released from protein within 30 min equals the total content of the free amino acid pool for most amino acids. This suggests that the rate of exchange of a free amino acid between the cerebral pool and plasma and its rate of incorporation into brain protein are of similar magnitude.

Amino acid uptake

When the concentration of amino acids in the plasma rises, the subsequent rise in the brain is usually much less than in most other organs. The restriction is not absolute, since some amino acids do accumulate more than others (Battistin *et al.* 1971). Aromatic amino acids belong to one group which is restricted (Guroff & Udenfriend 1962). Tracer experiments indicate that exchange of the essential amino acids is rapid and extensive while that of the non-essentials is small (Oldendorf 1971). With similar methods, Seta *et al.* (1972) found that the net uptake of the essential amino acids by the brains of both young and adult mice is much greater than the uptake of the non-essential ones. In general, the physiological amounts in brain of the non-essential amino acids (e.g. glutamate, aspartate and glycine) are much higher than those of the essential amino acids, and therefore it seems that the concentration of those compounds which are

TABLE 1

Comparison of amino acid concentrations *in vivo* with uptake by brain slices

Amino acid	Concentration in living brain[a] (µmol/g tissue)		Uptake by brain slices[b] (tissue/medium concentration ratio	
	Mouse	Hen	Mouse	Hen
Glu	9.9	12.2	26	21
Tau[c]	8.3	3.8	14	7.3
Gly	0.91	0.42	18	14
Lys	0.19	0.18	3.0	2.3
Leu	0.04	0.05	2.1	2.3
Phe	0.05	0.06	1.7	1.7

From Levi *et al.* (1967).

[a] Brain concentrations *in vivo* were measured with an amino acid analyser. In general the composition of the pool was remarkably similar in most species investigated.

[b] The uptake was measured after incubating the slice for 90 min in 2mM-amino acid and is expressed as the ratio of concentration in tissue water to that in the medium at the end of incubation. The value of the ratio is dependent on the concentration of amino acid in the medium because of saturation of uptake. Amino acids at high concentrations *in vivo* are taken up by brain slices more but there is no strict correlation between concentration and uptake.

[c] Taurine.

plentiful in the brain is under stronger control than is that of compounds which are scarce. Since many non-essential amino acids are physiologically active and may be neurotransmitters, it is not surprising that their distribution is carefully controlled. The fact that their presence is not easily increased *in vivo* does not mean that no uptake mechanism for these compounds exists. A comparison of uptake by brain slices with the amounts in brain (Table 1) shows that the plentiful compounds in the brain *in vivo* are accumulated to a greater degree by brain slices than compounds that are scarce in the living brain (Levi *et al.* 1967). Clearly the barriers and control are not passive, and compounds with a greater affinity for transport are better controlled. The fact that some of this transport is in glia and in postsynaptic sites indicates that one function of transport is to remove active compounds rapidly from sensitive sites (Balcar & Johnston 1973).

Exit and exchange

Uptake is not the only important process in cerebral amino acid flux. Convincing evidence indicates active exit mechanisms (including transport against a gradient) in the rat brain (Lajtha & Toth 1961) and heteroexchange, that is, flow of one amino acid driven by the counterflow of another amino acid

(Battistin *et al.* 1972). Uptake, efflux and exchange differ, for example in stereo-specificity (Lajtha & Toth 1963); these processes are sufficiently different to merit separate consideration. Furthermore, drugs and analogues of amino acids affect exit and uptake in different ways. A compound that inhibits uptake as well as exit need not affect cerebral concentrations greatly, since the two inhibitions may balance each other, whereas if a compound inhibits uptake and increases exit, the effect will be additive (Gaull *et al.* 1974). The first indication of active exit from the brain came from our observations (Lajtha & Toth 1961) that, after their intracerebral administration, amino acids leave the brain even against a concentration gradient (i.e. high plasma concentrations; Table 2). Since the efflux of the various amino acids differed, exit from the brain could not have been due mainly to removal by the bulk flow of the spinal fluid or diffusion. Snodgrass *et al.* (1969) have found mediated efflux of amino acids from the spinal fluid in ventriculocisternal perfusions in the cat. Since transport had been found both in choroid plexus (Lorenzo & Cutler 1969) and other brain areas, exit through spinal fluid seems to have an extra choroidal component, perhaps subarachnoid (Murray & Cutler 1970). While efflux from the spinal fluid could maintain cerebral amino acid concentrations, exit processes are likely to have other functions as well. Evidence that they operate within brain cells is the release of amino acids from spinal cord on electrical stimulation. This induced exit is specific for neurotransmitter-like amino acids (glutamate, glycine, 4-aminobutyric acid and taurine); it requires energy and is a mediated transport (Cutler *et al.* 1972). Specific release by efflux into, and inactivation by uptake from, sites of activity of neurotransmitter amino acids may represent a specific function of transport in neurotransmission.

TABLE 2

Efflux of leucine from the brain when plasma concentrations are elevated above those in brain

| Time (min) | Increase over control (μg leucine/g fresh tissue) | | | |
| | Lower dose | | Higher dose | |
	Plasma	Brain	Plasma	Brain
20	280	110		
30	180	56	590	82
45	250	24	980	73
90	260	23	710	54

From Lajtha & Toth (1961) and Lajtha (1973). Young male rats were injected subarachnoidally with 260 μg leucine in 0.02 ml of saline. Plasma concentrations were kept high by an initial intravenous dose 10 min after the subarachnoid injection and by repeated intraperitoneal injections at about 15 min intervals. Brain concentrations decrease although not completely to control values, even though plasma levels are elevated well above brain levels.

Diffusion seems to be unimportant in the cerebral flux of amino acids. Usually the saturable component of flux is taken to be carrier mediated and that not saturable as the diffusion part of flux. However, Cohen (1973) has found that in uptake by slices the non-saturable component may also be mediated, since it can result in accumulation against a concentration gradient.

Processes of exchange have been less frequently studied, and the function of exchange, especially heteroexchange is not clear. A label injected into blood appears in the brain when the physiological concentrations are unchanged; this shows rapid exchange *in vivo*. In continuous infusion in the rat, however, equilibrium was not reached, for the specific activity in the brain remained less than that in the plasma (Seta *et al.* 1973). This does not necessarily indicate that some amino acid is not exchangeable since protein turnover would tend to keep the specific activities of the brain less than those of plasma unless flux from plasma to brain is greatly above the flux from brain protein. During infusion, the specific activities of many amino acids reached high levels (Table 3), but those of glutamic acid and glycine remained far below the activity in plasma. Many factors seem to contribute to this lack of equilibration: slow exchange with blood, rapid metabolism in brain, binding and sequestration into slowly or non-exchanging compartments.

TABLE 3

Isotopic equilibrium of plasma and brain amino acids *in vivo*

Amino acid	Infusion time (h)	Specific activity in brain as % of that in plasma
Arg	3	69
	6	83
Val	3	70
	6	77
Lys	3	77
	6	86
Tyr	3	67
	6	90
Leu	3	47
	6	41
Gly	3	18
	6	23
Glu	6	7

From Seta *et al.* (1973). Tracer doses of a ^{14}C-labelled amino acid were administered by continuous intravenous infusion to rats. For calculations of specific activity (c.p.m./µmol). values had to be corrected to allow for the portion of label that was metabolized (about half the label in the tissue). Rapidly and slowly exchanging pools were found with incomplete exchange of glutamate and glycine.

When the concentrations of amino acids in slices of brain are greatly increased after incubation of the slices in media containing amino acids, those amino acids that were taken up exchange much faster than the amino acids originally in the slice (Shiu & Elliot 1973). Labelled endogenous glutamate, derived from [^{14}C]glucose, in slices exchanges more slowly than the glutamate that was taken up from the medium (Okamoto & Quastel 1972). These observations show differences in behaviour *in vitro* between the 'original' endogenous amino acid pool and the fraction that was taken up from the medium. There are some grounds for believing that most of the endogenous glutamate in the slice is in the neurons; uptake from the medium is mostly into glia (Okamoto & Quastel 1972). Large quantities of an amino acid may suppress the exchange of the medium with endogenous pool. We found that when slices are incubated in a medium containing only tracer amounts of amino acids at equilibrium (within 60 min), the specific activities of the slice and medium are similar, indicating only a small non-exchangeable pool (A. Neidle, J. Kandera & Lajtha, unpublished results). The endogenous amounts of amino acids did not grow under such conditions but proteolysis resulted in the release of some amino acid to the medium. The amino acid so produced could keep specific activities in slices lower by dilution; this effect should be the largest for amino acids with largest net increase in relation to pool size (Table 4).

TABLE 4

Isotopic equilibrium of amino acids in incubated slices of brain

Amino acid	Tissue concentration ($\mu mol/g$)	Amino acid increase in incubation ($\mu mol\ g^{-1}\ h^{-1}$)	Specific activity in slice as % of that in medium
Gly	1.2	0.58	102
Thr	0.61	0.57	94
Val	0.18	0.26	90
His	0.13	0.24	76
Leu	0.16	0.40	79
Lys	0.15	0.49	58
Tyr	0.07	0.24	83

From Neidle, Kandera & Lajtha (unpublished results). Slices of mouse brain were incubated in a medium containing tracer doses of ^{14}C-labelled amino acid for one hour. Initial tissue levels, and specific activities of the amino acids in the tissue and in the medium at the end of the incubation were determined with an amino acid analyser. Although during the incubation amino acid concentrations in the tissue did not change greatly, because of proteolysis, those of medium amino acids increased. The specific activities in the slice did not increase upon further incubation of one hour.

HETEROGENEITY OF TRANSPORT

The fact that the uptake, efflux and exchange differ indicates that we have to consider several mechanisms in amino acid fluxes. Amino acids may belong to more than one transport class and the participation of each class in uptake and exit might differ. Structural heterogeneity, differences of the outer and inner part of membranes, may add to the observed differences in behaviour.

Substrate specificity

The behaviour of amino acids showed that several transport classes can be distinguished within the brain. Structurally related amino acids have an affinity for the same carrier. We can distinguish, for example, small and large neutral amino acids, ω-amino acids, acidic amino acids, small and large basic amino acids among others (Blasberg & Lajtha 1966). Substrate specificity of these over-lapping transport classes in brain is similar to that in other tissues (Christensen 1973), although the details may not correspond. Aromatic amino acids probably belong to the large neutral class, although they may form a separate, somewhat different, class (Barbosa *et al.* 1970). The substrate specificity of efflux differs somewhat from that of uptake; competitive stimulation has been more often observed in efflux than in uptake, for example (Levi *et al.* 1966). The differences in specificity in uptake and efflux again indicate that the various transport classes participate to different extent.

Of particular interest for the brain are transport classes with high affinity. Highly specific transport and high affinity were first observed in synaptosomes, mostly with amino acids which function as neurotransmitters. The Michaelis constant, K_m, of the low affinity transport for the average amino acid, is usually 1 mM for the brain; the high affinity transport for glutamate, 4-aminobutyric acid and glycine is 10 μM. This is interesting since the concentrations of the compounds themselves are 1–10 mM in the brain; the concentrations at which the high affinity transport operates must therefore be a few orders of magnitude lower than average cerebral concentrations. The presence of high affinity systems in synaptosomes (Logan & Snyder 1972) suggested their role in synaptic trans-mission. The distinction between high affinity uptake and postsynaptic action (Balcar & Johnston 1973) in that, for example, synaptic antagonists do not affect uptake, and the presence of a high affinity uptake system in glia suggest that high affinity systems remove neurotransmitters from the synaptic environ-ment. The bulk of the low affinity systems are to be found in non-synaptosomal structures (Levi & Raiteri 1973).

Substrate specificity has been studied in greatest detail in brain slices (Blasberg & Lajtha 1966; Levi *et al.* 1966). The data so far available indicate that similar specificity exists in the living brain, as shown by experiments with tracer intravenous injections (Oldendorf 1971) or in the transport of amino acids from the spinal fluid in perfusions (Snodgrass *et al.* 1969). Separate transport classes are present in the brain for compounds other than amino acids, such as sugars and amines: we have found a class distinct from amino acids specific for diamines such as cadaverine (Piccoli & Lajtha 1971). The lack of absolute substrate specificity of the various transport classes (that is, a number of compounds share the same class) results in a competition between members of the same class, including compounds of related structure not present normally in the brain: transport of dinitrophenol is related to that of tyrosine and phenylalanine (Piccoli & Lajtha 1971).

Morphological compartments

Generally, the composition of the free amino acid pool is not homogeneous thoughout the brain. Distribution of putative neurotransmitters is especially inhomogeneous; 4-aminobutyric acid and glycine are plentiful in the region where they appear to act as neurotransmitters. The pattern of distribution varies with each amino acid (Kandera *et al.* 1968). Most likely, this distribution reflects the distribution of various structural elements, each containing a pool of different composition. Possibly the pool is different in neurons and glia, but present techniques are inadequate for determining this with accuracy. During isolation of neurons and glia in sucrose density gradients, much of the free amino acid leaks out from the pool into the sucrose. Incubation of cells isolated from rabbit cortex shows that glia accumulate amino acids to a greater extent than neurons do (Hamberger 1971), but this difference may, however, be due to the less-well-preserved membranes of neurons. Similar reasoning pertains to the subcellular elements. Isolated subcellular elements, especially synaptosomes, show amino acid uptake against a gradient (Navon & Lajtha 1969), but the free amino acid content of the pool in particulate fractions is low, perhaps having suffered leakage, and any difference between the various particulates may be due to differences in preservation of the membrane during preparation. The basic properties of transport (specificity and requirements) are similar in particulates to those observed in slices (Navon & Lajtha 1969; Grahame-Smith & Parfitt 1970); particulates have transport systems that can regulate the composition of the pool of each structure in a specific way. The composition of the pool can differ in cells and in subcellular elements, and extraneous metabolites or drugs may variously influence the several structures. That this structural ordering of amino acids into compartments results in metabolic compartmentation has been well established for glutamic acid (Balazs & Cremer 1973); therefore the metabolism of the amino acids differs in the various structural elements.

Heterogeneity was also observed in peripheral nerves in proximodistal concentration gradients in some amino acids; this varied from nerve to nerve (Marks et al. 1970) and might be due to differences in the rate of axonal flow of free amino acids (Csanyi et al. 1973), the rate of efflux from nerves or the rates of amino acid metabolism in peripheral nerves.

Changes in development

The composition of the amino acid pool alters during development; in general, the proportion of essential amino acids decreases while that of non-essential ones increases (Himwich & Agrawal 1969). The changes are complex, especially in the perinatal period (Table 5) ; the sudden, great change for some compounds either just before or just after birth indicates a change in function or metabolism (Lajtha & Toth 1973). It is generally assumed that barriers and homeostatic control mechanisms are absent in the immature brain. Recent evidence, however, supports the presence of barriers and transport mechanisms in the young brain; in newborn mice uptake is restricted (Seta et al. 1972) and in brain slices uptake is active (Levi et al. 1967). Barriers and transport mechanisms are an integral part of the functioning brain at all ages, although they are not as well developed

TABLE 5

Changes in the cerebral free amino acid pool during development of a mouse

| Amino acid | Amount (μmol amino acid/g fresh brain) | | | | |
| | Fetus | | Newborn animal | | Adult |
	15-day-old	19-day-old	0 h	24 h	
Tau[a]	14	12	14	16	6.6
Glu	7.5	5.7	4.8	5.0	10
Thr	4.3	0.90	0.93	0.90	0.70
Ala	5.1	3.0	4.3	0.80	0.51
Lys	0.86	0.88	0.91	0.41	0.22
Pro	0.89	0.52	0.66	0.57	0.08

From Lajtha & Toth (1973). Freshly excised mouse brain was extracted with perchloric acid and measured with an amino acid analyser. The developmental periods that showed the greatest change in amino acid levels are printed in bold type.
[a] Taurine.

in the immature brain; changes in mechanisms in the brain during development have not been studied in great detail. In the liver, transport increases markedly during the first 24 h after birth, presumably induced by the rapid diminution of plasma amino acids at birth. This sharp intensification of transport is repressed by an increase in the quantity of amino acids in the tissue (Christensen 1973). The effect of the amount of amino acid on its own transport, by influencing the synthesis of protein that participates in its transport, has been shown in chick-embryo heart cells (Franchi-Gazzola et al. 1973). Levi (1973) has noticed that the main characteristics of transport in brain (apparent K_m, substrate specificity, Na dependence) do not change greatly in development. The rate of efflux also increases during maturation; initial rates of uptake increase more than steady-state uptake values since exit processes also participate in the steady-state (Levi et al. 1967; Levi 1973). Changes in rates of transport, like changes in the composition of amino acid pools, are not linear but may have minimal or maximal values during development (Piccoli et al. 1971).

FACTORS INFLUENCING THE AMINO ACID POOL

The foregoing observations point to the dynamic state of the free pool wherein the amino acid flux is mediated by several mechanisms and shows structural heterogeneity. It is to be expected that such a complex system should react to a number of factors, the reactions being subordinated to the structure.

Requirements of transport

Knowledge of the requirements of transport will help us to find the conditions governing amino acid movements and to understand the possible indirect changes in brain composition. We have studied energy and ionic dependence, particularly from the point of view of specificity. We found no absolute substrate specificity for amino acid transport. This is puzzling, because the concentration and distribution for each compound is normally carefully maintained: for instance, aspartic acid is distributed differently from glutamic acid although they belong to the same transport class.

Sodium ions are vital for uptake of amino acids in slices of brain, for if all the sodium is removed from the tissue the concentrative uptake of amino acids ceases. Other ions do not replace sodium and the exclusion of other ions, including potassium, does not abolish concentrative uptake (Lahiri & Lajtha 1964). The diminution of uptake on lowering the quantity of sodium ions in the incubation medium of the brain slices varied from one amino acid to another (Margolis & Lajtha 1968). Changes in sodium concentrations induced by only a few compounds can increase selectivity of transport. In isolated cells, potassium concentrations had a greater effect on glial than neuronal transport, an observation which also shows selectivity (Hamberger 1971). Some of the ionic effects may be indirect—the effect on membrane structure, charge, leakiness etc.

Movement against a concentration gradient requires energy, but the primary source of the energy driving amino acid transport is not known. Significant discrepancies were found between the amount of ATP and concentrative uptake (Table 6), which indicate that either ATP is not used directly (and a reduction in ATP results in less transport only because ATP is a general measure of energy supply) or that ATP itself is compartmented (and the average ATP concentration is not a good measure of the local concentration of ATP and its variation). The consequence of lowering the amount of ATP varies from amino acid to amino acid (Banay-Schwartz et al. 1971), a specificity which indicates that availability of energy may also be one control through a process that increases selectivity. Movement of amino acids might be responsible for ionic movements and energy changes: Takagaki has proposed (1963) that both aspartic acid and glutamic acid aid transport of potassium ions. Acidic amino acids seem to lower ATP concentrations at least in preparations *in vitro*. In preliminary experiments (Banay-Schwartz, Teller and Lajtha, unpublished results), we found that most amino acid movements and ion movements were independent of each other.

TABLE 6

Inhibition of amino acid uptake and variation of ATP concentrations in slices of mouse brain

Compound	Concentration (mmol/l)	Percentage of control concentration			
		Concentrative uptake		ATP	
		Glutamate	Lysine	Glutamate	Lysine
Fluoroacetate	10	89	92	21	45
Sodium fluoride	2.0	92	53	19	32
Iodoacetate	0.5	6	53	4	3
Dinitrophenol	0.05	96	47	21	35
N-Ethylmaleimide	0.2	18	14	9	9
Valinomycin	0.05	44	48	10	10

From Banay-Schwartz et al. (1971) and Banay-Schwartz, Teller and Lajtha (unpublished results). Slices of mouse brain were first preincubated for 30 min (with or without inhibitor), then were incubated for 60 min with 1mM-glutamate or 2mM-lysine. The presence of glutamate decreases ATP levels. There seems to be no close correlation between the level of ATP and amino acid uptake in the slices. Concentrative uptake: slice uptake minus medium level (uptake above medium levels).

Differences between in vivo and in vitro preparations

Many properties, such as ionic requirements and energy, can be studied in greater detail in isolated preparations. The question to be answered is, to what extent can the properties of the living brain be deduced from *in vitro* preparations such as brain slices, isolated cell fractions or subcellular particulates? An important difference between slices and the living brain lies in the uptake of non-essential amino acids. On incubation, brain slices accumulate all amino acids against a concentration gradient until the quantity in the tissue is several times greater than endogenous levels (Levi *et al.* 1967). The endogenous levels of non-essential amino acids, unlike those in slices, cannot be greatly increased *in vivo* (Seta *et al.* 1972), and the tracer exchange of non-essential amino acids is much smaller than that of the essential amino acids (Oldendorf 1971). Compounds kept out of the brain to the greatest extent *in vivo* are accumulated to the highest degree *in vitro*. The effect of preparing slices can be shown by a difference in amino acid uptake between intact olfactory bulbs and slices therefrom (Table 7). Uptake of each amino acid by slices is several times higher than by the intact bulb. We found (Neidle *et al.* 1973) no great differences in the amounts of sodium and potassium ions, in the amount of ATP or in the rate of respiration between the two preparations; these parallels indicate that ionic or

TABLE 7

Uptake of amino acids by intact olfactory bulb and by slices of olfactory bulb

	Unit	*Intact bulb*	*Bulb slices*
Gly (0.1mM)	Tissue/medium	10	71
Gly (1.0mM)	Tissue/medium	4.0	36
Val (0.1mM)	Tissue/medium	3.4	20
Glu (0.1mM)	Tissue/medium	9.3	51
Glu (1.0mM)	Tissue/medium	4.3	34
ATP	μmol/ml	0.93	0.85
Na$^+$	μmol/ml	118	135
K$^+$	μmol/ml	94	89
Extracellular space	Inulin space, %	15	54
Val	K_m/mM	4.0	3.7
Glu	K_m/mM	1.5	1.4
Val	V_{max}/μmol/min	0.9	3.6
Glu	V_{max}/μmol/min	0.8	3.2

From Neidle *et al.* (1973). Olfactory bulbs (intact or slices) were incubated in a medium containing a ^{14}C-labelled amino acid at the concentrations noted for 30 min. Values of K_m and V_{max} were calculated from 3-min incubations at various concentrations. Uptake of all amino acids (at all concentrations and times) was much higher in slices than in the intact bulb, while ATP, Na$^+$, K$^+$ or the apparent K_m were not much different in the two preparations. The two main differences were in extracellular space and V_{max} of transport.

TABLE 8

Independence of amino acid uptake of swelling in incubated slices of brain

	Unit	Control	Control plus 4% Dextran 80
Swelling	Water increase (%)	48	2
Extracellular space	Inulin space (%)	43	39
Tissue Na$^+$	μmol/ml	111	102
Tissue K$^+$	μmol/ml	70	73
Gly uptake (2 mM)	μmol/ml	26	25
Glu uptake (1 mM)	μmol/ml	28	28
ATP	μmol/ml	2.0	2.0

From Banay-Schwartz et al. (1974).
Slices of mouse brain were incubated in the presence and absence of 4% Dextran 80 (molecular weight range 60 000–90 000) in a Krebs–Ringer medium containing either 2mM-[^{14}C]-glycine, 1mM-[^{14}C]glutamic acid, or 0.04% [^{14}C]inulin for the determination of amino acid uptake and extracellular space. Na$^+$ and K$^+$ were determined with a flame photospectrometer, swelling by dry weight. Dextran 80 abolishes the increase in water during incubation but has no effect on ion and ATP levels or on amino acid uptake. Since Dextran does not affect extracellular spaces, its main effect is on intracellular water.

energy changes are not responsible for the higher uptake by slices. Tissue damage seems to be only partly responsible for the difference. The uptake in cortical slices with one cut surface differs not greatly from that in slices with two cut surfaces, and the uptake of amino acids by thinner slices with much more tissue damage in most cases is not increased (Lajtha & H. Sershen, unpublished results), although concentrations at outer surfaces are higher than in the centre of the slices. One important distinction between intact and sliced olfactory lobes was the size of the extracellular space. When we measured the effect of swelling in slices, we found amino acid uptake was unrelated to uptake of water (Table 8). One possible reason for the greater uptake in slices may be that part of the transport system is damaged; active efflux is important in maintaining homeostasis in vivo and if it is damaged the result might be the excessive uptake by slices. Competitive inhibition of efflux from slices by amino acid analogues (Levi et al. 1966) indicates that all efflux is not cut off in slices.

Pathological changes

Great interest attends the question of whether transport processes participate in pathological alterations in the brain. The possible control of the synthesis of transport proteins by amino acid concentrations, as shown in other tissues (Christensen 1973; Franchi-Gazzola et al. 1973), indicates that alterations of these concentrations can result in long-term changes that persist even after the changes are restored. Diet can change brain protein metabolism; if a diet is deficient in proteins during development, a permanent deficit of cell numbers and brain proteins can result (Zamenhof *et al.* 1968). The extent of damage and of recovery depend on the period of dietary deficiency during development (Lajtha & Teller, unpublished results). A rise in amino acid levels in plasma can interfere with the uptake of amino acids by the brain. These effects have been studied especially in experimental phenylketonuria (Agrawal *et al.* 1970; Oldendorf 1973) and are complex, since not only uptake but also efflux can be inhibited or stimulated, and the amino acid increased in the brain can exchange with others. We found that alterations of plasma concentrations affect mostly the cerebral essential amino acid concentrations (Battistin *et al.* 1971). These changes could then have an influence on cerebral protein metabolism (Roberts 1968).

References

AGRAWAL, H. C., BONE, A. H. & DAVISON, A. H. (1970) Effect of phenylalanine on protein synthesis in the developing rat brain. *Biochem. J.* **117,** 325–331

BALAZS, R. & CREMER, J. E. (1973) (eds.) *Metabolic Compartmentation in the Brain*, Macmillan, London

BALCAR, V. J. & JOHNSTON, G. A. R. (1973) High affinity uptake of transmitters: studies on the uptake of L-aspartate, GABA, L-glutamate and glycine in cat spinal cord. *J. Neurochem.* **20,** 529–539

BANAY-SCHWARTZ, M., PIRO, L. & LAJTHA, A. (1971) Relationship of ATP levels to amino acid transport in slices of mouse brain. *Arch. Biochem. Biophys.* **145,** 199–210

BANAY-SCHWARTZ, M., GERGELY, A. & LAJTHA, A. (1974) Independence of amino acid uptake of tissue swelling in incubated slices of brain. *Brain Res.* **65,** 265–276

BARBOSA, E., JOANNY, P. & CORRIOL, J. (1970) Accumulation active du tryptophane dans le cortex cérébral isolé du rat. *C. R. Séances Soc. Biol. Fil.* **164,** 345–350

BATTISTIN, L., GRYNBAUM, A. & LAJTHA, A. (1971) The uptake of various amino acids by the mouse brain *in vivo. Brain Res.* **29,** 85–99

BATTISTIN, L., PICCOLI, F. & LAJTHA, A. (1972) Heteroexchange of amino acids in incubated slices of brain. *Arch. Biochem. Biophys.* **151,** 102–111

BLASBERG, R. & LAJTHA, A. (1966) Heterogeneity of the mediated transport systems of amino acid uptake in brain. *Brain Res.* **1,** 86–104

CHRISTENSEN, H. N. (1973) On the development of amino acid transport systems. *Fed. Proc.* **32,** 19–28

COHEN, S. R. (1973) Saturable and non-saturable transport of amino acids to and from brain slices. *Trans. Am. Soc. Neurochem.* **4,** 137

CSANYI, V., GERVAI, J. & LAJTHA, A. (1973) Axoplasmic transport of free amino acids. *Brain Res.* **56,** 271–284

CUTLER, R. W. P., MURRAY, J. E. & HAMMERSTAD, J. P. (1972) Role of mediated transport in the electrically-induced releases of [^{14}C]glycine from slices of rat spinal cord. *J. Neurochem.* **19,** 539–542

FRANCHI-GAZZOLA, R., GAZZOLA, G. C., RONCHI, P., SAIBENE, V. & GUIDOTTI, G. C. (1973) Regulation of amino acid transport in chick embryo heart cells. II. Adaptive control sites for the 'A' mediation. *Biochim. Biophys. Acta* **291**, 545–556

GRAHAME-SMITH, G. D. & PARFITT, A. G. (1970) Tryptophan transport across the synaptosomal membrane. *J. Neurochem.* **17**, 1339–1353

GAULL, G., TALLAN, H., RASSIN, D. & LAJTHA, A. (1974) in *Biology of Brain Dysfunction* (Gaull, G., ed.), Plenum Press, New York

GUROFF, G. & UDENFRIEND, S. (1962) Studies on aromatic amino acid uptake by rat brain *in vivo*. Uptake of phenylalanine and of tryptophan; inhibition and stereoselectivity in the uptake of tyrosine by brain and muscle. *J. Biol. Chem.* **237**, 803–806

HAMBERGER, A. (1971) Amino acid uptake in neuronal and glial cell fractions from rabbit cerebral cortex. *Brain Res.* **31**, 169–178

HIMWICH, W. A. & AGRAWAL, A. C. (1969) in *Handbook of Neurochemistry* (Lajtha, A., ed.), vol. 1, pp. 33–52, Plenum Press, New York

KANDERA, J., LEVI, G. & LAJTHA, A. (1968) Control of cerebral metabolite levels. II. Amino acid uptake and levels in various areas of the rat brain. *Arch. Biochem. Biophys.* **126**, 249–260

LAHIRI, S. & LAJTHA, A. (1964) Cerebral amino acid transport *in vitro*. I. Some requirements and properties of uptake. *J. Neurochem.* **11**, 77–86

LAJTHA, A. (1964) in *International Review of Neurobiology* (Pfeiffer, C.C. & Smythies, J. R., eds.), vol. 6, pp. 1–98, Academic Press, New York

LAJTHA, A. (1973) in *Problems of Brain Biochemistry* (Buniatian, H.Ch., ed.), vol. 9, Armenian Academy of Sciences, Yerevan

LAJTHA, A. & TOTH, J. (1961) The brain barrier system. II. Uptake and transport of amino acids by the brain. *J. Neurochem.* **8**, 216–225

LAJTHA, A. & TOTH, J. (1963) The brain barrier system. V. Stereospecificity of amino acid uptake, exchange and efflux. *J. Neurochem.* **10**, 909–920

LAJTHA, A. & TOTH, J. (1966) Instability of cerebral proteins. *Biochem. Biophys. Res. Commun.* **23**, 294–298

LAJTHA, A. & TOTH, J. (1973) Perinatal changes in the free amino acid pool of the brain in mice. *Brain Res.* **55**, 238–241

LEVI, G. (1973) in *Biochemistry of the Developing Brain* (Himwich, W. A., ed.), Dekker, New York

LEVI, G. & RAITERI, M. (1973) Detectability of high and low affinity uptake systems for GABA and glutamate in rat brain slices and synaptosomes. *Life Sci.* **12**, 81–88

LEVI, G., BLASBERG, R. & LAJTHA, A. (1966) Substrate specificity of cerebral amino acid exit *in vitro*. *Arch. Biochem. Biophys.* **114**, 339–351

LEVI, G., KANDERA, J. & LAJTHA, A. (1967) Control of cerebral metabolite levels. I. Amino acid uptake and levels in various species. *Arch. Biochem. Biophys.* **119**, 303–311

LOGAN, W. J. & SNYDER, S. H. (1972) High affinity uptake systems for glycine, glutamic and aspartic acids in synaptosomes of rat central nervous tissues. *Brain Res.* **42**, 413–431

LORENZO, A. V. & CUTLER, R. W. P. (1969) Amino acid transport by choroid plexus *in vitro*. *J. Neurochem.* **16**, 577–585

MARGOLIS, R. & LAJTHA, A. (1968) Ion dependence of amino acid uptake in brain slices. *Biochim. Biophys. Acta* **163**, 374–385

MARKS, N., DATTA, R. K. & LAJTHA, A. (1970) Distribution of amino acids and of exo and endopeptidases along vertebrate and invertebrate nerves. *J. Neurochem.* **17**, 53–63

MURRAY, J. E. & CUTLER, R. W. P. (1970) Transport of glycine from the cerebrospinal fluid. *Arch. Neurol.* **23**, 23–31

NAVON, S. & LAJTHA, A. (1969) The uptake of amino acids by particulate fractions from the brain. *Biochim. Biophys. Acta* **173**, 516–531

NEIDLE, A., KANDERA, J. & LAJTHA, A. (1973) The uptake of amino acids by the intact olfactory bulb of the mouse. A comparison with tissue slice preparations. *J. Neurochem.* **20**, 1181–1193

OKAMOTO, K. & QUASTEL, H. J. (1972) Uptake and release of glutamate in cerebral-cortex slices from the rat. *Biochem. J.* **128**, 1117–1124

OLDENDORF, W. H. (1971). Brain uptake of radiolabeled amino acids, amines, and hexoses after arterial injection. *Am. J. Physiol.* **221**, 1629–1639

OLDENDORF, W. H. (1973) Saturation of blood brain barrier transport of amino acids in phenylketonuria. *Arch. Neurol.* **28**, 45–48

PERRY, T. L., HANSEN, S., BERRY, K., MOK, C. & LESK, D. (1971) Free amino acids and related compounds in biopsies of human brain. *J. Neurochem.* **18**, 521–528

PICCOLI, F. & LAJTHA, A. (1971) Some aspects of uptake of non-metabolites in slices of mouse brain. *Biochim. Biophys. Acta* **225**, 356–369

PICCOLI, F., GRYNBAUM, A. & LAJTHA, A. (1971) Developmental changes in Na^+, K^+ and ATP and in the levels and transport of amino acids in incubated slices of rat brain. *J. Neurochem.* **18**, 1135–1148

ROBERTS, S. (1968) in *Brain Barrier Systems* (Lajtha, A. & Ford, D. H., eds.), pp. 235–243, Elsevier, New York

SETA, K., SERSHEN, H. & LAJTHA, A. (1972) Cerebral amino acid uptake *in vivo* in newborn mice. *Brain Res.* **47**, 415–425

SETA, K., SANSUR, M. & LAJTHA, A. (1973) The rate of incorporation of amino acids into brain proteins during infusion in the rat. *Biochim. Biophys. Acta* **294**, 472–480

SHIU, P. C. & ELLIOTT, K. A. C. (1973) Binding and uptake of amino acids by brain tissue. *Can. J. Biochem.* **51**, 121–128

SNODGRASS, S. R., CUTLER, R. W. P., KANG, E. S. & LORENZO, A. V. (1969) Transport of neutral amino acids from feline cerebrospinal fluid. *Am. J. Physiol.* **217**, 974–980

TAKAGAKI, G. (1963) Aspartic acid and the accumulation of potassium ions in cerebral tissue. *Life Sci.* **10**, 759–764

ZAMENHOF, S., VAN MARTHENS, E. & MARGOLIS, F. L. (1968) DNA (cell number) and protein in neonatal brain: alteration by maternal dietary protein restrictions. *Science (Wash. D.C.)* **160**, 322–323

Discussion

Felix: In synaptic transmission in the brain, uptake is possibly the mechanism whereby the function of a transmitter is terminated. Is the uptake mechanism *in vitro* similar to the mechanism that terminates the synaptic transmission within the intact brain?

Lajtha: The present evidence suggests that the activity of most transmitters including transmitter amino acids is terminated not by enzymic metabolism but by uptake. There is a high-affinity uptake mechanism with high substrate specificity for transmitter amino acids, which, curiously, is present not only in synaptosomes but also in glia. Since drugs that inhibit excitatory activity do not inhibit high-affinity transport and transport inhibitors do not always inhibit excitation, it seems that this high-affinity transport is not an excitatory release mechanism but a postsynaptic removal mechanism. It is interesting that the K_m values (the affinity constants) of the high-affinity uptake systems are a few orders of magnitude lower than the average concentration of the amino acids (K_m for glutamate is 10 μM while glutamate concentration is 10 mM in the brain). This indicates the efficiency of the removal mechanism and also that small quantities of transmitter amino acids could be present at active synaptic sites.

Wurtman: Is a physiologically significant fraction of the tyrosine, tryptophan or phenylalanine in brain used to make neurotransmitters? It certainly seems true that in terms of its total brain or total body pools, the proportion of tryptophan used to make 5-hydroxytryptamine is much less than the proportion used for protein synthesis. However, that is probably not the physiologically relevant question. Rather we should ask what fraction of the amino acid within these cells that make both protein and 5-hydroxytryptamine goes to make each? This can be answered for the cells of the pineal. We incubated rat pineals in organ culture with [^{14}C]tryptophan and measured the accumulation of 5-hydroxytryptamine, *N*-acetyl-5-methoxytryptamine (melatonin), 5-hydroxyindoleacetic acid and protein. We found that more than 100 times as much tryptophan is used for the synthesis of 5-hydroxytryptamine and its products than for making protein (Wurtman *et al.* 1969). Perhaps in this situation, there is sufficient competition for tryptophan to limit protein synthesis.

Roberts: This is an exceptional circumstance, because the pineal gland is so active in synthesizing 5-hydroxytryptamine and may be relatively inactive in protein synthesis!

Gál: I agree; the pineal is a specific situation. Studying the cerebral metabolism of free L-tryptophan, we found that 20% of the amino acid was consumed in the synthesis of the neurotransmitter but that much less than 1% was converted into tryptamine. The rest may have gone through other pathways (for example, to kynurenine) and into protein synthesis. However, as Dr Lajtha said, we must define whether we are talking about functional or structural protein when considering the rate of amino acid used for protein synthesis such as the hydroxylating enzymes—which usually have a half-life of roughly 2.5–3.0 days in the brain. These values, of course, depend greatly on the method of measurement.

Wurtman: Do the cells which use phenylalanine, tyrosine and tryptophan for neurotransmitters have specialized uptake systems for them?

Glowinski: We can partially answer this question from the results of our tryptophan uptake studies. Two uptake processes for this amino acid have been found in synaptosomes of the rat cortex, in cultured glial cells, in fibroblasts and in neuroblastoma clones of the noradrenergic and cholinergic types (Bauman *et al.* 1974). These data suggest that high affinity uptake of tryptophan is not specific to 5-hydroxytryptaminergic neurons.

Wurtman: If cells are consuming tryptophan to make 5-hydroxytryptamine, how do they maintain the appropriate amino acid pattern that allows them also to make protein?

Carlsson: Most 5-hydroxytryptamine is made in the nerve terminals. Perhaps there is not so much protein synthesis there.

Wurtman: To what extent do other specific amino acids affect either the uptake or efflux of tryptophan?

Lajtha: Although the transport classes overlap, we can clearly differentiate the following classes of amino acids: (1) acidic (glutamic acid, aspartic acid); (2) large basic (lysine, arginine); (3) small basic (diaminobutyric acid, diamino-propionic acid); (4) small neutral (glycine, alanine, 3-amino-2-methylpropionic acid); (5) ω-amino (4-aminobutyric acid, β-alanine, 5-aminovaleric acid); (6) proline-type (hydroxyproline); and (7) large neutral (most other amino acids). Additional transport classes exist in brain, such as for diamines, like cadaverine, while compounds related to amino acids can interfere with transport; for example, dinitrophenol competes with tyrosine and phenylalanine (see p. 33). In the large neutral group, the introduction of a hydroxy group, sulphur atom or aromatic group had little apparent effect but a heterocyclic side chain did cause a difference in uptake. Therefore, at least one subclass exists for histidine and possibly tryptophan. In uptake, we observed principally competitive inhibition, while in efflux we found not only inhibition but also competitive stimulation; we noticed that the large neutral amino acids, leucine and phenylalanine, stimulated the exit of other amino acids.

Wurtman: How do the large neutral amino acids affect exchange?

Lajtha: We have demonstrated exchange in the living brain and in brain slices (Battistin *et al.* 1972; see p. 28). Although we did not study the specificity of exchange in detail, we found evidence that heteroexchange is not a 1:1 exchange: a flow of one mole of amino acid in one direction is accompanied by a counterflow of more than one (up to four) moles of another amino acid. Basic amino acids exchange most actively, but the general pattern of exchange indicated substrate specificity similar to that of uptake. Since leucine exchanged with valine and arginine but not with glycine, I expect that phenylalanine and tyrosine also exchange predominantly with large neutral and large basic amino acids. Therefore, when the concentration of phenylalanine or tryptophan in the plasma is raised, their influx might cause the efflux of leucine, valine, tyrosine and arginine in heteroexchange.

Kaufman: Oldendorf *et al.* (1971) elegantly took advantage of a natural phenomenon by showing the severe competition between the uptake of phenylalanine, present in high concentrations in the plasma, and of [^{75}Se]selenomethionine in the brain of a patient with phenylketonuria. Do phenylalanine and methionine share a common transport system? If so, is this consistent with the *in vitro* results?

Lajtha: Yes.

Grahame-Smith: We have been interested in the phenomenon of counter-transport in synaptosomes where it is easily demonstrable (see Parfitt & Grahame-Smith, this volume pp. 175–192). We were curious about the acute effect of an enormous intravenous dose of phenylalanine on concentrations of tryptophan in the brain. If the behaviour was similar to that in slices and in synaptosomes, within half an hour there should have been a big fall in the amount of brain tryptophan. There was hardly any change (Grahame-Smith, unpublished observations). So there is some *in vivo* protection against this particular phenomenon, and I do not think all is explained by what is found in slices and synaptosomes.

Wurtman: Some of my associates have recently obtained evidence that the brain regenerates methionine from homocysteine (Ordonez & Wurtman 1973). Moreover, this apparently is a major source of brain methionine: rats, in which cerebral S-adenosylmethionine has been acutely depleted by a single dose of L-dopa, do not exhibit a significant reduction in the amount of methionine in the brain. However, if the capacity of the brain to regenerate methionine from homocysteine is impaired by, for example, prior induction of folate deficiency, a single dose of L-dopa will profoundly depress brain methionine (Ordonez & Wurtman 1974).

Is there any evidence that hormones which influence amino acid uptake into other tissues—such as insulin, growth hormone, cortisol etc.—affect amino acid uptake in the brain?

Lajtha: Although the available evidence is somewhat controversial, most experiments show no such significant effect of the hormones tried so far.

Wurtman: So the main determinant for brain uptake of an amino acid is the concentration of it and its competing amino acids in the extracellular fluid.

Roberts: Although hormones may not generally affect amino acid uptake by brain directly, they may do so indirectly by affecting the plasma concentrations of amino acids. For example, we have observed that injected adrenocortical steroids increase the uptake of amino acids by the brain by elevating plasma concentrations (S. Roberts & B. S. Morelos, unpublished findings, 1968).

In addition to transport phenomena, the extent to which the brain uses an amino acid is important in determining the cerebral concentrations of amino acids. The formation of neurotransmitters constitutes only a small percentage of the total consumption of aromatic amino acids within the brain in most circumstances. I wonder whether the measurements of amino acid uptake or efflux *in vitro* aren't complicated by several factors, one of them being that degradation of protein contributes to the free amino acid pool within the slice. In addition, amino acids probably enter the brain *in vivo* primarily through astroglial cells, whereas *in vitro*, surfaces of neurons directly confront the incubation medium.

Lajtha: In other words, can we really derive information about *in vivo* control mechanisms from experiments on *in vitro* preparations? Of course, this question is an eternal one in biochemistry and physiology; it was asked about the elucidation of mitochondrial oxidative phosphorylation in isolated mitochondrial preparations and of mechanisms of protein synthesis in isolated ribosomal systems. Brain slices are useful; they give us information about the *in vivo* brain transport mechanism in that some properties such as substrate specificity of transport can be tested both *in vivo* and in slices and seem to be similar *in vivo* and *in vitro*. Other properties, such as energy and ion dependence, cannot be studied well *in vivo*, and therefore most experiments have to be performed *in vitro*. The qualitative aspects of specificity, energy and ion dependence, must be the same whether *in vivo* or *in vitro*, but preparation of the slices leads to important quantitative changes. The fact that the slices are altered, however, can be put to good purpose: we are studying the differences between slices and the living brain in the hope that the alteration of some control mechanism upon preparation of the slices will give us some information about the control mechanisms themselves.

One important difference in slices is pertinent to our discussion. During incubation, there is a net breakdown of protein. The composition of the free amino acid pool in the slice remains similar to that in the living brain. The breakdown results in the net production of free amino acids which leak out to the incubation medium.

Roberts: Wouldn't the increased efflux indirectly influence the uptake *per se*?

Lajtha: This would depend on the experimental conditions, but if the amino acid concentrations in the medium are high, the net production of small amounts of amino acids by the slice would not affect the uptake from the medium; if, however, only tracer amounts of labelled amino acids are added, their specific activity will be greatly decreased by the dilution of the endogenously produced amino acid.

Gessa: In explaining some *in vitro* and *in vivo* differences, we should remember the possible effect on the transport across the blood–brain barrier of enzymes, such as decarboxylase and monoamine oxidase, in the wall of the brain capillaries, which limit the penetration of monoamines and amino acids such as L-dopa and 5-hydroxytryptophan.

Lajtha: Yes, even if the brain were permeable to amines, they could not penetrate the organ because of the rapid oxidation by monoamine oxidase. I do not consider this as a permeability barrier. We must keep in mind that several mechanisms can prevent the entrance of a compound into the brain, for example rapid removal of the compound from the circulation by the liver or rapid metabolism of the compound outside the brain. Metabolic degradation does not seem to block the entry of amino acids since flux between plasma and brain is more rapid for amino acids than the rate of metabolism.

Wurtman: I suspect that exogenous L-dopa and 5-hydroxytryptophan, which are not normally found in circulation but which are unusually good substrates for aromatic L-amino acid decarboxylase, are much more affected by the enzyme in capillary walls than are phenylalanine, tyrosine and tryptophan.

Mandel: What is the ratio of the influx of amino acids to the spinal fluid to the efflux into plasma?

Lajtha: We have only indirect evidence about the route of amino acids into and out of the brain in normal conditions. The general consensus is that removal by the spinal fluid is not significant. The short half-life of the free amino acid pool in the brain and the differences in half-life among the various compounds are two pieces of evidence that militate against removal of several amino acids by the bulk flow process of the spinal fluid. As far as influx from the spinal fluid into the brain is concerned, the present evidence only shows transport in the other direction—from the choroid plexus to the blood—and uptake by spinal perfusion into brain tissue seems to be rather slow. Certainly, the same dose causes a greater increase in the brain after intraspinal administration than after intravenous administration, but this is not due to easier access from the spinal fluid but to the smaller volume and slower circulation of the spinal fluid.

Curzon: You mentioned Oldendorf's tracer studies on rats (Oldendorf 1971) (p. 27). Two points come out of this: first, he was able to show neatly that other amino acids (leucine, valine and histidine) interfere with the transport of tryptophan and, secondly, that the rate of transport of essential amino acids was much greater than that of non-essential ones. What is the mechanism by which this occurs?

Lajtha: The injection of tracer amounts of amino acids and measurement of uptake within a single circulation through the brain, although an elegant technique, is less physiological than it appears. Oldendorf found that when the injected material contained quantities comparable to those in the plasma, removal of the injected amino acids by the brain was 50–80 % less than with tracer amounts. Curiously, he found that only the non-essential amino acids are removed at high rate from the circulation. If one increases blood concentrations *in vivo*, in general a much greater increase occurs in the essential amino acids in the brain than in the non-essential ones. Perhaps the latter (already in high concentration in the brain) are under more careful control than the former.

Wurtman: By killing the animal after 15 s, Oldendorf is measuring tracer uptakes but not net flux. Data on isolated perfused brain preparations in which one could determine the steady-state losses of particular amino acids into brain would be more useful.

Weiner: Dr Lajtha, the finding of a relatively larger concentration of an amino acid in a certain particulate fraction (see Table 1) does not necessarily mean that the amino acid is actively accumulated there. This measurement reflects the total amount of substance and not necessarily the concentration of free amino acid or (in more thermodynamic terms) the activity of the material in the brain slice compared with that in the extracellular compartment. The accumulation could be due to simple diffusion into the slice and subsequent binding to all types of macromolecules, such as macromolecular components in the nucleus—that may be a non-specific interaction. What is the effect of different sodium concentrations in the extracellular fluid on tyrosine uptake in brain slices?

Lajtha: We found that brain slices have an absolute requirement of sodium for the accumulation of all amino acids.

Sourkes: We found that 40mM-sodium in the medium bathing slices of rat-brain cortex is sufficient to achieve the same uptake (i.e. transport ratio) during 15 min incubation under 100% oxygen as with 145mM-sodium. In these conditions the exclusion of sodium ions from the medium resulted in a 30% decrease of the transport ratio, but even without added sodium ions this ratio (concentration of tryptophan in cell water to that in medium) was well above the value expected from simple diffusion phenomena (Kiely & Sourkes 1972).

Lajtha: A concentration of 40 mM might be at the limit. We found that 20mM-sodium ions produce 30–40% inhibition of the uptake of most amino acids. In 10mM-sodium ions, uptake is further decreased.

Munro: Dr Lajtha, what evidence is there of compartmentation in free amino acid pools in the brain? Also, I wonder how good is the evidence that, for example, charging of tRNA is related to certain pools and not to others, if there is more than one pool.

Lajtha: The compartmentation of glutamic acid in the brain is well established (for a review see Balazs & Cremer 1973). Glutamate and perhaps enzymes metabolizing it are in several morphologically separate pools. Other amino acids have been studied much less, but it seems likely that further compartmentation will be found. We found compartmentation of glycine in hippurate formation (Garfinkel & Lajtha 1963).

Roberts: In addition to the evidence cited by Dr Lajtha, several other investigations strongly suggest the compartmentation of amino acids for protein synthesis. I shall discuss this point in some detail in my paper (pp. 299–318). Briefly, it is possible to show such compartmentation by administering a tracer amount of an amino acid either intracisternally or intravenously at the same time that an excess of the same or another amino acid is given intravenously. When the labelled amino acid is administered intravenously, its transport and entry into the brain cells are inhibited and we see what appears to be a concurrent inhibition of protein synthesis. In contrast, when the labelled amino acid is given intracisternally, the concurrent intravenous administration of another amino acid in excess does not seem to inhibit the incorporation of the label into brain protein. The most obvious explanation for these results is compartmentation of the amino acid precursor for protein synthesis.

Munro: Dr Lajtha, your nuclear and mitochondrial studies suggest independence in their concentrating capacity. This, in the case of the nucleus, is curious, because all the recent evidence on nuclei suggests that they are virtually unable to synthesize protein, so what is the biological advantage of having this particular compartmental concentration gradient?

Lajtha: I don't know.

Wurtman: For what does the nucleus use amino acids?

Lajtha: Most structures in the nervous system, including nuclei, are able to control their metabolite composition—their internal environment—possibly transport. The precise maintenance of the amino acid pool with concentration ratios very different in the free pool from that in a protein-bound pool and with high concentrations of the non-essential amino acids indicates many roles for the amino acids in addition to that for protein synthesis.

Gál: Some of the amino acids in the nuclei might contribute to the stabilization by 'charge transfer' to DNA, although we don't know how. We have evidence for this *in vitro*.

Munro: In the liver, RNA nucleotidyltransferase (EC 2.7.7.6) is apparently quickly stimulated by tryptophan, according to Henderson (1970).

Wurtman: If the route by which the amino acids normally enter or leave the brain is vascular and not via the cerebrospinal fluid (p. 46), how are we to interpret the physiological significance of studies in which those compounds are given intracisternally?

Roberts: The incorporation rate is much greater when the amino acid is given intracisternally than intravenously, but the dilution is also much less.

Lajtha: Can we distinguish between transport compartments and precursor compartments?

Roberts: No, I don't think we can.

Mandel: The amount of ATP in incubated slices is rather low, so the energy available is limited. If ATP is not the source of energy, what is the source?

Lajtha: The amount of ATP is not low in incubated slices, although in freshly prepared slices there is a lot of ADP and AMP; the nucleotides do not seem to leak out from slices but are rapidly converted into ATP upon incubation in glucose-containing medium. The calculated intracellular concentration of ATP in brain slices, after a 30 min recovery period, is about 2 mM, which is almost that present in the living brain. When the amount of ATP is lowered by metabolic inhibitors, one can observe high ATP with low uptake or low ATP with high uptake, which indicates sources of energy for uptake other than ATP. The present evidence is also against ion gradients as the primary driving force.

References

BALAZS, R. & CREMER, J. E. (1973) (eds.) *Metabolic Compartmentation in the Brain*, Macmillan, London

BATTISTIN, L., PICCOLI, F. & LAJTHA, A. (1972) Heteroexchange of amino acids in incubated slices of brain. *Arch. Biochem. Biophys.* **151**, 102–111

BAUMAN, A., BOURGOIN, S., BENDA, P., GLOWINSKI, J. & HAMON, M. (1974) *Brain Res.*, in press

GARFINKEL, D. & LAJTHA, A. (1963) A metabolic inhomogeneity of glycine *in vivo*. 1. Experimental determination. *J. Biol. Chem.* **238**, 2429–2434

HENDERSON, A. R. (1970) The effect of feeding with a tryptophan-free amino acid mixture on rat liver magnesium ion-activated deoxyribonucleic acid-dependent ribonucleic acid polymerase. *Biochem. J.* **120** 205–214

KIELY, M. & SOURKES, T. L. (1972) Transport of L-tryptophan into slices of rat cerebral cortex. *J. Neurochem.* **19**, 2863–2872

OLDENDORF, W. H. (1971) Brain uptake of radiolabeled amino acids, amines and hexoses after arterial injection. *Am. J. Physiol.* **221**, 1629–1639

OLDENDORF, W. H., SISSON, W. B. & SILVERSTEIN, A. (1971) Brain uptake of selenomethionine Se75: II. Reduced brain uptake of selenomethionine Se75 in phenylketonuria. *Arch. Neurol.* **24**, 524–528

ORDONEZ, L. A. & WURTMAN, R. J. (1973) Enzymes catalyzing the *de novo* synthesis of methyl groups in the brain and other tissues of the rat. *J. Neurochem.* **21**, 1447–1456

ORDONEZ, L. A. & WURTMAN, R. J. (1974) Folic acid deficiency and methyl group metabolism in rat brain: effects of L-dopa. *Arch. Biochem. Biophys.*, in press

PARFITT, A. & GRAHAME-SMITH, D. G. (1974) The transfer of tryptophan across the synaptosome membrane, *This Volume*, pp. 175–192

ROBERTS, S. (1974) Effects of amino acid imbalance on amino acid utilization, protein synthesis and polyribosome function in cerebral cortex, *This Volume*, pp. 299–318

WURTMAN, R. J., SHEIN, H. M., AXELROD, J. & LARIN, F. (1969) Incorporation of ^{14}C-tryptophan into ^{14}C-protein by cultured rat pineals: stimulation by L-norepinephrine. *Proc. Natl. Acad. Sci. U.S.A.* **62**, 749–755

Octopamine, phenylethanolamine, phenylethylamine and tryptamine in the brain†

JULIUS AXELROD and JUAN M. SAAVEDRA

Laboratory of Clinical Science, National Institute of Mental Health, Bethesda, Maryland

Abstract Sensitive enzymic assays have been developed for picogram quantities of octopamine, phenylethylamine, phenylethanolamine and tryptamine in tissues. With these assays we have found that these biogenic amines are normally present in the brain and are formed by the following pathways: tyrosine → tyramine → octopamine; phenylalanine → phenylethylamine → phenylethanolamine; tryptophan → tryptamine. The decarboxylation of the amino acids is catalysed by aromatic amino acid decarboxylase and β-hydroxylation by dopamine β-hydroxylase. Octopamine can also be formed by decarboxylation, dehydroxylation and β-hydroxylation of dopa. Inhibition of monoamine oxidase causes a rise in the concentrations of all the biogenic amines in the brain. Inhibition of tryptophan hydroxylase or phenylalanine hydroxylase by *p*-chlorophenylalanine results in an elevation of the concentrations of tryptamine, phenylethylamine and phenylethanolamine in the brain whereas inhibition of dopamine β-hydroxylase causes a fall in the cerebral amount of octopamine and phenylethanolamine. After the sympathetic nerves have been destroyed, the amount of octopamine and phenylethanolamine in tissue decreases, which suggests that these compounds are made and stored in adrenergic nerves.

The amino acids phenylalanine, tyrosine and tryptophan can be decarboxylated to form the corresponding biogenic amines (Lovenberg *et al.* 1962). However, the favoured path for the metabolism of these amino acids is hydroxylation. Thus, if any direct decarboxylation takes place, the resulting amines would be found in tissues only in exceedingly small concentrations. We have developed sensitive assays for octopamine* (Molinoff *et al.* 1969), phenylethanolamine* (Saavedra & Axelrod 1973*a*), phenylethylamine* (Saavedra & Axelrod, unpublished results) and tryptamine (Saavedra & Axelrod 1972*a*), the decarboxylated products of aromatic amino acids. These assays depend on the ability of trans-

---

† Read for Dr Axelrod in his absence by Dr Wurtman.
* Structures are shown in Figs. 1 and 3 (see also Editors' Note, p. IX).

methylating enzymes to transfer the radioactive methyl group of S-adenosyl-$[Me$-^{14}C]methionine to an amine acceptor group. Thus, phenylethanolamine N-methyltransferase can methylate β-hydroxyphenylethylamine derivatives, such as octopamine and phenylethanolamine, present in tissues (Axelrod 1962a). The [^{14}C]- or [^{3}H]-methylated metabolites of these β-hydroxylated amines are then separated from tissues and other interfering compounds by extraction into solvents of varying degrees of polarity. Amines such as phenylethylamine can be β-hydroxylated enzymically with dopamine hydroxylase and then methylated with phenylethanolamine N-methyltransferase. Tryptamine is assayed by methylation with non-specific N-methyltransferase from rabbit lung (Axelrod 1962b; Saavedra & Axelrod 1972a). Specificity is achieved by extraction of radioactive N-methyltryptamine from tissues by a non-polar organic solvent. With these enzyme assays, as little as 100 pg of octopamine and phenylethanolamine and 2–3 ng of phenylethylamine and tryptamine can be measured. The sensitivity of these assays has made it possible to examine the normal occurrence of these biogenic amines in brain and other tissues and to study their disposition after the administration of amino acid precursors, enzyme inhibitors and drugs.

OCTOPAMINE

Octopamine was first found in the posterior salivary gland of *Octopus vulgaris* (Erspamer & Boretti 1951). It is also present in nervous system of the octopus in relatively large amounts (0.5–5 µg/g) (Juorio & Molinoff 1971). Until recently this amine was detected in mammalian tissues only after its metabolism had been inhibited *in vivo* by monoamine oxidase (Armstrong *et al.* 1956; Kopin *et al.* 1964). By using specific enzyme assays (Molinoff *et al.* 1969), we found that octopamine occurred normally in mammalian brain (Table 1), heart (14–30 ng/g), salivary gland (50 ng/g), spleen (14 ng/g), vas deferens (30 ng/g) and pineal (500 ng/g) (Molinoff & Axelrod 1969, 1972). Large amounts of octopamine are normally present in lobster thoracic nerve cord (800 ng/g) and abdominal nerve cord (45 ng/g) (Molinoff & Axelrod 1972) and in the octopus nerve (10 µg/g) (Juorio & Molinoff 1971). The various regions of the rat brain where the highest concentrations were found are the cord, midbrain and hypothalamus (10 ng/g) and lowest concentrations in the cerebellum (4 ng/g) (Molinoff & Axelrod 1972). Subcellular studies indicated that octopamine, like noradrenaline, is localized in the sympathetic nerve endings. Destruction of sympathetic nerves either surgically or chemically with 6-hydroxydopamine results in a marked fall in the amount of octopamine in the tissue (Molinoff & Axelrod 1969, 1972).

We can measure the turnover of octopamine in tissues by following the dis-

TABLE 1

Minor biogenic amines in the brain of the rat

Biogenic amine	Concentration (ng/g) (±s.e.m.)
Octopamine	4.7 ± 0.9
Phenylethanolamine	6.2 ± 0.2
Phenylethylamine	2 ± 1
Tryptamine	23 ± 2

appearance of this biogenic amine after inhibition of dopamine β-hydroxylase, the enzyme that converts tyramine into octopamine (Molinoff & Axelrod 1972). Using such inhibitors, we found that the turnover of octopamine in the rat heart was six times as rapid as that of noradrenaline. The half-life of octopamine in the rat brain was twice that in the heart. The rapid turnover of octopamine is supported by the high rate of excretion of its deaminated metabolite, p-hydroxy-mandelic acid (Armstrong et al. 1956).

Biosynthesis and metabolism

The administration of an inhibitor of monoamine oxidase causes a fivefold elevation of octopamine in the brain and heart (Table 2); this indicates that the main pathway for its metabolism is deamination by monoamine oxidase. The

TABLE 2

Effect of amino acids and metabolic inhibitors on octopamine in the brain of the rat (results quoted with s.e.m.)

Treatment	Brain octopamine (ng/g)	
		With MAOI
None	6.0 ±1.5	28 ± 24
L-Phe	6.6 ±1.5	60 ± 6[b]
L-Tyr	1.6 ± 0.5[b]	37 ± 1.3[b]
L-Dopa	< 1.5[b]	78 ± 5[b]
Tyramine	8.5 ± 0.6[a]	260 ± 19[b]
Disulfiram	1.8 ± 0.2[b]	

L-Phenylalanine (800 mg/kg i.p.), L-tyrosine (500 mg/kg), L-dopa (500 mg/kg i.p.) and tyramine (500 mg/kg i.p.) were given 3 h before the rats were killed. The monoamine oxidase inhibitor (MAOI; pheniprazine, 10 mg/kg i.p.) was given 24 and 3 h before rats were killed. Disulfiram, a dopamine hydroxylase inhibitor (400 mg/kg i.p). was given 13 h before rats were killed (Brandau & Axelrod 1972).

[a] $P < 0.05$, compared with control value.

[b] $P < 0.01$, compared with control rats treated with an inhibitor of monoamine oxidase.

administration of phenylalanine, tyrosine or tyramine together with an inhibitor of monoamine oxidase results in a marked elevation of the concentration of octopamine in the brain compared to the action of the inhibitor alone (Table 2) (Brandau & Axelrod 1972). When dopamine hydroxylase is inhibited, the amounts of octopamine in tissues fall. These results indicate that octopamine is formed and metabolized as shown in Fig. 1. Administration of L-dopa with a monoamine oxidase inhibitor also raises the concentration of octopamine in the brain in normal and germ-free animals (Table 2) (Brandau & Axelrod 1972). This suggests an alternative pathway for the formation of octopamine from catecholamines. A minor pathway for metabolism of octopamine *in vivo* (Creveling *et al.* 1962) and *in vitro* (Axelrod 1963) is hydroxylation to noradrenaline.

The physiological role of octopamine in mammals is still obscure. Previously, Kopin *et al.* (1964) postulated that octopamine served as a false neurotransmitter since it accumulated in sympathetic nerves after inhibition of monoamine oxidase. Exogenously administered [^{3}H]octopamine is taken up in sympathetic neurons of an isolated, perfused cat spleen (Kopin *et al.* 1964) and released

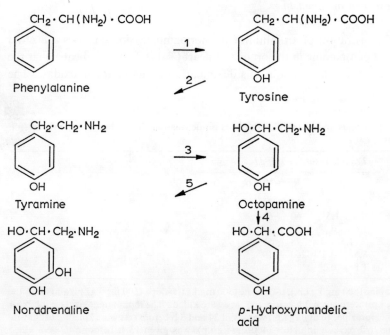

FIG. 1. Formation and metabolism of octopamine: 1, phenylalanine 4-hydroxylase (EC 1.99.1.2); 2, aromatic amino acid decarboxylase (EC 4.1.1.26); 3, dopamine hydroxylase (EC 1.14.2.1); 4, monoamine oxidase (EC 1.4.3.4); 5, mixed-function oxidase.

upon stimulation of the adrenergic nerves. Normally occurring octopamine can be released with noradrenaline when the cat spleen is stimulated (P. B. Molinoff & I. J. Kopin, unpublished results). Thus, octopamine is synthesized in, stored in and released from nerves and thus can be considered as a putative neurotransmitter. It might modulate the actions of its co-transmitter, noradrenaline, or it might conceivably have actions of its own, especially as octopamine has powerful stimulatory effects on insect nerve cord phosphorylase (Robertson & Steele 1972) and on the light organ of fireflies (Carlsson 1968). The considerable presence of octopamine in the nerves of the octopus (Juorio & Molinoff 1971) and lobster (Molinoff & Axelrod 1972) indicates that this biogenic amine may be a primary transmitter at some synapses in non-mammalian species.

PHENYLETHANOLAMINE

Phenylethanolamine is widely distributed in rat tissue, the highest concentrations being in the pineal (300 ng/g) and lowest in the brain (6 ng/g) (Table 1) (Saavedra & Axelrod 1973a). It is also present in heart, spleen, salivary gland and vas deferens (50 ng/g) and lung (90 ng/g). The distribution of phenylethanolamine in the brain of the rat is unequal, the highest concentration being in the hypothalamus and midbrain (25 ng/g) and the lowest in the cerebral cortex and cerebellum (2.5 ng/g).

The administration of phenylethylamine roughly trebles the amount of phenylethanolamine in the brain (Fig. 2) (Saavedra & Axelrod 1973a), but when phenylethylamine is given together with a monoamine oxidase inhibitor, the increase is almost twentyfold. Monoamine oxidase inhibitors alone cause about a tenfold elevation of phenylethanolamine in the brain. After the administration of an inhibitor of dopamine hydroxylase, sodium diethyldithiocarbamate, together with phenylethylamine and iproniazid (a monoamine oxidase inhibitor), the formation of phenylethanolamine in the brain is considerably reduced in comparison with the result of giving the precursor and monoamine oxidase inhibitor alone (Fig. 2). The administration of phenylalanine, the presumed amino acid precursor of phenylethanolamine, results in a negligible increase in the amount of the biogenic amine in the brain. However, when a phenylalanine hydroxylase inhibitor, p-chlorophenylalanine, is administered with phenylalanine, phenylethanolamine concentrations in the brain are roughly quadrupled (Fig. 2). These observations indicate that the formation and metabolism of phenylethanolamine is as shown in Fig. 3.

To determine whether phenylethanolamine is made and stored in sympathetic nerves, these nerves to the rat salivary gland were destroyed unilaterally by

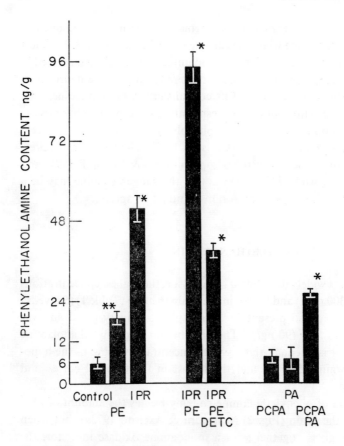

FIG. 2. Effect of enzyme inhibitors and precursors on phenylethanolamine in the brain of the rat. Compounds were dissolved in water and administered intraperitoneally as follows: 2-phenylethylamine (PE), 50 mg/kg, 90 min; iproniazid (IPR), 150 mg/kg, 18 h; sodium diethyl-dithiocarbamate (DETC), 500 mg/kg, 30 min; p-chlorophenylalanine methyl ester (PCPA), 375 mg/kg, 48 and 24 h; L-phenylalanine methyl ester hydrochloride (PA), 800 mg/kg, 90 min before rats were killed. * Statistically significant, $P < 0.01$. ** Statistically significant, $P < 0.05$.

CH$_2$·CH$_2$·NH$_2$ HO·CH·CH$_2$·NH$_2$ HO·CH·COOH

Phe $\xrightarrow{1}$ $\xrightarrow{2}$ $\xrightarrow{3}$

Phenylethylamine   Phenylethanolamine   Mandelic acid

FIG. 3. Formation and metabolism of phenylethanolamine: 1, aromatic amino acid de-carboxylase; 2, dopamine hydroxylase; 3, monoamine oxidase.

removal of a superior cervical ganglion. This resulted in a 40% fall in phenylethanolamine content in the denervated salivary gland (Saavedra & Axelrod
1973a). We concluded that a fraction of phenylethanolamine is stored in sympathetic nerves while some is present outside these nerves. The uptake of exogenous phenylethanolamine is five times greater on the innervated side of the
pineal gland than on the denervated side; this suggests that the amine is selectively taken up by sympathetic nerves.

PHENYLETHYLAMINE

An enzymic assay for phenylethylamine was recently developed and we are
currently studying the distribution, formation and metabolism of the amine
(Saavedra & Axelrod, unpublished results). We found that phenylethylamine is
normally present in the brain in small amounts (about 2 ng/g), however, the
administration of a monoamine oxidase inhibitor raised the level of this biogenic
amine about 20 times. The presumed precursor of phenylethylamine, phenylalanine, did not change the concentration of the amine in the brain. Administration of phenylalanine with a monoamine oxidase inhibitor raised this concentration about 200-fold to 700 ng/g. When rats are pretreated with a decarboxylase inhibitor, this elevation is almost completely blocked. Inhibition of phenylalanine hydroxylase with p-chlorophenylalanine resulted in considerable (30-
fold) rise in brain phenylethylamine.

TRYPTAMINE

Though tryptamine has been found in urine (Rodnight 1956), its normal occurrence in the tissues could not be demonstrated until recently. The development of
a specific and sensitive enzymic assay for tryptamine made it possible to examine
the tissue distribution of this amine and to study changes in its concentration
after the administration of drugs and amino acid precursors (Saavedra & Axelrod
1972a). Tryptamine is present in both rat (Table 3) and human brain (40 ng/g),
and in rat lung (50 ng/g), heart (30 ng/g), duodenum (40 ng/g), liver (60 ng/g),
salivary gland (60 ng/g) and kidney (30 ng/g) (Saavedra & Axelrod 1973b).
    The administration of tryptophan to rats causes a small but significant eleva-
tion of tryptamine in the brain (Table 3) (Saavedra & Axelrod 1973b). A large
(fivefold) increase in brain tryptamine is observed after giving tryptophan together with a monoamine oxidase inhibitor. Inhibition of aromatic amino acid
decarboxylase lowers the amount of tryptamine while inhibition of the decarboxylase peripherally, but not in the brain (thus permitting more tryptophan to

TABLE 3

Effect of amino acids and inhibitors on tryptamine in the brain of the rat (results quoted with
S.E.M.)

Treatment	Brain tryptamine (ng/g)	
		with MAOI
None	23 ± 2	35 ± 5
L-Trp	34 ± 8^a	111 ± 18^b
p-Chlorophenylalanine	38 ± 4^a	84 ± 11^b
p-Chlorophenylalanine + L-Trp	47 ± 7^a	156 ± 33^b
L-Trp + NSD-1055	——	10 ± 2^b
NSD-1055	13 ± 15^a	——
MK-486	46 ± 5^a	——

Iproniazid, a monoamine oxidase inhibitor (150 mg/kg), L-tryptophan (400 mg/kg), MK-486
(100 mg/kg) and NSD-1055 (200 mg/kg) were injected intraperitoneally 16 h, 1 h and 1.5 h,
respectively, before the rats were killed. p-Chlorophenylalanine (300 mg/kg) was given
intraperitoneally 48 and 24 h before killing. MK-486, L-3-(3,4-dihydroxyphenyl)-2-hydrazino-
2-methylpropionic acid, is a peripheral decarboxylase inhibitor. NSD-1055, 5-aminooxy-
methyl-2-bromophenol, inhibits decarboxylase activity peripherally and centrally (Saavedra &
Axelrod 1973b).
^{a} $P < 0.5$, compared with untreated controls.
^{b} $P < 0.001$, compared with controls treated with a monoamine oxidase inhibitor.

enter the brain), results in a doubling of the quantity of endogenous brain
tryptamine. p-Chlorophenylalanine, which inhibits tryptophan hydroxylase also,
elevates the brain concentration of tryptamine by about 50% but when given
together with tryptophan and a monoamine oxidase inhibitor, brain tryptamine
concentrations are increased sevenfold.

We have found an enzyme in rabbit lung that can N-methylate both tryptamine and N-methyltryptamine to form NN-dimethyltryptamine, a psychotomimetic compound. We have recently uncovered a similar enzyme in the brain
of rats and humans (Saavedra & Axelrod 1972b). The intracisternal injection of
tryptamine in the brain of a rat previously treated with monoamine oxidase inhibitors results in the formation of N-methyl- and NN-dimethyl-tryptamine.

From the foregoing results, we can describe the formation and metabolism of
tryptamine as follows: tryptophan → tryptamine → indoleacetic acid. A minor
pathway is the N-methylation of tryptamine.

SUMMARY

The development of sensitive enzymic assays has made it possible to establish
the normal occurrence of octopamine, phenylethanolamine, phenylethylamine

and tryptamine in the brain of the rat. The administration of precursors of amino acids together with inhibitors of monoamine oxidase results in a marked elevation of these amines in the brain.

References

ARMSTRONG, M. D., SHAW, K. N. F. & WALL, P. E. (1956) Phenolic acids of human urine. *J. Biol. Chem.* **218**, 293–303

AXELROD, J. (1962*a*) Purification and properties of phenylethanolamine *N*-methyltransferase. *J. Biol. Chem.* **237**, 1657–1660

AXELROD, J. (1962*b*) The enzymatic *N*-methylation of serotonin and other amines. *J. Pharmacol. Exp. Ther.* **132**, 28–33

AXELROD, J. (1963) Enzymatic formation of adrenaline and other catechols from monophenols. *Science (Wash. D.C.)* **140**, 499–500

BRANDAU, K. & AXELROD, J. (1972) The biosynthesis of octopamine. *Naunyn-Schmiedeberg's Arch. Pharmakol.* **273**, 123–133

CARLSSON, A. D. (1968) Effect of drugs on luminescence in larval fireflies. *J. Exp. Biol.* **49**, 195–199

CREVELING, C. R., LEVITT, M. & UDENFRIEND, S. (1962) An alternative route for the biosynthesis of norepinephrine. *Life Sci.* **10**, 523–526

ERSPAMER, V. & BORETTI, G. (1951) Identification and characterisation by paper chromatography of enteramine, octopamine, tyramine, histamine and allied substances in extracts of posterior salivary glands of octopoda and in other tissue extracts of substrates. *Arch. Int. Pharmacodyn. Ther.* **88**, 296–332

JUORIO, A. V. & MOLINOFF, P. B. (1971) Distribution of octopamine in nervous tissues of *Octopus vulgaris*. *Br. J. Pharmacol.* **43**, 438–439P

KOPIN, I. J., FISCHER, J. E., MUSACCHIO, J. & HORST, W. D. (1964) Evidence for a false neurochemical transmitter as a mechanism for the hypotensive effect of monoamine oxidase inhibitors. *Proc. Natl. Acad. Sci. U.S.A.* **52**, 716–721

LOVENBERG, W., WEISSBACH, H. & UDENFRIEND, S. (1962) Aromatic L-amino acid decarboxylase. *J. Biol. Chem.* **237**, 89–93

MOLINOFF, P. B. & AXELROD, J. (1969) Octopamine: normal occurrence in sympathetic nerves of rats. *Science (Wash. D.C.)* **164**, 428–429

MOLINOFF, P. B. & AXELROD, J. (1972) Distribution and turnover of octopamine in tissues. *J. Neurochem.* **19**, 157–163

MOLINOFF, P. B., LANDSBERG, L. & AXELROD, J. (1969) An enzymatic assay for octopamine and other β-hydroxylated phenylethylamines. *J. Pharmacol. Exp. Ther.* **179**, 253–261

ROBERTSON, H. A. & STEELE, J. E. (1972) Activation of insect nerve cord phosphorylase by octopamine and adenosine 3′,5′-monophosphate. *J. Neurochem.* **19**, 1603–1606

RODNIGHT, A. (1956) Separation and characterization of urinary indoles resembling 5-hydroxytryptamine and tryptamine. *Biochem. J.* **64**, 621–626

SAAVEDRA, J. M. & AXELROD, J. (1972*a*) A specific and sensitive assay for tryptamine in tissues. *J. Pharmacol. Exp. Ther.* **182**, 363–369

SAAVEDRA, J. M. & AXELROD, J. (1972*b*) Psychotomimetic *N*-methylated tryptamines: formation in brain *in vivo* and *in vitro*. *Science (Wash. D.C.)* **172**, 1365–1366

SAAVEDRA, J. M. & AXELROD, J. (1973*a*) The demonstration and distribution of phenylethanolamine in the brain and other tissues. *Proc. Natl. Acad. Sci. U.S.A.* **70**, 769–772

SAAVEDRA, J. M. & AXELROD, J. (1973*b*) Effect of drugs on the tryptamine content of rat tissues. *J. Pharmacol. Exp. Ther.* **185**, 523–529

Discussion

Sandler: Oxidative deamination of octopamine gives an aldehyde which may be either oxidized further to *p*-hydroxymandelic acid or reduced to an alcohol, *p*-hydroxyphenylethane-1,2-diol (*p*-hydroxyphenylglycol) (see Fig. 1). Since noradrenaline is predominantly metabolized in the rat by reduction to an alcohol, wouldn't you expect octopamine also to be metabolized preferentially to its alcohol rather than to the acid?

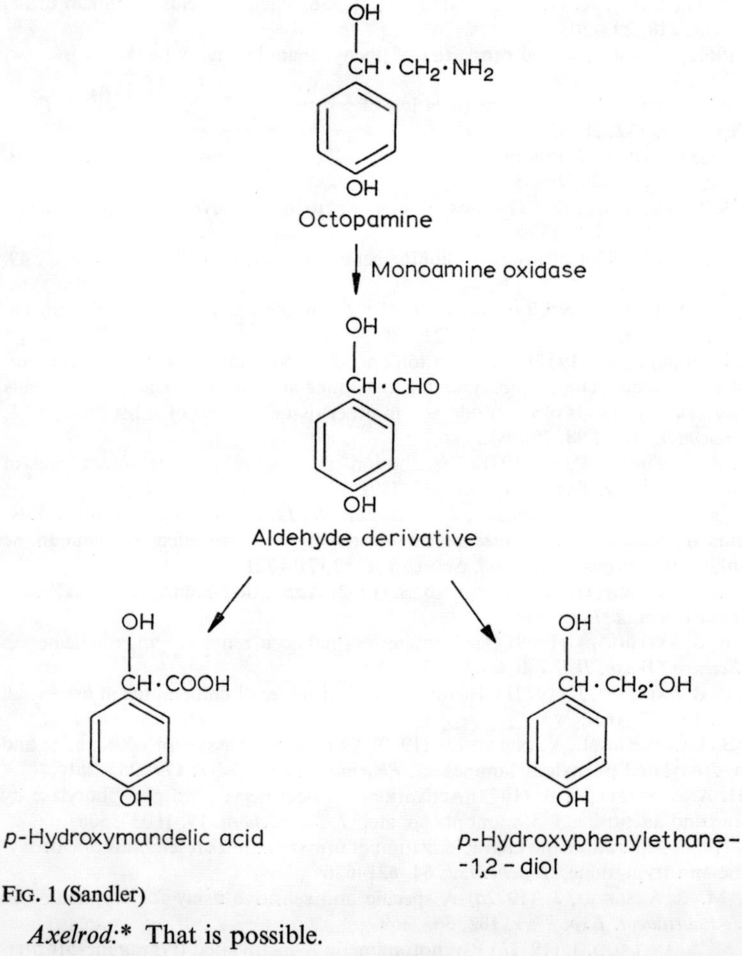

FIG. 1 (Sandler)

*Axelrod:** That is possible.

Fernstrom: Isn't it rather strange that phenylalanine increases the amount of phenylethanolamine but tyrosine depresses the amount of octopamine in the brain?

* These comments were added as the result of postal exchanges.

Axelrod: Administration of phenylalanine does not increase the concentrations of phenylethanolamine. When administered alone, tyrosine decreases the amount of octopamine (Brandau & Axelrod 1972) probably because an increase in the concentrations of catecholamines triggers a release of octopamine, which is then rapidly metabolized by monoamine oxidase. Octopamine concentrations rose after the simultaneous administration of tyrosine and a monoamine oxidase inhibitor.

Fernstrom: You administered *p*-chlorophenylalanine with tryptophan and phenylalanine to inhibit the hydroxylation but you didn't try to inhibit the conversion of tyrosine into octopamine. Have you used α-methyltyrosine?

Axelrod: Yes. This compound reduced the tissue concentrations of octopamine, possibly by direct action, causing the release of octopamine, or by a secondary increase of activity of the sympathetic nervous system.

Wurtman: The dose of tryptophan (400 mg/kg) you use is massive: it is about 40 times that necessary to induce a major rise in brain 5-hydroxytryptamine.

Axelrod: Lower tryptophan doses (100 mg/kg) also raise the amount of tryptamine (Saavedra & Axelrod 1973). Ichiyama *et al.* (1970) found that mammalian brain supernatant fractions were able to decarboxylate L-tryptophan to tryptamine in significant amounts at high L-tryptophan concentrations, whereas at low concentrations of L-tryptophan 5-hydroxytryptamine alone was the end product. This could be explained by the different affinity of aromatic L-amino acid decarboxylase toward L-tryptophan and 5-hydroxytryptophan (K_m values of 14 mM and 5.4 μM, respectively) (Ichiyama *et al.* 1970).

Sandler: You mentioned that the highest concentration of phenylethanolamine in the brain was in the pineal. Can the approach you used measure the amount of 2-phenylethylamine there as well, or is it too insensitive?

Axelrod: We have recently improved the sensitivity of the method for determining 2-phenylethylamine, and we hope to be able to demonstrate the amine's presence in the pineal gland although its endogenous levels might be too low to be detected. We may have to resort to blockade of the specific monoamine oxidase in the pineal.

Sandler: The possibility of high concentrations of phenylethylamine in the pineal is intriguing. Yang *et al.* (1972) have noted high activity in this site of their monoamine oxidase fraction B which appears to be highly specific for phenylethylamine.

Sourkes: Dr A. Boulton has been stressing that some of these β-hydroxylated compounds are formed by dehydroxylation of the corresponding catecholamines (personal communication).

Wurtman: One problem to explain is how administration of dopa elevates the concentration of octopamine. Is there an enzyme that will catalyse the removal of a *m*-hydroxy group from dopa or dopamine?

Kaufman: Dopa is a potent inhibitor of tyrosine hydroxylase. An alternative mechanism for this elevation is that dopa might raise the tissue concentration of tyrosine and phenylalanine and thus allow octopamine to be formed. Dehydroxylation need not be invoked.

Wurtman: But the acute administration of dopa does not seem to have any effect on the amount of tyrosine in the brain (Weiss *et al.* 1971).

Axelrod: After inhibition of the monoamine oxidase, administration of L-dopa does increase octopamine formation. The dehydroxylation of L-dopa to 3-hydroxy- and 4-hydroxy-phenylethylamine (tyramine) has recently been reported (Sandler *et al.* 1971; Boulton & Quan 1970; Brandau & Axelrod 1972).

Sourkes: A controversy over tryptamine has been running between Drs Axelrod and Costa on the one hand and Dr Boulton on the other. Dr Boulton has been finding higher values for this monoamine in the brain than the other investigators (Wu & Boulton 1973). [*Note added in proof:* Dr Boulton has since reported that his values for tryptamine are now much more in agreement with those of Axelrod (June 1973, unpublished results).]

Maître: When you stimulated the sympathetic nerve, could you measure the release of phenylethylamine and phenylethanolamine as well as that of octopamine?

Axelrod: At that time we had no sensitive methods to measure these compounds. We are now in a position to try that.

Glowinski: [^{3}H]Tryptamine is formed from [^{3}H]tryptophan in some structures in the rat on incubation of the slices with the labelled amino acid. [^{3}H]-Tryptamine accumulated only in presence of a monoamine oxidase inhibitor, particularly in the striatum (Hamon *et al.* 1973*a*). In the absence of such an inhibitor, small quantities of the [^{3}H]indole acid, the [^{3}H]tryptamine metabolite, were found in the striatum (Hamon *et al.* 1973*a*) and to a lesser extent in the brain stem and the hypothalamus (Hery *et al.* 1972) but not in the hippocampus (Hamon *et al.* 1973*b*).

Axelrod: We found that endogenous levels of tryptamine are higher in hypothalamus, spinal cord and striatum of the rat brain (Saavedra & Axelrod 1973); these results have recently been confirmed by Snodgrass & Iversen (1973).

Fernstrom: The measurement of 20 pg of 5-hydroxytryptamine is impressive. How does the assay work?

Axelrod: It is based on a two-step enzymic conversion of the amine, first into *N*-acetyl-5-hydroxytryptamine by the rat liver *N*-acetyltransferase and then into *N*-acetyl-5-methoxytryptamine (melatonin) with *S*-adenosyl[*Me*-^{3}H]methionine as methyl donor and the hydroxyindole *O*-methyltransferase from bovine pineal gland. The radioactive melatonin is easily separated from the S-adenosyl-[Me-^{3}H]methionine by solvent extraction. As little as 25 pg of the amine can be detected.

Curzon: How does p-chlorophenylalanine increase the amount of tryptamine formed? Such an increase suggests formation in 5-hydroxytryptaminergic neurons, but we find raphe lesions only slightly decrease rat-brain tryptamine (Knott *et al.* 1974), indicating that the amine is mostly formed elsewhere. Similarly, why does brain tryptamine increase so little with increasing brain tryptophan? If it is assumed that endogenous brain tryptophan concentration is well below the K_m of the decarboxylase then proportionality might be expected.

Wurtman: When tryptophan is present in large concentrations and cannot be metabolized elsewhere, endothelial decarboxylases might be sufficient to generate some tryptamine.

Axelrod: It is possible that at least part of the endogenous brain tryptamine is formed in non-5-hydroxytryptaminergic neurons, but we do not have any data to confirm this hypothesis. Reserpine treatment does not deplete tryptamine in brain tissue, and this may indicate the existence of different storage mechanisms for tryptamine and 5-hydroxytryptamine.

Wurtman: Significant amounts of L-dopa are formed within the gut from dietary cereals by a tyrosine-hydroxylating enzyme present in cereals that differs from mammalian tyrosine hydroxylase (Hoeldtke *et al.* 1972). After a rat eats Purina chow (or other standard lab chows), dopa continues to be formed for considerable periods of time while the food is working its way through the gut (Hoeldtke & Wurtman 1973a). The pH optimum for the enzyme is low, and so the enzyme is relatively unaffected by gastric hydrochloric acid. We further measured the amounts of conjugated dopamine and dihydroxyphenyl-acetic acid, and free noradrenaline in the urine of MIT students on cereal-free or control diets for two days; here also there were major reductions from normal values (Hoeldtke & Wurtman 1973b). Dietary dopa might thus be a major source of urinary and even tissue monoamines. The dangerous effects of tyramine-containing foods, such as blue cheese and beer, on people taking mono-amine oxidase inhibitors are well known. The tyramine in these and other foods might be responsible for the tyramine and octopamine found in certain tissues. In any study on the sources of tissue octopamine, tyramine, and related mono-amines, it is vital that potentially-contributing nutritional sources be controlled. Any food containing tyrosine and the necessary plant enzymes could generate virtually all these monoamines.

Gál: If tryptophan was decarboxylated to tryptamine elsewhere, we would find a much higher concentration of tryptamine in the brain than we do. From studies with p-chlorophenylalanine, I conclude that in the absence of 5-hydroxytryptophan more tryptophan is available for the decarboxylase, so that more tryptamine could be formed.

Wurtman: The K_m for dopa decarboxylase is much more favourable for decarboxylation of the hydroxylated products, dopa and 5-hydroxytryptophan, than it is for decarboxylation of tyrosine and tryptophan. Dopa and 5-hydroxytryptophan are barely detectable in tissues of untreated animals. Therefore, even though the K_m values of dopa decarboxylase for these substrates are low, the enzyme is almost certainly largely unsaturated *in vivo*; and the decarboxylations might be first-order. If the K_m is thus not the determining factor, why is 5-hydroxytryptophan decarboxylated in preference to tryptophan? Perhaps this has to do with the intracellular locations of tryptophan hydroxylase and dopa decarboxylase.

Weiner: Another point is that the concentrations of these amines in tissues are more a function of storage mechanisms than of rates of synthesis. As Dr Axelrod mentioned, the turnover of these amines is high, and measurements of tissue concentration provide no reliable indication of the rate of synthesis. Furthermore, the preferential storage of an amine in one region of the brain does not necessarily mean that it is synthesized there, as Dr Wurtman mentioned.

Gál: Are you suggesting that *p*-chlorophenylalanine will increase the ability of the brain to store tryptamine? That could be tested.

Weiner: No, I am saying that *p*-chlorophenylalanine could favour tryptamine synthesis by inhibition of tryptophan hydroxylase. This synthesis may not necessarily occur in the regions where the tryptamine ultimately accumulates.

Moir: We have some evidence of an interaction between the tryptophan and dopa pathways. One possibility was that this interaction was through tryptamine. Unfortunately we did not have an assay sensitive enough for tryptamine. Is there any evidence to show that perhaps the caudate nucleus is one of the regions in which tryptamine is made preferentially?

Glowinski: During our studies on the formation of tryptamine in discrete areas of the rat brain, we discovered that the striatum was one structure in which higher concentrations of [^{3}H]tryptamine or of 5-hydroxy[^{3}H]indoleacetic acid were found when slices were incubated with or without a monoamine oxidase inhibitor respectively. Since these were studies on slices *in vitro*, there is no doubt that tryptamine is synthesized locally.

Wurtman: Dr Axelrod, what is the source of these monoamines in the brain? Are they synthesized in brain or derived from diet?

Axelrod: In germ-free rats, the cerebral concentration of octopamine is the same as in normal matched controls (Brandau & Axelrod 1972). The same is true for tryptamine (Saavedra & Axelrod, unpublished observations). These results argue in favour of octopamine being formed in brain tissue. Tryptamine could also be formed in the periphery because it can cross the blood–brain barrier, but then we have to postulate specific uptake or storage mechanisms to explain the uneven distribution of this amine in the brain.

Wurtman: Is it possible that these compounds have physiological functions, either as false transmitters thus modifying the storage of true transmitters or because they act postsynaptically?

Axelrod: The possible physiological functions of these compounds are presently unknown. They can, of course, act as secondary transmitters or false neurotransmitters, or by modulating the neuronal function by other mechanisms. We do not yet have any evidence to show that they can also be released in the central nervous system and then act postsynaptically.

Wurtman: How does the administration of dopa affect tissue octopamine concentrations?

Axelrod: Injection of large amounts of L-dopa can lead to a decrease in the amounts of octopamine by increasing the catecholamine concentrations and provoking octopamine release. The increase in octopamine concentration after administration of monoamine oxidase and L-dopa indicates the possibility of a secondary pathway from dopamine to tyramine by ring dehydroxylation (Brandau & Axelrod 1972).

Wurtman: Are the rates of synthesis of these compounds (if they are formed *in vivo*) sufficient to alter the quantities of neuronal tyrosine, tryptophan or phenylalanine which are available for protein synthesis or for *bona fide* neurotransmitter synthesis?

Axelrod: This is possible, although unlikely, owing to the low concentrations of these compounds. Only a very small fraction of the amino acids are used to form biogenic amines.

Weiner: Studies with aromatic amino acid decarboxylases from various tissues are consistent with Dr Axelrod's observations regarding the presence of relatively high concentrations of tryptamine and smaller amounts of phenylethylamine derivatives in tissues. The coupled decarboxylase assay for tyrosine hydroxylase works well in various tissues, with high substrate concentrations, because there is little direct decarboxylation of tyrosine to interfere with the assay. In contrast, Hayaishi and his colleagues (Ichiyama *et al.* 1970) have reported (and we have confirmed [unpublished observations]) that tryptophan is more readily directly decarboxylated, particularly with high tryptophan concentrations. Thus, the coupled decarboxylase assay for tryptophan hydroxylase is much more difficult to perform because the direct decarboxylation of tryptophan occurs at a much higher rate than that of tyrosine and, in order to apply the assay successfully to tissues, the tryptophan concentration must be very low, well below its K_m for the enzyme.

References

BOULTON, A. A. & QUAN, L. (1970) Formation of p-tyramine from DOPA and dopamine in rat brain. *Can. J. Biochem.* **48**, 1287–1291

BRANDAU, K. & AXELROD, J. (1972) The biosynthesis of octopamine. *Naunyn-Schmiedeberg's Arch. Pharmakol.* **273**, 123–133

HAMON, M., BOURGOIN, S. & GLOWINSKI, J. (1973a) Feed-back regulation of serotonin synthesis in rat striatal slices. *J. Neurochem.* **20**, 1727–1746

HAMON, M., BOURGOIN, S., JAGGER, J. & GLOWINSKI, J. (1973b) Effect of LSD on synthesis and release of 5-HT in rat brain slices. *Brain Res.*, in press

HERY, F., ROUER, R. & GLOWINSKI, J. (1972) Daily variations of serotonin metabolism in the rat brain. *Brain Res.* **43**, 445–465

HOELDTKE, R. & WURTMAN, R. J. (1973a) Synthesis of dopa in rat stomach following ingestion of cereals. *Metabolism*, in press

HOELDTKE, R. D. & WURTMAN, R. J. (1973b) Cereal ingestion and catecholamine excretion. *Metabolism*, in press

HOELDTKE, R., BALIGA, B. S., ISSENBERG, P. & WURTMAN, R. J. (1972) Dihydroxyphenyl alanine in rat food containing wheat and oats. *Science (Wash. D.C.)* **175**, 761–762

ICHIYAMA, A., NAKAMURA, S., NISHIZUKA, Y. & HAYAISHI, O. (1970) Enzymic studies on the biosynthesis of serotonin in mammalian brain. *J. Biol. Chem.* **245**, 1699–1709

KNOTT, P. J., MARSDEN, C. A. & CURZON, G. (1974) Comparative studies of brain 5-hydroxy-tryptamine and tryptamine. *Adv. Biochem. Psychopharmacol.*, in press

SAAVEDRA, J. M. & AXELROD, J. (1973) Brain tryptamine and the effects of drugs in *Proceedings of the International Symposium on 5-Hydroxytryptamine and other Indolealkylamines in Brain* (Forte Village, St. Margherita [Cagliari], Italy, May 27 – June 2), Raven Press, New York, in press

SANDLER, M., GOODWIN, B. L., RUTHVEN, C. R. J. & CALNE, D. B. (1971) Therapeutic implications in Parkinsonism of m-tyramine from L-dopa in man. *Nature (Lond.)* **229**, 414–416

SNODGRASS, S. R. & IVERSEN, L. L. (1973) Occurrence and release of 5 HT and tryptamine from rat spinal cord slices in *Proceedings of the International Symposium on 5-Hydroxytryptamine and other Indolealkylamines in Brain* (Forte Village, St. Margherita [Cagliari]. Italy, May 27 – June 2), Raven Press, New York, in press

WEISS, B. F., MUNRO, H. N. & WURTMAN, R. J. (1971) L-Dopa: disaggregation of brain polysomes and elevation of brain tryptophan. *Science (Wash. D.C.)* **173**, 833–835

WU, P. H. & BOULTON, A. A. (1973) Metabolism, distribution and turnover of tryptamine in rat brain. *Can. J. Biochem.*, in press

YANG, H.-Y. T., GORIDIS, C. & NEFF, N. H. (1972) Properties of monoamine oxidases in sympathetic nerve and pineal gland. *J. Neurochem.* **19**, 1241–1250

Tyrosine aminotransferase in the rat brain

P. MANDEL and D. AUNIS*

Centre de Neurochimie du CNRS and Institut de Chimie Biologique, Faculté de Médecine, Strasbourg

Abstract   L-Tyrosine:2-oxoglutarate aminotransferase (EC 2.6.1.5) is localized in the rat brain mainly on the inner mitochondrial membrane. The enzyme has been purified 2500 times. Only one band of molecular weight 100 000 was detected by disc electrophoresis on urea and sodium dodecanesulphate. After treatment with 2-mercaptoethanol the enzyme yielded subunits of molecular weight 25 000. Serine and alanine were identified by dansylation as the *N*-terminal amino acids, present in equimolar amounts. We can assume that each molecule of enzyme contains two molecules of serine and two of alanine. These results suggest that tyrosine aminotransferase in the rat brain has a tetrameric structure of the $\alpha_2\beta_2$ type. As for all other aminotransferases, pyridoxal phosphate is the cofactor of tyrosine aminotransferase. The association of two molecules of pyridoxal phosphate with each molecule of enzyme indicates that two subunits are catalytic; the two others have a still unexplained role. The enzymic mechanism is Ping-Pong Bi-Bi in Cleland's nomenclature. The rate equation and the kinetic constants have been determined. The real Michaelis constant for tyrosine is 20 mM and that for 2-oxoglutarate 0.1 mM. Tyrosine and 3-iodo-tyrosine appeared to be the most active substrates.

The maximum velocity of tyrosine aminotransferase is 400 µmol/h of *p*-hydroxyphenylpyruvate formed by 1 mg of enzyme. Homogentisate, fumarate and maleate, from the catabolism of tyrosine, inhibit tyrosine aminotransferase activity. Complete inhibition was obtained with 0.66mM-homogentisate. We find that homogentisate might regulate transamination.

Noradrenaline was a competitive inhibitor: at a concentration of 1.75 mM, the inhibition was 66.6%. Tyrosine aminotransferase in the liver, in contrast to that in the brain, is primarily a soluble enzyme. There are four easily dissociable molecules of pyridoxal phosphate per molecule of liver enzyme whereas tyrosine aminotransferase in the brain contains only two tightly bound pyridoxal phosphates per enzyme molecule. The inhibition characteristics of the enzyme in the brain and liver are also different. The different properties of the brain and liver isoenzymes might reflect the different roles of the enzymes in the two tissues.

* D. Aunis is attaché de recherche à l'INSERM.

The metabolism of tyrosine is of particular interest since several transmitters and hormones are derived from this amino acid. However, only a small fraction of brain tyrosine is converted into catecholamines (Mark *et al.* 1970). The major pathway of tyrosine metabolism is degradation in which several inborn errors of metabolism have been observed. The first step in this pathway is reversible transamination with 2-oxoglutarate (α-ketoglutarate).

LOCALIZATION OF TYROSINE AMINOTRANSFERASE IN RAT BRAIN

The enzyme activities detected in 0.32M-sucrose homogenates were low but increased if 0.2M-phosphate or Triton X-100 1% was added. The enzyme activity was similar in the cortex, caudate nucleus, mesencephalon, hypothalamus, cerebellum and brain stem (Mark *et al.* 1970); no marked regional differences were observed. At least 60% of tyrosine aminotransferase (TAT)* activity was localized in the crude mitochondrial fraction (Table 1). Further subfractionation of the crude mitochondrial fraction showed a strong parallel between the distribu-

TABLE 1

Subcellular distribution of tyrosine aminotransferase (TAT) in the rat brain

Fraction	TAT activity ± s.e.m.	TAT activity (% of total activity recovered)
Cytoplasmic	0.53 ± 0.16 (6)	18.3
Nuclear	0.95 ± 0.21 (6)	13.5
Microsomal	0.15 ± 0.04 (6)	5.2
Crude mitochondrial	1.25 ± 0.37 (6)	63

tion of succinate dehydrogenase (EC 1.3.99.1) and TAT in saccharose (Table 2). From this parallelism and from previous data (Mark *et al.* 1970) it can be concluded that about 75% of the enzymic activity is localized in the nerve-end mitochondria and 12% in the free mitochondria. TAT activity is much lower in brain than in liver, but after addition of phosphate and Triton X-100 the enzyme activity in liver increases only about 18%, whereas in brain there is a 10-fold increase (Table 3). The latter data are in accord with the suggestion that hepatic TAT is mainly soluble and the enzyme in brain is particulate (Mark *et al.* 1970). Similar results were reported by Gibb & Webb (1969) and Miller & Litwack (1969). Phosphate buffer detached some of the TAT activity from the mitochondrial membranes (Table 2) as was shown by Waksman & Rendon (1969) for aspartate aminotransferase (EC 2.6.1.1). Although the techniques for mito-

* L-Tyrosine: 2-oxoglutarate aminotransferase (EC 2.6.1.5).

TABLE 2

Distribution of tyrosine aminotransferase in submitochondrial fractions as a function of extraction medium

Activity (%)[a]	Succinate dehydrogenase	TAT			
			Phosphate buffer		
	0.32M-saccharose	0.32M-saccharose	1 mM	2.5 mM	0.2 M
Outer-mitochondrial space	1	8	8	8	28
Membranes	95[b]	74	64	45	7
Matrices	4	19	28	46	65

[a] Expressed as a percentage of recovered total activity.
[b] Essentially in the internal membrane.

chondrial subfractionation are controversial, there is little doubt that the enzyme is detached from the internal mitochondrial membranes by phosphate buffer as a function of ionic strength. Such changes of enzyme localization after modifications of the microenvironment may play a role in metabolic regulation.

TAT was purified by conventional methods over 2 500-fold compared to the initial homogenate (Aunis *et al.* 1971). Using the dansyl method to determine free and *N*-terminal amino acids, we found no significant contamination (less than 10^{-11}M of the purified enzyme molecule). At low ionic strength, the enzyme was relatively stable and could be maintained at $-20\,^{\circ}$C for at least three weeks with less than 10% loss of activity. Inactivation at $0\,^{\circ}$C could be prevented by addition of 1 mg/ml of albumin. The specific activity increased slightly as a function of protein concentration, as has been observed for some other enzyme activities. The mechanism of this phenomenon has to be clarified, but it might be a result of interactions between protein molecules or between the enzyme and the aqueous environment.

TABLE 3

Tyrosine aminotransferase activity in different media

Medium	Tissue	
	Brain[a]	Liver[a]
0.32M-sucrose	0.31	39.0
0.32M-sucrose + 2mM-EDTA	0.15	28.0
0.2M-phosphate buffer	0.83	48.0
0.2M-phosphate buffer (pH 7.3) + Triton X-100 1%	2.99	46.0

[a] Values are expressed as μmol *p*-hydroxyphenylpyruvate h^{-1} (g wet wt.)$^{-1}$.

MOLECULAR WEIGHT—SUBUNITS

By gel filtration, a single peak with a molecular weight of 110 000 daltons was obtained. By disc-gel electrophoresis in 8M-urea and sodium dodecanesulphate (1%) only one band was detected. The molecular weight was estimated to be 100 000 ± 5 000 (Fig. 1) by the method of Dunker & Rueckert (1969). This value agrees with that obtained by amino acid analysis. Gel electrophoresis after 2-mercaptoethanol treatment gave four bands with molecular weights of 100 000, 75 000, 50 000 and 25 000. Thus the native TAT protein was probably dissociated by 1% mercaptoethanol into four subunits each with molecular weight 25 000, and it would thus be a tetramer. The trimer and dimers were probably due to incomplete dissociation in the presence of mercaptoethanol. The absence of dissociation in the presence of sodium dodecanesulphate and urea and the effect of 2-mercaptoethanol suggest that the subunits are held together by disulphide bounds.

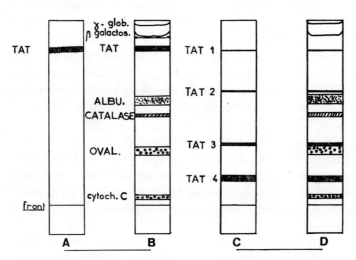

FIG. 1. Disc-gel electrophoresis of tyrosine aminotransferase (TAT): γ-glob., gammaglobulin; β galactos., β-galactosidase; ALBU., albumin; OVAL., ovalbumin; cytoch.C, cytochrome oxidase (EC 1.9.3.1): A,B, electrophoresis of native TAT; C,D, electrophoresis of 2-mercaptoethanol-treated TAT.

AMINO ACID COMPOSITION

The amino acid composition of rat brain TAT is different from that of liver TAT (Table 4). It is relevant to note that brain TAT contains a high percentage of proline residues which may be involved in the attachment of the enzyme to

TABLE 4

Amino acid composition of brain TAT compared with that of liver TAT (Valeriote *et al.* 1969).

Amino acid	Number/mol brain TAT	Number/mol liver TAT
Cysteine	14	28
Methionine	21	25
Aspartate	81	95
Threonine	40	34
Serine	62	68
Glutamate	103	115
Proline	92	6
Glycine	86	62
Alanine	69	70
Valine	43	60
Isoleucine	31	55
Leucine	71	94
Tyrosine	16	27
Phenylalanine	31	33
Lysine	102	49
Histidine	25	18
Arginine	31	46
Tryptophan	?	9
Total	908	897

the mitochondrial membrane. There is also a clear-cut difference in the ratio of basic amino acids to dicarboxylic amino acids in the two enzymes. The liver enzyme is acidic while the brain enzyme is slightly basic.

Serine and alanine, identified as the N-terminal amino acids, are present in equimolecular amounts. Moreover, for one mole of enzyme there were 2 alanine and 2 serine as judged by the ratio of N^6-dansyllysine to N-dansylalanine and N-dansylserine. This fact, together with the probable tetrameric structure, suggests an $\alpha_2\beta_2$ structure for brain TAT. Aurichio *et al.* (1970) have shown that hepatic TAT is composed of four similar or identical subunits, each of them with a molecule of pyridoxal 5'-phosphate.

PYRIDOXAL PHOSPHATE CONTENT

To estimate the amount of pyridoxal phosphate bound to the enzyme we used the method of Adams (1969). Two molecules of pyridoxal phosphate were found for each molecule of TAT. The pyridoxal phosphate could not be dissociated from the brain enzyme by the usual methods of dialysis nor by dialysis in the presence of tyrosine as is the case for liver TAT (Hayashi *et al.* 1967). This suggests that for brain TAT there is complementary ionic, hydrophobic or π–π

binding of pyridoxal phosphate. In addition, after dissociation of pyridoxal phosphate by dialysis, liver TAT became temperature-sensitive if the coenzyme was not replaced (Hayashi *et al.* 1967). The sensitivity of brain TAT did not change after dialysis, an observation which suggests that the irreversible fixation of pyridoxal phosphate to brain TAT stabilizes it (Aunis & Mandel, unpublished results).

ABSORPTION SPECTRA

In a phosphate buffer (0.1 M, pH 7.6), the main absorption peaks (Fig. 2) were observed at 280 nm, an absorption corresponding to aromatic amino acids, and at 410 nm due to the presence of pyridoxaldimine, the Schiff base resulting from the condensation of the 6-amino group of lysine with the aldehyde group of pyridoxal phosphate.

Upon addition of tyrosine, the peak at 410 nm virtually disappeared and a peak at 340 nm, which corresponds to pyridoxamine phosphate, appeared. Although pyridoxamine phosphate absorbs at 325 nm, the slight shift might be due to structural strain or changes resulting from the position of the substrate in the enzyme molecule. Addition of 2-oxoglutarate does not change the absorption spectrum of the native enzyme.

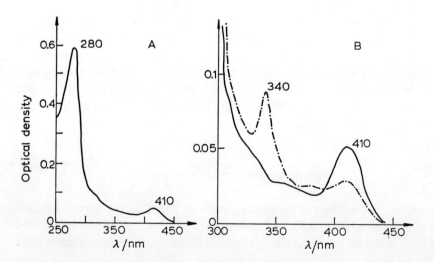

Fig. 2. Ultraviolet absorption spectra of tyrosine aminotransferase: A, native proteins in phosphate buffer, 0.1 M, pH 7.6; B, ————, native proteins, –·–·–·–, native proteins with 0.1M-tyrosine.

ACETYLATION BY IODOACETIC ACID

Acetylation of brain TAT at pH 5.5 inhibited enzyme activity, suggesting that the imidazole residues (one or more) are affected. Also, the inactivation of the enzyme by acetylation at pH 8.0 suggests that thiol groups contribute to the enzymic activity (Aunis & Mandel, unpublished results).

ENZYMIC MECHANISMS

Transaminations are multistep processes. Their complete kinetic analysis necessitates a study of the reaction velocity–substrate relationship for one substrate at several concentrations of the other substrates. Thus, the enzyme activity was measured at various concentrations of tyrosine at constant concentrations of 2-oxoglutarate or with tyrosine at constant concentrations with varying concentrations of 2-oxoglutarate. For every concentration of 2-oxoglutarate when the concentration of tyrosine is the variable, an apparent Michaelis constant, K_m, can be calculated for tyrosine. Similarly, different apparent Michaelis constants can be calculated for 2-oxoglutarate, when its concentration is variable

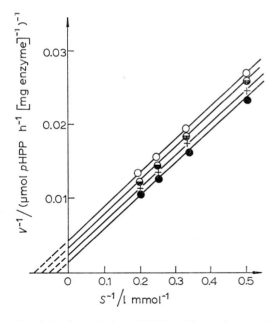

FIG. 3. Reciprocal plots of TAT activity against tyrosine concentration (S) at four different concentrations of 2-oxoglutarate: ○, 0.5 mM; ◓, 1.0 mM; +, 2.0 mM; ●, 2.5 mM. The activity is recorded in units of μmol p-hydroxyphenylpyruvate (pHPP) formed each hour by 1 mg enzyme.

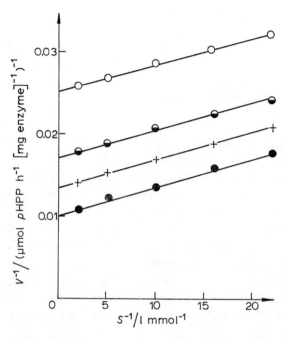

FIG. 4. Reciprocal plots of TAT activity against 2-oxoglutarate concentration (S), at four different concentrations of tyrosine: $\circ$, 2 mM; $\ominus$, 3 mM; $+$, 4 mM; $\bullet$, 5 mM. Units of activity are explained in Fig. 3.

and different constant concentrations of tyrosine are used. Figs. 3 and 4 show that in both cases parallel lines were obtained by the double reciprocal plotting method of Lineweaver & Burk (1934). Thus, the affinity of the enzyme for a given substrate varies as a function of the concentration of the other substrate. There is a compensation phenomenon: when the concentration of one substrate decreases, the affinity of the enzyme to the second substrate increases and *vice versa*. When the values of the apparent K_m for one substrate as a function of the other substrate were plotted, a straight line was obtained. The intersection of this line with the ordinate gives the limit K_m for the lowest concentration of the substrate (close to zero). This limit K_m is 7.5 mM for tyrosine, 0.01 mM for 2-oxoglutarate (Aunis & Mandel, unpublished results).

The real Michaelis constant can be evaluated for both tyrosine and 2-oxo-glutarate according to Cleland (1963) by a replot of the reciprocal of the apparent maximum velocity for 2-oxoglutarate ($1/V_{max(og)}$) as a function of the reciprocal of the tyrosine concentration and of the reciprocal of the apparent maximum velocity for tyrosine ($1/V_{max(tyr)}$) as a function of the reciprocal of the 2-oxoglutarate concentration (Fig. 5). The values obtained are given in Table 5.

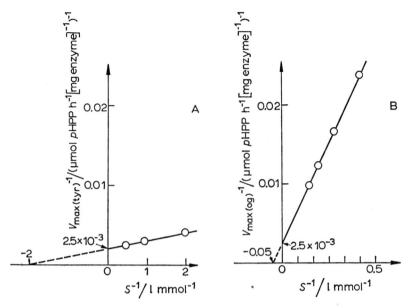

FIG. 5 Reciprocal plots of apparent maximum velocity for 2-oxoglutarate ($V_{max(og)}$) and tyrosine ($V_{max(tyr)}$) against the concentration of (A) 2-oxoglutarate and (B) tyrosine. The intercept with the X-axis gives the real Michaelis constants and that with the Y-axis the real maximum velocities.

According to Cleland (1963), the fact that conventional plots (see Figs. 3 and 4) give parallel lines suggests a Ping-Pong Bi-Bi mechanism. As tyrosine bound to the native enzyme and 2-oxoglutarate did not do so, the proposed mechanism, where the enzyme oscillates between two forms, is shown in Fig. 6.

In order to investigate whether conformational changes and cooperative processes occurred during transamination, we used the Hill equation. When log-$[V/(V_{max}-V)]$ was plotted against the logarithm of the substrate concentration (log[S]) for tyrosine or 2-oxoglutarate, the slope of the line was unity, whatever the concentration of the second substrate (Aunis & Mandel, unpublished re-

TABLE 5

Determination of real Michaelis constants for 2-oxoglutarate and tyrosine

Substrate	Apparent K_m (mM)	Real constant (mM)	Limit K_m (mM)	Real V_{max}[c]
Tyrosine	20[a]	20	7.5	400
2-Oxoglutarate	0.04[b]	0.5	0.01	400

[a] Determined with 2mM-2-oxoglutarate.
[b] Determined with 5mM-tyrosine.
[c] In μmol p-hydroxyphenylpyruvate h^{-1} (mg enzyme)$^{-1}$.

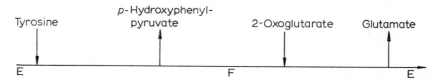

FIG. 6. The proposed mechanism for transamination, in which the enzyme oscillates between two forms: E is the Schiff base (pyridoxaldimine) from lysine and pyridoxal and F is the pyridoxamine.

sults). This means that the Hill number is 1.0 for each substrate, and that there was no cooperativity during transamination.

ENZYME REACTION WITH ANALOGUES

Table 6 shows data for the reactions of TAT towards analogues of tyrosine. Note that dopa is a substrate for the enzyme, as are phenylalanine and 3-iodo-L-tyrosine.

TABLE 6

The effect of structural analogues of tyrosine on the activity of TAT

Substrate	Apparent $K_m{}^a$	Apparent $V_{max}{}^b$
D-Tyrosine	0	0
L-Tyrosine	20	400
Phenylalanine	8	120
p-Chlorophenylalanine	20	100
p-Fluorophenylalanine	33	80
3-Iodo-L-tyrosine	11	500
p-Aminophenylalanine	0	0
α-Methyltyrosine	0	0
Dopa	20	400

[a] Apparent values of K_m were determined with 2mM-2-oxoglutarate.
[b] V_{max} values are expressed in μmol p-hydroxyphenylpyruvate h^{-1} (mg protein)$^{-1}$.

INHIBITION OF TAT

Both noradrenaline and 2,5-dihydroxyphenylacetic acid (homogentisic acid) inhibited TAT (see Table 7). When tyrosine was the variable substrate, the inhibition of TAT by noradrenaline was competitive. This inhibition differs from that found with TAT from the rat liver (Black & Axelrod 1969). Homo-

TABLE 7

Inhibition of TAT by noradrenaline, 2,5-dihydroxyphenylacetic acid and some analogues

Compound	Concentration (μM)	Inhibition (%)	Inhibition type
Noradrenaline	10	12	Competitive
	50	50	
	300	66	Inhibitor constant, $K_i = 125$ μM
2,5-Dihydroxyphenyl-acetic acid	60	25	Non-competitive
	160	60	Inhibitor constant, K_i, variable —
	300	86	a function of tyrosine
	660	100	concentration (cooperativity)
(3,4-Dihydroxyphenyl)-hydroxyacetic acid	60	25	Inhibition resembles that of 2,5-
	160	60	dihydroxyphenylacetic acid
(3,4-Dihydroxyphenyl)-ethane-1,2-diol	300	86	Inhibition resembles that of 2,5-
	660	100	dihydroxyphenylacetic acid

gentisic acid inhibited TAT non-competively with either tyrosine or 2-oxoglu-tarate as the variable substrate. It seems likely that this acid plays a regulatory role in the enzymic mechanism. The inhibition of TAT by homogentisate when tyrosine was the variable substrate appeared to be parabolic for the inhibitor but linear for tyrosine (Aunis & Mandel, unpublished results). This type of curve is due to the formation of a complex (enzyme)$_1$–(inhibitor)$_1$, followed by a complex (enzyme)$_1$–(inhibitor)$_2$ when the inhibitor concentration rises. When the Hill number was investigated for this interaction, we found a value of 0.8 for the complex (enzyme)$_1$–(inhibitor)$_1$ and 1.6 for the complex (enzyme)$_1$–(inhibitor)$_2$. As this number depends on the tyrosine–inhibitor ratio, some cooperativity can be assumed.

DISCUSSION

Comparison of brain and liver TAT

Table 8 lists some differences between cerebral and hepatic TAT. Whereas TAT in brain is mainly insoluble and localized in the mitochondria, TAT in liver is mainly soluble and localized in the cytosol. The distribution of amino acids and especially the proline content show considerable differences. The ratio of basic and dicarboxylic amino acids is also divergent. The coenzyme requirement differs: two molecules of pyridoxal phosphate are bound to the cerebral apo-enzyme whereas four molecules are bound to the liver apoenzyme. The ultra-violet absorption spectra differ. The interaction between pyridoxal phosphate and the apoenzyme suggests a stronger attachment of the coenzyme to the brain

TABLE 8

Comparison of cerebral TAT and hepatic TAT

	TAT in brain	TAT in liver
Localization	mitochondrial	cytoplasmic
Molecular weight	$100\ 000 \pm 5\ 000^a$	$111\ 000 \pm 5\ 000^a$
N-Terminal amino acids	2 Ala + 2 Ser	?
Proline (%)	10.0	0.7
Ratio of acidic to basic amino acids	1.16	1.8
Number of molecules of pyridoxal phosphate	2	4
Absorption maxima, λ_{max}/nm	280, 410 → 340	280, 425 → 327
Structure type	$\alpha_2\beta_2$	α_4
Number of subunits	4	4
Mechanism	Ping-Pong with two steps	Ping-Pong with three steps (fixation of pyridoxal phosphate)

^{a} $\pm$ S.E.

enzyme due to complementary binding forces. Taking into account the N-terminal amino acids and the pyridoxal phosphate content, we propose an $\alpha_2\beta_2$ structure for the cerebral enzyme and an α_4 type for the enzyme in the liver. Further, it is relevant to note that cerebral TAT is not induced by corticosteroids as is the hepatic enzyme (Aunis & Mandel, unpublished results). These divergent properties may be explained by the adaptation of TAT to its role in the tissue and to tyrosine uptake. It seems reasonable to suppose that in liver, where the input of tyrosine is rather high and where the amino acid has to be degraded, the enzyme is localized in the cytosol. In contrast, in brain where the input is much lower and it is necessary to preserve tyrosine molecules, the enzyme is mainly attached to membranes in the nerve endings where it could have a physiological role. Similarly, the strong binding of pyridoxal phosphate might also be due to a phenomenon of adaptation since the pyridoxal phosphate availability is low in the brain and high in liver. According to these speculations, TAT in the brain could be a tissue-adapted enzyme.

Role of TAT in the brain

Since the affinity of tyrosine hydroxylase for tyrosine is much higher than that of TAT, it seems unlikely that the two enzymes compete for the tyrosine substrate in brain. But the uptake of catecholamines which occurs after the release of noradrenaline should be followed by synthesis to compensate for losses. The noradrenaline which is taken up could inhibit tyrosine hydroxylase, thus blocking synthesis of noradrenaline. But the noradrenaline which is taken up could also inhibit TAT, thereby increasing the pool of tyrosine and the ratio of tyrosine to noradrenaline. In this way, the inhibition of tyrosine hydroxylase could be overcome. A role for TAT in the regulation of catecholamine synthesis is thus not to be excluded.

References

ADAMS, E. (1969). *Anal. Biochem.* **31**, 118–122
AUNIS, D., MARK, J. & MANDEL, P. (1971). *Life Sci.* **10**, 617–625
AURICCHIO, F., VALERIOTE, F., TOMKINS, G. & RILEY, W. D. (1970). *Biochim. Biophys. Acta* **221**, 307–313
BLACK, I. R. & AXELROD, J. (1969). *J. Biol. Chem.* **224**, 6124–6129
CLELAND, W. (1963). *Biochim. Biophys. Acta* **67**, 104–137
DUNKER, A. K. & RUECKERT, R. (1969). *J. Biol. Chem.* **244**, 5074–5080
GIBB, J. W. & WEBB, J. G. (1969). *Proc. Natl. Acad. Sci. U.S.A.* **63**, 364–369
HAYASHI, S., GRANNER, D. K. & TOMKINS, G. M. (1967). *J. Biol. Chem.* **242**, 3998–4006
LINEWEAVER, L. & BURK, D. (1934). *J. Am. Chem. Soc.* **56**, 658–666
MARK, J., PUGGE, H. & MANDEL, P. (1970). *J. Neurochem.* **17**, 1393–1401
MILLER, J. E. & LITWACK, G. (1969). *Arch. Biochem. Biophys.* **134**, 149–159
VALERIOTE, F. A., AURICCHIO, F., TOMKINS, G. M. & RILEY, D. (1969). *J. Biol. Chem.* **244**, 3618–3624
WAKSMAN, A. & RENDON, A. (1969). *Biochem. Biophys. Res. Commun.* **36**, 324–330

Discussion

Grahame-Smith: Professor Mandel, is there any tyrosine aminotransferase in the adrenal medulla?

Mandel: Certainly, but to my knowledge the enzyme has not been investigated.

Barondes: Have you ever lysed the synaptosomes and then shown that the enzymes are associated with the mitochondria from the lysed synaptosomes?

Mandel: Yes. They are still associated with mitochondria.

Munro: The hepatic aminotransferases in the soluble and mitochondrial fractions are isozymes. Have you compared these fractions in the brain?

Mandel: After the usual preparation of mitochondria, about 85% of the aminotransferase is in the cytosol and the remaining 15% is in mitochondria in the liver.

Noradrenaline is only competitive for the aldehyde of pyridoxal phosphate as Black & Axelrod (1969) showed for hepatic tyrosine aminotransferase. Dopamine is not a substrate.

Weiner: Could dopamine be oxidatively deaminated?

Sourkes: Oxidative deamination of dopamine yields an aldehyde. Since the aldehyde dehydrogenase is probably in the soluble portion of the nerve terminals, a certain amount of the aldehyde dehydrogenase should also be present in synaptosomes to convert the aldehyde into 3,4-dihydroxyphenylacetic acid. If O-methylation takes place, the final product would be homovanillic acid.

Sandler: Another item for the comparison of brain and liver tyrosine aminotransferase (Table 8) is the effect of noradrenaline. Whereas noradrenaline inhibits the brain enzyme, like dopamine it potentiates the hepatic enzyme. So, for example, oral administration of L-dopa (which is largely metabolized to dopamine peripherally) leads to an over-production of the tyrosine transamination product, p-hydroxyphenylpyruvic acid.

I was fascinated by your speculation about the role of tyrosine aminotransferase in the brain. The presence of so much enzyme with no obvious function is most puzzling. A few years ago, Gey (1965) performed a tantalizing experiment. He gave p-hydroxyphenylpyruvate to a rat and found that, even after tyrosine hydroxylase was blocked, dopa production was better than with tyrosine as precursor. Is it possible that another route to dopa is by hydroxylation of p-hydroxyphenylpyruvate by some as yet undefined mechanism, followed by transamination?

Mandel: p-Hydroxyphenylpyruvate is a strong inhibitor of tyrosine aminotransferase. In this way, the concentration of tyrosine builds up and so the ratio of tyrosine to noradrenaline increases. This might help to remove the inhibition of tyrosine hydroxylase by competition.

Sourkes: Inhibition is not the only possible explanation. p-Hydroxyphenylpyruvate could be forming 2,5-dihydroxyphenylpyruvate by a mechanism similar to that of ring oxidation in the formation of 2,5-dihydroxyphenylacetic acid. The 2,5-dihydroxyphenylpyruvate could then be transaminated to 2,5-dihydroxyphenylalanine, a positional isomer of the 3,4-dihydroxy compound, dopa, with which we are commonly concerned. This is, admittedly, hypothetical but if such a reaction took place, the formation of the 2,5-isomer might burden transamination mechanisms (e.g. through engagement of available pyridoxal phosphate coenzyme) to the detriment of action on tyrosine.

Weiner: We have evidence (Cloutier & Weiner 1973) that the increase in tyrosine hydroxylase activity during nerve stimulation does not seem to be related to end-product feedback inhibition (see our paper, pp. 135–147). Have you administered 6-hydroxydopamine to see whether the amount of tyrosine aminotransferase decreases in the brain? In other words, is there some selective localization of this enzyme in adrenergic neurons?

Mandel: We don't know. It would be interesting to perform such an experiment.

Barondes: Several groups (e.g. De Robertis *et al.* 1962) reported separation of amine-rich and amine-poor synaptosome fractions by sucrose density gradient centrifugation. Was this enzyme localized in the amine-rich synaptosome fraction?

Mandel: No.

Wurtman: Zigmond & Wilson (1973) have found that intracisternal administration of 6-hydroxydopamine in concentrations that lower the concentrations of noradrenaline and dopamine in the brain does not lower brain tyrosine aminotransferase activity.

Mandel: That does not seem the best control if the regulation at an enzymic level exists *in vivo.* We are investigating the metabolism of noradrenaline in the adrenergic neurons of neuroblastoma cells in culture by following the transamination of [^{14}C]tyrosine and noradrenaline synthesis.

Wurtman: One would like to find evidence that *p*-hydroxyphenylpyruvate is actually formed in brain, or that synaptosomes which are able to synthesize catecholamines from tyrosine can also synthesize them from *p*-hydroxyphenylpyruvate. It would also be important to obtain some evidence that the destruction of intraneuronal tyrosine by tyrosine aminotransferase could depress concentrations of the amino acid to the point of limiting the rate of catecholamine synthesis. Can tyrosine availability control the rate at which it is hydroxylated to dopa?

Mandel: It has been suggested that the increase in the amount of tyrosine stimulates the synthesis of noradrenaline. We have good evidence with an adrenergic clone of neuroblastoma cells that when the tyrosine concentration in the medium is raised, the synthesis of noradrenaline increases. We rigorously purified the tyrosine, because commercial material contains a considerable amount of dopa.

Weiner: From our studies on aminotransferases, amino acids and acceptors, we discovered that two of the best acceptors for transamination of tryptophan are *p*-hydroxyphenylpyruvate and phenylpyruvate (Lees & Weiner 1973). The aromatic ketoacids seem to be much better acceptors than 2-oxoglutarate for the brain tryptophan aminotransferase.

Wurtman: Has *p*-hydroxyphenylpyruvate been found in the brain?

Sharman: *p*-Hydroxyphenyllactate, a reduction product of *p*-hydroxyphenylpyruvate, has been found in the cerebrospinal fluid (Sjöquist & Änggård 1972).

Gál: The chlorine analogue, *p*-chlorophenylpyruvic acid, is almost instantaneously transaminated into *p*-chlorophenylalanine in the brain (Gál *et al.* 1970. The brain favours the production of the amino acid (*in vivo* or *in vitro*). I presume the preferred ketoacid acceptor in that reaction is glutamate.

Roberts: It is a mistake to suppose that amino acid metabolism in the brain is directed primarily towards neurotransmitter formation. There is enough evidence in the literature to indicate that, in certain circumstances at least, amino acids may serve as a source of energy in brain. The activities of enzymes facilitating the early steps in amino acid utilization may reflect the latter processes to a greater extent than the former.

Mandel: It is curious that in the liver where the input and degradation of tyrosine both proceed at a high rate, tyrosine aminotransferase is soluble. In contrast, in brain where the input of tyrosine is much lower and the brain must retain its tyrosine, the enzyme is particulate. In this way, degradation will automatically be lower.

Roberts: What I said does not eliminate the possibility that there is compartmentation of degradative enzymes within the brain that favours certain mechanisms taking place in specific cell structures.

Gál: Will your explanation serve, Professor Mandel? In the liver, hexokinase is more cytoplasmic while in the brain it is particulate.

Mandel: Maybe if it can be considered as an adaptive process as we think that it is for tyrosine aminotransferase.

Wurtman: Another enzyme in brain that catabolizes an aromatic amino acid is, of course, tryptophan oxygenase (Hirata & Hayaishi 1972). This enzyme can catalyse the destruction of 5-hydroxytryptophan, 5-hydroxytryptamine and N-acetyl-5-methoxytryptamine as well as L-tryptophan. Hence, there is some precedent for an enzyme catabolizing both amino acids and monoamines. This enzyme is widely distributed but its physiological significance is not known.

Gál: For the record, Professor Wurtman, I have described formation of kynurenine in brain homogenates from D- and L-tryptophan several years ago (Gál *et al.* 1966) but it has obviously escaped notice.

Barondes: Tyrosine aminotransferase is a classic inducible enzyme in the liver. Have you induced it (e.g. with steroids) in the brain?

Mandel: No; it is not inducible in the brain.

Wurtman: I don't know of any evidence that the physiological regulation of the tyrosine aminotransferase in liver involved steroids. Normally, the hepatic enzyme exhibits a five-fold daily rhythm (Wurtman & Axelrod 1967); this rhythm persists after the pituitary or adrenals are removed. It seems to be generated by the diet (Wurtman *et al.* 1968), that is the cyclic consumption of dietary protein causes a corresponding rhythm in the aggregation of hepatic polyribosomes (Fishman *et al.* 1969), and thus in the synthesis of this enzyme and other proteins.

References

BLACK, I. B. & AXELROD, J. (1969) Inhibition of tyrosine transaminase activity by norepinephrine. *J. Biol. Chem.* **244**, 6124–6129

CLOUTIER, G. & WEINER, N. (1973) Further studies on the increased synthesis of norepinephrine during nerve stimulation of guinea-pig vas deferens preparation: effect of tyrosine and 6,7-dimethyltetrahydropterin. *J. Pharmacol. Exp. Ther.* **186**, 75–85

DE ROBERTIS, R., PELLEGRINO DE IRALDI, A., RODRIGUEZ DE LORES GARNAIZ, G. & SALGANICOFF, L. (1962) Cholinergic and non-cholinergic nerve endings in rat brain. I. Isolation and subcellular distribution of acetylcholine and acetylcholinesterase. *J. Neurochem.* **9**, 23–25

FISHMAN, B., WURTMAN, R. J. & MUNRO, H. N. (1969) Daily rhythms in hepatic polysome profiles and tyrosine transaminase activity: role of dietary protein. *Proc. Natl. Acad. Sci. U.S.A.* **64**, 667–682

GÁL, E. M., GINSBERG, B. & ARMSTRONG, J. (1966) The nature of *in vitro* hydroxylation of L-tryptophan by brain tissue. *J. Neurochem.* **13**, 643–653

GÁL, E. M., ROGGEVEEN, A. E. & MILLARD, S. A. (1970) DL-[2-^{14}C]*p*-Chlorophenylalanine as an inhibitor of tryptophan 5-hydroxylase. *J. Neurochem.* **17**, 1221–1235

GEY, K. F. (1965) 4-Hydroxyphenylpyruvat und 3,4-Dihydroxyphenylpyruvat als Noradrenalin-Vorstufen. *Helv. Physiol. Pharmacol. Acta* **23**, C 89

HIRATA, F. & HAYAISHI, O. (1972) New degradative routes of 5-hydroxytryptophan and serotonin by intestinal tryptophan 2,3-dioxygenase. *Biochem. Biophys. Res. Commun.* **47**, 1112–1119

LEES, G. J. & WEINER, N. (1973) Transaminations between amino acids and keto acids elevated in phenylketonuria and maple syrup urine disease. *J. Neurochem.* **20**, 389–403

SJÖQUIST, B. & ÄNGGÅRD, E. (1972) Gas chromatographic determination of homovanillic acid in human cerebrospinal fluid by electron capture detection and by mass fragmentography with a deuterated internal standard. *Anal. Chem.* **44**, 2297–2301

WEINER, N., LEE, F.-L., WAYMIRE, J. C. & POSIVIATA, M. (1974) The regulation of tyrosine hydroxylase activity in adrenergic nervous tissue, *This Volume*, pp. 135–147

WURTMAN, R. J. & AXELROD, J. (1967) Daily rhythmic changes in tyrosine transaminase activity of the rat liver. *Proc. Natl. Acad. Sci. U.S.A.* **57**, 1594–1598

WURTMAN, R. J., SHOEMAKER, W. J. & LARIN, F. (1968) Mechanism of the daily rhythm in hepatic tyrosine transaminase activity: role of dietary tryptophan. *Proc. Natl. Acad. Sci. U.S.A.* **59**, 800–807

ZIGMOND, M. J. & WILSON, S. P. (1973) Studies on the interaction between catecholamines and tyrosine aminotransferase in brain. *Biochem. Pharmacol.* **17**, 2151–2164

Properties of the pterin-dependent aromatic amino acid hydroxylases

SEYMOUR KAUFMAN

Laboratory of Neurochemistry, National Institute of Mental Health, Bethesda, Maryland

Abstract The three pterin-dependent aromatic amino acid hydroxylases—phenylalanine, tyrosine and tryptophan hydroxylase—show remarkable similarities in their kinetic properties. This paper focuses on those properties that may be relevant to the *in vivo* regulation of the enzymes.

One property shared by all the enzymes is that their behaviour varies dramatically with the nature of the pterin cofactor used. In many instances, regulatory properties are revealed only in the presence of the naturally occurring cofactor, tetrahydrobiopterin. Not only is the apparent affinity of the enzymes for this pterin much greater than it is for synthetic analogues, such as 6,7-dimethyltetrahydropterin, but the apparent affinity for the other substrates is also increased. Thus, in the case of cerebral tryptophan hydroxylase, the K_m for tryptophan determined in the presence of dimethyltetrahydropterin, has been reported to be 300 μM, a value that is far higher than the probable concentration of the amino acid in the brain. This puzzling discrepancy was resolved by the finding that in the presence of tetrahydrobiopterin, K_m for tryptophan is 50 μM, a value that is close to the estimated concentrations of this amino acid in the brain.

The substitution of tetrahydrobiopterin for dimethyltetrahydropterin not only decreases K_m for the other substrates, but it also reveals a new regulatory property of the enzymes, that is, inhibition by high concentrations of the amino acid substrate and, in most cases, by a high concentration of oxygen.

Another property of this group of enzymes that can be altered in the presence of tetrahydrobiopterin is the substrate specificity. Thus, in the presence of tetrahydrobiopterin but not 6,7-dimethyltetrahydropterin, phenylalanine is an excellent substrate for tyrosine hydroxylase.

An understanding of certain important aspects of the oxidative metabolism of aromatic amino acids in the brain started with a study of the enzymic conversion of phenylalanine into tyrosine in the liver. Several years ago, I reported that a nonprotein cofactor isolated from rat liver extracts was an essential component of the hepatic phenylalanine hydroxylating system (Kaufman 1958). Subsequent

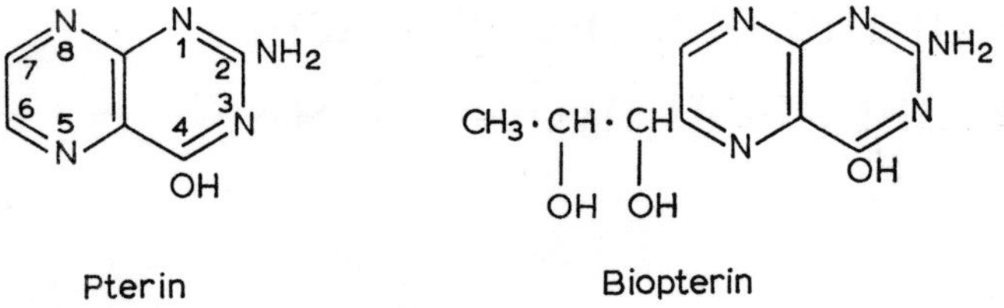

FIG. 1. Structure of pterin and biopterin.

results indicated that the cofactor might be a pteridine. Furthermore, certain synthetic, unconjugated derivatives of pterin (2-amino-4-hydroxypteridine), such as 6-methyl- and 6,7-dimethyl-tetrahydropterin ($6MPH_4$ and $DMPH_4$, respectively), were found to have high cofactor activity with the phenylalanine hydroxylating system (Kaufman 1959; Kaufman & Levenberg 1959). Our earlier suppositions that the naturally occurring cofactor was a pterin were substantiated with the report that the hydroxylation cofactor from rat liver was the reduced form of the unconjugated pterin, biopterin, 2-amino-6-(1,2-dihydroxypropyl)-4-hydroxypteridine (see Fig. 1) (Kaufman 1963). The work on phenylalanine 4-hydroxylase (EC 1.14.3.1) established the first metabolic role for this class of compound.

Mainly through the use of the 6,7-dimethyl analogue of the cofactor, the general manner in which the pterin acts as cofactor in hydroxylations was elucidated. The conversion of phenylalanine into tyrosine was shown to proceed according to equations (1) and (2), where RH stands for phenylalanine, ROH for tyrosine, XH_4 for the tetrahydropterin and XH_2 for the quinonoid dihydropterin (Kaufman 1964a). Reaction (1) is catalysed by phenylalanine 4-hydrox-

$$RH + O_2 + XH_4 \rightarrow ROH + H_2O + XH_2 \qquad (1)$$
$$XH_2 + NADPH + H^+ \rightarrow XH_4 + NADP^+ \qquad (2)$$

ylase and reaction (2) by a separate enzyme, dihydropteridine reductase. Although both NADPH and NADH can function with the latter enzyme, the value of K_m is lower and V_{max} is higher with NADH (Nielsen et al. 1969; Scrimgeour & Cheema 1971; Craine et al. 1972). In the absence of the reductase, reducing agents such as thiols and ascorbate can also reduce the quinonoid dihydropterin and allow the pterin to function catalytically (Kaufman 1959).

I have reviewed this background material because phenylalanine 4-hydroxylase is the prototype of a group of enzymes—the aromatic amino acid hydroxylases—with remarkably similar properties. On account of these similarities, the phenylalanine hydroxylating system has served as a model which on the one hand has greatly facilitated studies on tyrosine and tryptophan hydroxylases and on the other allows predictions to be made about certain properties of these related enzymes. Since our knowledge about hepatic phenylalanine 4-hydroxylase is somewhat more advanced than that about the other two hydroxylases, we are not yet in a position to test some of the more interesting predictions.

Rather than attempt to review all that is known about these hydroxylases I shall focus attention on the dramatic dependence of the properties of these hydroxylases on the particular pterin cofactor with which they are assayed. I shall also discuss how the properties of these enzymes possibly regulate their activities *in vivo*.

TYROSINE HYDROXYLASE

Most of our knowledge about tyrosine hydroxylase is limited to the enzyme from adrenal medulla. It seems likely, however, that the cerebral enzyme shares many of the properties of the adrenal enzyme. This likelihood is strengthened by our recent finding that antibodies to highly purified, bovine adrenal tyrosine hydroxylase cross-react with the enzyme from the caudate of bovine brain (Lloyd & Kaufman 1973).

Partially purified preparations of tyrosine hydroxylase from adrenal medulla have an absolute requirement for a tetrahydropterin (Nagatsu *et al.* 1964; Brenneman & Kaufman 1964; Kaufman 1964*b*). This established the enzyme as the second pterin-dependent aromatic amino acid hydroxylase. It appeared likely that the role of the tetrahydropterin in this hydroxylating system would prove to be the same as the one previously established in the phenylalanine hydroxylating system, an expectation that was fulfilled by subsequent work.

I must introduce a note of caution about the relationship between pterins and enzymes. Since tetrahydropterins are excellent reducing agents, and may, by virtue of this property, stimulate certain enzymes in a non-specific manner (Zannoni *et al.* 1963), the mere demonstration that an enzyme-catalysed reaction is stimulated by a tetrahydropterin is not adequate to support the conclusion that the pterin is functioning as a coenzyme. As I have outlined previously (Kaufman 1964*b*), the following criteria must be met before such a conclusion is justified: the pterin should (*a*) be specific, (*b*) function catalytically, and (*c*) be used during the hydroxylation and this use should be substrate dependent (see reaction [1]). There are claims for the discovery of new pterin-dependent reactions that float in scientific limbo because these criteria have been ignored.

Not only have these criteria been met with tyrosine hydroxylase, but it was also shown that the hydroxylase can be coupled with dihydropteridine reductase (Brenneman & Kaufman 1964). This coupling strongly indicates that the tetrahydropterin is oxidized to the quinonoid dihydropterin in this hydroxylation just as it is during the hydroxylation of phenylalanine.

The stoichiometry of the reaction catalysed by bovine adrenal tyrosine hydroxylase was only established fairly recently (Shiman *et al.* 1971). In the presence of tetrahydrobiopterin, the reaction proceeds as shown in equation (3) (dopa is 3,4-dihydroxyphenylalanine). A different aspect of the stoichiometry will be discussed later.

$$\text{L-Tyrosine} + O_2 + \text{Tetrahydrobiopterin} \rightarrow \text{L-Dopa} \qquad (3)$$
$$+ H_2O + \text{Quinonoid dihydrobiopterin}$$

Because soluble tyrosine hydroxylase has never been purified to any extent, almost all our knowledge about the properties of the enzyme from bovine adrenal medulla has come from studies on partially purified preparations that were solubilized by limited proteolysis with either trypsin (Petrack *et al.* 1968) or chymotrypsin (Shiman *et al.* 1971). I shall refer to the latter preparation simply as the 'solubilized' enzyme.

This method of preparing the enzyme raises some problems. It is possible, for example, that the properties, both physical and kinetic, of the solubilized enzyme bear little or no relation to the properties of the native enzyme. Indeed, since the molecular weight of the native adrenal enzyme is 135 000–155 000 whereas that of the trypsin-solubilized enzyme is 34 000 (Musacchio *et al.* 1971), Musacchio has concluded that the trypsin-solubilized enzyme is "only a fragment of the native enzyme" and, therefore, has expressed the concern that studies of the kinetic properties of the solubilized enzyme would be "highly misleading" (Musacchio *et al.* 1971).

It is worth examining both these statements. Although the question of a change in physical properties cannot be answered conclusively until the native enzyme has been purified by a method that does not resort to treatment with a protease, the question of whether the protease treatment alters the kinetic properties of the enzyme can be examined simply by comparison of the properties of the native and the solubilized enzymes. We have carried out a detailed comparison of this kind and can find no support for Musacchio's concern. Where the kinetic properties of the particulate and solubilized adrenal enzyme have been compared, they have been shown to be essentially the same (Shiman *et al.* 1971). Some of these data will be presented later.

Even the evidence in favour of the conclusion that trypsin treatment yields an enzyme that is only a fragment of the native enzyme is not compelling. It is just as likely that treatment of the enzyme with either trypsin or chymotrypsin (the molecular weight of the hydroxylase purified after chymotrypsin treatment is about 40 000 [Shiman *et al.* 1971]) under carefully controlled conditions leads to only a minor decrease in the size of the peptide chain(s) of the hydroxyl-ase, but that the protease-modified enzyme no longer aggregates. Attempts to induce aggregation of the trypsin-treated enzyme have been unsuccessful (Musacchio *et al.* 1971). Furthermore, our recent finding that the chymotrypsin-solubilized, highly purified hydroxylase is antigenically indistinguishable from the undigested enzyme (Lloyd & Kaufman 1973) provides no support for the idea that treatment with chymotrypsin has led to extensive fragmentation of the hydroxylase.

On the dramatic variation in the behaviour of these hydroxylases with the pterin cofactor used, the first point worthy of note is that even the stoichiometry of the reaction catalysed by tyrosine hydroxylase varies with the cofactor. As I mentioned, in the presence of tetrahydrobiopterin there is tight coupling between tetrahydropterin oxidized and dopa formed, that is, the ratio of these two quantities is close to unity. However, in the presence of 6,7-dimethyl-tetrahydropterin ($DMPH_4$) (the commercially available, commonly used analogue of the naturally occurring cofactor) the hydroxylation is only loosely coupled to tetrahydropterin oxidation and the ratio of tetrahydropterin oxidized to dopa formed is about 1.75 (Shiman *et al.* 1971). The uncoupling induced by the analogue and first discovered with phenylalanine hydroxylase (Kaufman 1961; Storm & Kaufman 1968) leads to the formation of an abnormal product in the hydroxylation, namely hydrogen peroxide, H_2O_2 (Storm & Kaufman 1968), which could alter the characteristics of the enzyme. Table 1 summarizes the stoichiometry of the hydroxylation in the presence of various model cofactors (Shiman *et al.* 1971). Loose coupling with the 7-methyltetrahydropterin is observed with both tyrosine and phenylalanine hydroxylases (Storm & Kaufman 1968), but tight coupling is observed with phenylalanine hydroxylase in the presence of the dimethylpterin.

For both the solubilized and the particulate enzyme, the hydroxylation is about twice as fast in the presence of tetrahydrobiopterin as it is in the presence of $DMPH_4$ (Shiman *et al.* 1971). However, there is partial uncoupling in the

TABLE 1

Stoichiometry of tyrosine hydroxylation with model cofactors

Tetrahydropterin	Ratio of dihydropterin to dopa (± s.d.)	Range	Number of determinations
$6MPH_4$	1.10 ± 0.15	(1.05–1.12)	6
$7MPH_4$	1.60 ± 0.05	(1.56–1.62)	4
$PH_4{}^a$	1.15 ± 0.05	(1.12–1.15)	4
$DMPH_4$	1.75 ± 0.15	(1.61–2.0)	10

[a] 2-Amino-4-hydroxytetrahydropteridine.

presence of the latter compound; the activity of the enzyme, therefore, is the same in the presence of either pterin when this activity is measured as tyrosine-dependent oxidation of the tetrahydropterin.

TABLE 2

Michaelis constants for pterins with tyrosine hydroxylase from bovine adrenal

Pterin	Solubilized enzyme (mM)	Particulate enzyme (mM)
DMPH$_4$	0.30	0.30
6MPH$_4$	0.30	0.30
Tetrahydrobiopterin	0.10	0.10

As can be seen in Table 2, the Michaelis constant, K_m, of both the solubilized and the particulate enzyme for tetrahydrobiopterin is lower than it is for DMPH$_4$ (Shiman & Kaufman 1970). Although it is not at all surprising that the K_m for the naturally occurring cofactor is lower than that of the dimethyl compound, a much more interesting and somewhat unexpected finding is that for all three pterin-dependent hydroxylases, the apparent K_m values for their substrates (both the amino acid substrate and oxygen) also vary with the pterin used (Kaufman 1970). The K_m values (see Table 3) illustrate this point for tyrosine hydroxylase (Shiman & Kaufman 1970). For both the purified, solubilized and the particulate enzyme, the K_m value for tyrosine is much lower in the presence of tetrahydrobiopterin than it is in the presence of the dimethyl compound.

The apparent K_m value for oxygen is also markedly lower in the presence of tetrahydrobiopterin than it is in the presence of DMPH$_4$. With the dimethyl compound, a value of 6% has been reported (Ikeda et al. 1966), whereas in the presence of tetrahydrobiopterin the K_m for oxygen is about 1% (Fisher & Kaufman 1972). With cerebral tyrosine hydroxylase, the K_m for oxygen is about 1% in the presence of DMPH$_4$ but less than 1% in the presence of tetrahydrobiopterin (Fisher & Kaufman 1972).

One of the most dramatic examples of the influence of the nature of the pterin cofactor on the enzyme's properties is the specificity of the enzyme for its amino acid substrates. Although the adrenal enzyme was originally thought to be absolutely specific for L-tyrosine (Nagatsu et al. 1964), it was later shown to

TABLE 3

Michaelis constants for tyrosine with tyrosine hydroxylase from bovine adrenal

Pterin	Solubilized enzyme (10^{-5} M)	Particulate enzyme (10^{-5} M)
DMPH$_4$	20	10
6MPH$_4$	7	4
Tetrahydrobiopterin	1.5	0.4

use L-phenylalanine also as a substrate (Ikeda *et al.* 1965). The low rate of phenylalanine hydroxylation by tyrosine hydroxylase in the presence of DMPH$_4$ indicated that this reaction was probably of minor physiological significance.

A more detailed study of the phenylalanine hydroxylating activity of tyrosine hydroxylase, however, led to the surprising finding that in the presence of tetrahydrobiopterin, the rate of phenylalanine hydroxylation by both the highly purified, solubilized enzyme and the particulate enzyme is equal to, or greater than, the rate of tyrosine hydroxylation (Shiman *et al.* 1971). Furthermore, as will be discussed in greater detail later, whereas under these conditions the enzyme is sensitive to inhibition by an excess of tyrosine, there is no evidence for inhibition by an excess of phenylalanine. These results made it highly likely that phenylalanine could serve as an important precursor of noradrenaline *in vivo*, a prediction that has since been verified (Karobath & Baldessarini 1972).

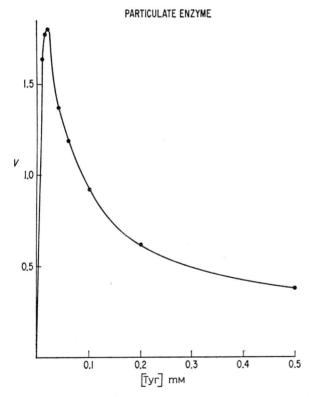

FIG. 2. Substrate inhibition by an excess of tyrosine with particulate tyrosine hydroxylase from bovine adrenal medulla. The tetrahydrobiopterin concentration was 0.1 mM.

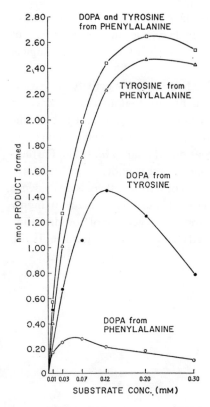

FIG. 3. Activity of tyrosine hydroxylase isolated from the rat brain as a function of either phenylalanine or tyrosine concentration in the presence of 0.126mM-tetrahydrobiopterin. Incubation time was 30 min at 37 °C.

The final pterin-dependent property of tyrosine hydroxylase that I shall discuss is its inhibition by substrates and products. This enzyme shares with hepatic phenylalanine hydroxylase the property of being inhibited by excessive concentrations of either of its substrates, tyrosine (Shiman *et al.* 1971) and oxygen (Fisher & Kaufman 1972). With both enzymes, marked inhibition is only apparent in the presence of tetrahydrobiopterin. The inhibition of adrenal tyrosine hydroxylase by tyrosine has been observed with both the particulate and the solubilized enzyme (Shiman *et al.* 1971) (see Fig. 2). The brain enzyme is also inhibited by an excess of tyrosine in the presence of tetrahydrobiopterin but it is less sensitive than the adrenal enzyme: 50% inhibition is observed with about 0.35mM-tyrosine (Lloyd & Kaufman, unpublished results) (see Fig. 3). Inhibition by the other substrate, oxygen, has been reported for both the solubilized adrenal enzyme and the brain enzyme (Fisher & Kaufman 1972).

The product of the tyrosine hydroxylase-catalysed reaction, dopa, as well as a variety of other catechols inhibit tyrosine hydroxylase (Ikeda *et al.* 1966; Udenfriend *et al.* 1965; Carlsson *et al.* 1963); the inhibition is competitive with the tetrahydropterin (Ikeda *et al.* 1966).

We have found that the inhibition by dopa, observed with both the solubilized and particulate adrenal enzyme, is non-competitive with respect to both tyrosine and tetrahydropterin (Shiman & Kaufman, unpublished results). The data for the solubilized enzyme with tetrahydrobiopterin and the particulate enzyme with $DMPH_4$ are shown in Figs. 4 and 5. I should explain that the extent of inhibition decreases as the concentration of tetrahydropterin increases, but even at infinite concentrations of tetrahydropterin, inhibition is still manifest; in the accepted kinetic terminology, this pattern of inhibition is 'non-competitive'. The same conclusion applies to the effect of tyrosine on the extent of inhibition.

The data in Figs. 4 and 5 also show that the enzyme is much more sensitive to inhibition by dopa in the presence of tetrahydrobiopterin than in the presence of $DMPH_4$. This difference is shown by both the solubilized and the particulate enzyme. The latter form of the enzyme is even more sensitive to dopa inhibition

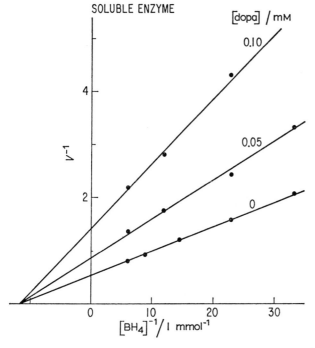

FIG. 4. Inhibition of solubilized bovine adrenal tyrosine hydroxylase by dopa in the presence of tetrahydrobiopterin (BH_4).

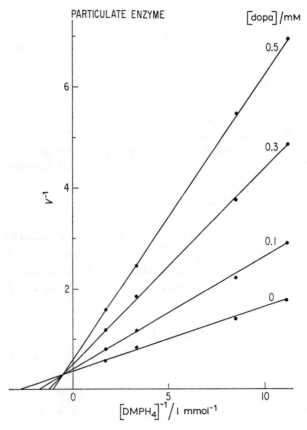

FIG. 5. Inhibition of particulate bovine adrenal tyrosine hydroxylase by dopa in the presence of DMPH₄.

than is the former. Inhibition data for both forms of the enzyme in the presence of dopa are summarized in Table 4.

To recapitulate: the apparent affinity of the enzyme for its substrates and its product (as measured by K_m and K_i values) is higher in the presence of the naturally occurring cofactor, tetrahydrobiopterin, than it is in the presence of the synthetic model cofactor, DMPH₄. In other words, substrates are better substrates and inhibitors are better inhibitors (including the substrates in high concentrations) in the presence of tetrahydrobiopterin than in the presence of DMPH₄.

I shall now consider some properties that might be relevant to the *in vivo* regulation of the enzyme's activity, limiting my comments to those aspects that do not involve alterations in the tissue concentrations of the enzyme, that is,

TABLE 4

Inhibition of tyrosine hydroxylase by L-dopa

Tetrahydropterin	Particulate enzyme K_i (slope)/μM	Soluble enzyme K_i (slope)/μM
Tetrahydrobiopterin	40	60
DMPH$_4$	150	500

Summary of K_i (slope) values obtained with two different pterins with dopa as the inhibitor. The K_i (slope) values were obtained from a replot of slope against [L-dopa]. The K_i values obtained were identical (within experimental error) for inhibition against either pterin or L-tyrosine. The concentrations of the non-varied substrates were at their K_m values.

ignoring mechanisms of regulation that appear to involve synthesis or degradation of the enzyme.

Is the *in vivo* activity of tyrosine hydroxylase normally limited by availability of its substrates and coenzyme? For an answer to this question, the K_m values of these compounds must be compared with their concentrations in the tissues under consideration. There are two obvious limitations to any firm conclusions that can be reached: (*a*) the K_m values have been determined with partially purified preparations of the enzyme and might not accurately reflect the properties of the enzyme *in vivo* and (*b*) in most cases, the concentration of the substrate in the intracellular compartment in which the enzyme might be located is not known. The only estimates of tissue concentration that can be made are inadequate, based on the weak assumption of uniform intracellular distribution of the substance in question. A final caveat is that I am assuming that the naturally occurring pterin cofactor in adrenal medulla (Lloyd & Weiner 1971) and in brain tissue is either tetrahydrobiopterin or a closely related pterin.

A key to the evaluation of the foregoing question is the fact that with tyrosine hydroxylase, just as with phenylalanine hydroxylase, the K_m for tetrahydrobiopterin is not only lower than it is with DMPH$_4$ but that the K_m values of the other two substrates, tyrosine and oxygen, are also lower in the presence of the naturally occurring pterin than in the presence of the synthetic analogue.

The K_m of the adrenal enzyme for tetrahydrobiopterin is about 0.1 mM (Shiman & Kaufman 1970) (Table 2) whereas the concentration of biopterin in bovine adrenal medulla (assuming uniform intracellular distribution) is estimated to be about 20 μM (Lloyd & Weiner 1971). The same value has been reported for whole rat adrenal glands (Rembold 1964).

Thus, the activity of the adrenal enzyme appears to be limited by availability of the cofactor, but the limitation is not nearly as severe as might have been estimated from the K_m of DMPH$_4$. On the other hand, the activity of brain enzyme would appear to be absurdly limited by low levels of the cofactor. The K_m for tetrahydrobiopterin assayed with a partially purified preparation of the enzyme from brain is about 0.25 mM (Lloyd & Kaufman, unpublished results), whereas the concentration of biopterin in rat brain is about 1 μM (Rembold 1964). This discrepancy suggests that, in brain tissue, one of the assumptions stated earlier is almost certainly invalid, in other words, (a) the K_m measured *in vitro* is not a true reflection of the value *in vivo* or (b) there is a much higher concentration of tetrahydrobiopterin in some intracellular compartment in which the hydroxylase functions or (c) the cofactor in brain is not tetrahydrobiopterin.

The K_m of the adrenal enzyme for tyrosine in the presence of tetrahydrobiopterin lies between 4 μM (particulate enzyme) and 15 μM (solubilized enzyme) (Shiman & Kaufman 1970) (see Table 3), whereas the concentration of tyrosine in bovine adrenal medulla is about 50–100 μM (Hall *et al.* 1961). It is unlikely, therefore, that the activity of the enzyme is limited by low concentrations of tyrosine.

The variation of the K_m of tyrosine with the pterin cofactor used provides a dramatic example of the importance of knowing the K_m in the presence of the naturally occurring pterin cofactor. The K_m for tyrosine in the presence of $DMPH_4$ is 13–25 times higher than it is in the presence of tetrahydrobiopterin. Based on these K_m values for tyrosine in the presence of $DMPH_4$, which are close to the tissue concentrations of the amino acid, one would conclude that tyrosine hydroxylase is limited by low concentrations of tyrosine.

We know that in the presence of tetrahydrobiopterin the K_m of the other substrate of tyrosine hydroxylase, oxygen, is about 1 % and for the brain enzyme it is somewhat less than 1 % (Fisher & Kaufman 1972).

The brain concentration of oxygen in rats breathing air at atmospheric pressure is 37 mM or 5 % (Jamieson & Van den Brenk 1965). Therefore, *in vivo*, the brain tyrosine hydroxylase is probably saturated with oxygen. The same conclusion is probably valid for the adrenal enzyme.

Tyrosine hydroxylase shares with phenylalanine hydroxylase the important property of being inhibited by excessive concentrations of either of its substrates tyrosine (Shiman *et al.* 1971) and oxygen (Fisher & Kaufman 1972). With both enzymes, marked inhibition is apparent only in the presence of tetrahydrobiopterin. The inhibition by tyrosine has been observed with both the particulate (see Fig. 2) and the solubilized adrenal enzyme (Shiman *et al.* 1971). Inhibition by oxygen has been reported for both the solubilized adrenal enzyme and the brain enzyme (Fisher & Kaufman 1972).

The inhibition by tyrosine may be physiologically significant since the enzyme is about 30 % inhibited (Shiman *et al.* 1971) at only twice the normal tissue concentration of tyrosine (Olten & Putnam 1966). In contrast to tyrosine, phenylalanine (up to 0.5 mM) does not inhibit its own hydroxylation (Shiman *et al.* 1971), although it will compete with tyrosine when both substrates are presented simultaneously to the enzyme (Ikeda *et al.* 1965).

There has been one attempt to test the possibility that an excess of tyrosine can inhibit tyrosine hydroxylase *in vivo*. Dairman (1972) reported that the concentration of tyrosine in the brains and hearts of rats roughly doubles on injection of tyrosine. In another approach, tissue concentrations of tyrosine were more than quadrupled by administration of cycloheximide to the animals. Neither treatment led to a decrease in the amount of noradrenaline in the two tissues that were examined. Dairman concluded that tyrosine hydroxylase is probably not inhibited by an excess of tyrosine *in vivo*.

Both the approach that was used and the conclusion drawn can be questioned. First, the amount of noradrenaline in the tissue might not be a sensitive measure of the activity of the hydroxylase. Second, the enzyme might not show the same sensitivity to inhibition by an excess of tyrosine in all tissues. We know that the brain hydroxylase is considerably less sensitive than is the adrenal enzyme. Indeed, it can be calculated that even if a sensitive assay of the hydroxylase had been used, the brain tyrosine concentrations after injection of tyrosine would not have led to significant inhibition. As far as the results obtained with cycloheximide are concerned the many possible effects of this drug preclude any meaningful conclusion from being made.

On the sensitivity of the enzyme to an excess of oxygen, it is known that hyperbaric oxygen (3–6 atm of pure oxygen) leads to a decrease in the amount of noradrenaline (Faiman & Heble 1966) and dopamine (Häggendal 1967) in the brain as well as noradrenaline in the heart (Buckingham *et al.* 1966). At these cerebral concentrations of oxygen (the amount of oxygen in the brains of rats breathing 4 atm of pure oxygen is 820 mM or 108% [Jamieson & Van den Brenk 1965]), 50% inhibition of bovine tyrosine hydroxylase would be expected, an inhibition that could account for the decreased amounts of noradrenaline and dopamine that have been observed after hyperbaric treatment with oxygen. From these results, we suggested (Fisher & Kaufman 1972) that the inhibition of brain tyrosine hydroxylase by oxygen could account for some of the neurotoxic effects of the treatment of newborn animals with high concentrations of oxygen (Haugaard 1968).

Finally, is there any evidence that the enzyme's sensitivity to product inhibition (and to inhibition by catechols in general), a sensitivity that is much greater in the presence of tetrahydrobiopterin than in the presence of $DMPH_4$, operates *in vivo*? The answer to this is clear. When the monoamine oxidase inhibitor, pargyline, was administered to guinea pigs, the brain and heart concentrations of noradrenaline increased 2–3 fold, whereas the rate of incorporation of $[^{14}C]$-tyrosine into noradrenaline decreased 4–5 fold (Spector *et al.* 1967).

Although the elevated amounts of noradrenaline might have inhibited one of the other enzymes in the biosynthetic pathway, the most likely explanation is that tyrosine hydroxylase, believed to catalyse the rate-limiting step in the pathway, is the site of inhibition.

We conclude that two important regulatory properties of tyrosine hydroxylase are lost or attenuated when $DMPH_4$ is substituted for tetrahydrobiopterin: inhibition by an excess of tyrosine and inhibition by dopa, the product of the reaction. It would be interesting to test the consequent prediction that $DMPH_4$ *in vivo* might lead to synthesis of noradrenaline that is relatively uncontrolled by these two restraints.

TRYPTOPHAN HYDROXYLASE

Considerably less is known about the third pterin-dependent aromatic amino acid hydroxylase, tryptophan hydroxylase.

The enzyme is believed to catalyse the rate-limiting step in the biosynthesis of the putative neurotransmitter, 5-hydroxytryptamine (Grahame-Smith 1964; Jequier et al. 1967). Since the pathway involved in the amine's synthesis consists of only two consecutive steps (hydroxylation of tryptophan to 5-hydroxytryptophan followed by decarboxylation) and the decarboxylase has a much higher activity *in vitro* than does the hydroxylase (Ichiyama et al. 1968), the conclusion that the hydroxylation is rate-limiting is reasonable if largely untested.

Grahame-Smith (1964) reported the first evidence for a specific tryptophan hydroxylase in a normal mammalian tissue. The enzyme was detected in homogenates of whole brains, but not in the high-speed supernatant fraction (100 000 g for one hour) from the centrifugate of the brain stem from dogs and rabbits. The anatomical distribution of the enzyme closely parallels the content of 5-hydroxytryptamine in these areas. Tryptophan hydroxylation was not inhibited by 50μM-aminopterin, 0.2mM-5-hydroxytryptamine or 0.2mM-5-hydroxytryptophan. No other properties of the enzyme were reported.

After Grahame-Smith's demonstration that a specific tryptophan hydroxylase exists in brain, conditions were described for its ready solubilization and the enzyme was partially purified from this tissue and from pineal glands. I shall now deal with the properties of the tryptophan hydroxylase system from these two tissues.

There has been considerable disagreement about the requirement of a tetrahydropterin for this hydroxylation. The reason for the disagreement is that the particulate preparations of the enzyme used by the earlier workers catalysed the hydroxylation without any supplementation.

In contrast to the particulate enzyme, the soluble enzyme from both brain and the pineal shows an absolute requirement for a pterin (Lovenberg et al. 1967), and shows the by-now-familiar characteristics of a pterin-dependent hydroxylase that were originally elucidated for hepatic phenylalanine hydroxylase (Kaufman 1971). Lovenberg et al. (1967) showed that both the brain and pineal enzymes required oxygen and 2-mercaptoethanol in addition to the pterin. The activity of the enzyme from the rat brain stem is 1.7 times higher with tetrahydrobiopterin than with $DMPH_4$; the K_m values reported are 30 and 5 μM for $DMPH_4$ and tetrahydrobiopterin, respectively (Jequier et al. 1969). Ichiyama et al. (1970) reported a K_m value of 60 μM for $DMPH_4$ for the brain enzyme from guinea pigs.

A more detailed study of the pterin specificity of a partially purified preparation of the soluble enzyme from rabbit hindbrain has recently been reported (Friedman et al. 1972) (see Table 5). We confirmed the finding of Jequier et al. (1969) that the K_m for tetrahydrobiopterin is much lower than it is for $DMPH_4$. The values that we reported for both pterins, however, are five times greater than those reported for the rat brain enzyme; the value for $DMPH_4$ is twice as high as that reported for the enzyme from guinea pig brain enzyme (Ichiyama et al. 1970). Whether these differences are characteristic of the enzyme in different species or whether they reflect differences in the assay conditions used still remains unknown.

The V_{max} for the enzyme from rabbit brain in the presence of tetrahydrobiopterin is slightly (but perhaps not significantly) higher than it is with $DMPH_4$ (Friedman et al. 1972) in contrast to the results of Jequier et al. (1969) with the enzyme from rat brain. As already mentioned, they reported that the rate of hydroxylation with tetrahydrobiopterin is almost 70% faster than it is with

TABLE 5

Summary of apparent K_m values for various substrates for tryptophan hydroxylase from rabbit hindbrain

Substrate	Cofactor		
	Tetrahydrobiopterin	DMPH₄	6MPH₄
L-Tryptophan	50 μM	290 μM	78 μM
Pterin	31 μM	130 μM	67 μM
Oxygen	2.5%	20%	——

All values were calculated from double reciprocal plots of initial velocity against variable substrate concentrations. The K_m for L-tryptophan with tetrahydrobiopterin as cofactor was obtained by extrapolation to the abscissa of those points in the plot that fell on a straight line (tryptophan concentrations were 0.1mM or less). When 6MPH₄ concentration was varied, the concentration of L-tryptophan was 0.7 mM. When L-tryptophan concentration was varied in the presence of 6MPH₄, the concentration of the latter was 0.29 mM.

$DMPH_4$, although it is not clear whether their rates are V_{max} values. The results obtained by Friedman *et al.* with tryptophan hydroxylase in the presence of $DMPH_4$ and tetrahydrobiopterin are consistent with those previously obtained with both phenylalanine and tyrosine hydroxylase and may be summarized as follows:—with all three pterin-dependent hydroxylases, the major difference between the naturally occurring cofactor and the cofactor analogue, $DMPH_4$, is that the K_m is much lower for the former, whereas the V_{max} with tetrahydrobiopterin is either about the same (for tryptophan hydroxylase and tyrosine hydroxylase [Friedman *et al.* 1972; Shiman *et al.* 1971]) or somewhat lower (for phenylalanine hydroxylase [Kaufman 1970]) than it is with $DMPH_4$.

Although it was reasonable to assume that tetrahydropterins function with tryptophan hydroxylase in the same way that they were first shown to function with phenylalanine hydroxylase (Kaufman 1971), this assumption has not been tested until recently. Indeed, it had never been shown that the pterin functions as a cofactor, in other words, that it functions catalytically during tryptophan hydroxylation. Friedman *et al.* (1972) proved this point for the first time when they showed that with 8.5 nmol of tetrahydrobiopterin, 40 nmol of 5-hydroxytryptophan were formed.

The demonstration of the catalytic role for the pterin implies that the tetrahydropterin must be regenerated during the hydroxylation. Unfortunately, the use of 2-mercaptoethanol in some assays (Lovenberg *et al.* 1968) has made it difficult to demonstrate the need for a tetrahydropterin-generating system. Thiols can stimulate hydroxylases dependent on pterins not only by reducing quinonoid dihydropterins (Kaufman 1959, 1964a) but perhaps also by stabilizing one of the enzymes in the system.

The first demonstration that tryptophan hydroxylation can be stimulated by highly purified dihydropteridine reductase, in addition to NADPH and tetrahydropterin, was reported recently by Friedman *et al.* (1972). Since I had previously established that the substrate for dihydropteridine reductase is a quinonoid dihydropterin (Kaufman 1964a), the stimulation of tryptophan hydroxylation by the reductase provides strong support for the conclusion that, just as is the case with phenylalanine hydroxylase, the quinonoid dihydropterin is the product of tetrahydropterin oxidation during the enzymic hydroxylation of tryptophan.

It is surprising that until recently, the stoichiometry of the reaction catalysed by tryptophan hydroxylase had not been determined: it was tacitly assumed to be the same as that previously established for the phenylalanine (Kaufman 1957) and tyrosine (Shiman *et al.* 1971) hydroxylases.

Friedman *et al.* (1972) showed that with tetrahydrobiopterin this assumption is correct. They observed that one mole of 5-hydroxytryptophan is formed for each mole of tetrahydrobiopterin oxidized. This result, taken together with their demonstration that the hydroxylation is stimulated by dihydropteridine reductase, led to the reaction being formulated as in equation (4) where BH_4 stands for tetrahydrobiopterin and BH_2 for the quinonoid dihydrobiopterin.

$$L\text{-Tryptophan} + BH_4 + O_2 \rightarrow 5\text{-Hydroxy-L-tryptophan}$$
$$+ BH_2 + H_2O \tag{4}$$

It is not yet known whether the stoichiometry with $DMPH_4$ is the same as that shown in equation (4) or whether the analogue leads to partial uncoupling, as it does with tyrosine hydroxylase (Shiman *et al.* 1971).

Just as we have already shown for phenylalanine and tyrosine hydroxylase, the K_m of tryptophan hydroxylase from rabbit brain (Friedman *et al.* 1972) for tryptophan varies with the tetrahydropterin used (see Table 5); with tetrahydrobiopterin it is only about one-sixth of what it is in the presence of $DMPH_4$. Also, the K_m of the enzyme for oxygen is dramatically sensitive to the pterin used; the value in the presence of the naturally occurring pterin is one-eighth of that with $DMPH_4$.

For a discussion of the regulatory properties of tryptophan hydroxylase, the same restrictions that limited the discussion of the regulatory properties of tyrosine hydroxylase apply; I shall not consider those phenomena that appear to involve alterations in the rates of synthesis or degradation of the hydroxylase. In trying to correlate *in vivo* concentrations of substrates for the enzyme with the K_m values for these substrates, I shall adopt the same assumptions as I did for tyrosine hydroxylase.

The early values of 300 μM reported for the K_m of this enzyme for tryptophan, all determined in the presence of $DMPH_4$, generated considerable speculative interest because this value is far above the plasma or the brain concentration of tryptophan, both of which are about 30 μM for rats (Grahame-Smith 1971; Fernstrom & Wurtman 1971). (The brain concentration was calculated from the reported value of 5 μg/g by assuming uniform distribution of the amino acid in the tissue.) These considerations led to the conclusion that tryptophan hydroxylase activity in brain is severely limited by availability of its amino acid substrate (Fernstrom & Wurtman 1971 and references therein). As Table 5 shows, however, the K_m of the enzyme for tryptophan varies significantly with the pterin cofactor used. The value in the presence of tetrahydrobiopterin is 50 μM, a figure that is close to the estimated brain concentration of this amino acid.

There is little doubt that a K_m of 50 μM is more consistent with *in vivo* observations than is the previously accepted value of 300 μM. Fernstrom & Wurtman (1971) showed that, for example, the concentration of 5-hydroxytryptamine in the brain increases when the concentration of tryptophan in the brain is increased from 5 μg/g (about 30 μM) to 15 μg/g (about 90 μM), but further increases of tryptophan (up to 45 μg/g) do not lead to further increases in the amount of 5-hydroxytryptamine. The hydroxylating system in brain, therefore, appears to be saturated with tryptophan at about 90 μM. These *in vivo* observations agree well with the apparent K_m value of 50 μM (Friedman *et al.* 1972), but would be difficult to explain if one accepted the higher value of 300 μM.

Thus, it does not appear that this hydroxylase operates *in vivo* under any unique disadvantage of substrate limitation but rather it appears to function, as probably most other enzymes do (Cleland 1967), with tissue concentrations of its substrate in the region of the K_m value of the substrate.

Tryptophan hydroxylase resembles tyrosine and phenylalanine hydroxylase in that the K_m for tetrahydrobiopterin is significantly lower than the K_m for DMPH$_4$ (Table 5). Even with the naturally occurring pterin, however, the K_m (31 μM) is considerably higher than the estimated tetrahydrobiopterin concentration in brain of 1 μM (see p. 96). This discrepancy suggests that either brain tryptophan hydroxylase is severely limited by its cofactor concentration or that the estimated brain concentration, based on the assumption of uniform distribution, is incorrect.

Just as with tryptophan, the K_m of the enzyme for oxygen is also strongly influenced by the pterin cofactor used (Table 5). The K_m for oxygen in the presence of tetrahydrobiopterin is 2.5%, a value somewhat less than the oxygen concentrations of 5% in the brain of animals breathing air (Jamieson & Van den Brenk 1965). It seems likely, therefore, that the enzyme is not normally limited by oxygen supply. This conclusion contrasts with the one that would have been reached if one relied on the oxygen K_m value of 20% determined in the presence of the artificial cofactor, DMPH$_4$.

Another property that brain tryptophan hydroxylase shares with the phenylalanine and tyrosine hydroxylases is its sensitivity to inhibition by high concentrations of its substrate (Friedman *et al.* 1972). Just as with the other two hydroxylases, this inhibition (see Figs. 6 and 7) is apparent with tetrahydrobiopterin as the cofactor but not with DMPH$_4$ (up to 2mM-tryptophan).

Evidence suggests that inhibition of brain tryptophan hydroxylase by high concentrations of tryptophan can occur *in vivo*. The administration of large amounts of tryptophan to rats increases the brain content of tryptophan (Grahame-Smith 1971; Fernstrom & Wurtman 1971). In one study (Grahame-Smith 1971) the tryptophan concentration achieved in the brain (about 0.5 mM) was high enough to inhibit the hydroxylase if the hydroxylating system *in vivo* showed the same sensitivity to an excess of tryptophan as it does *in vitro* in the presence of tetrahydrobiopterin (see Fig. 7). Although the variation of the accumulation of 5-hydroxytryptamine as a function of tryptophan concentra-

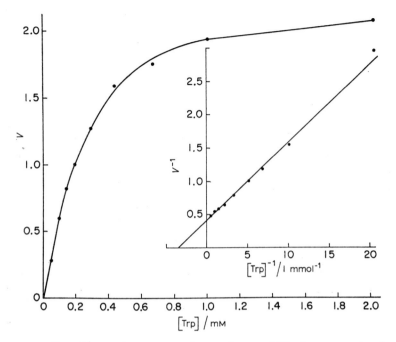

FIG. 6. Effect of tryptophan concentration on the rate of 5-hydroxytryptophan formation in
the presence of 0.33mM-DMPH$_4$. Velocity (V) is expressed as nmol 5-hydroxytryptophan
formed in 25 min at 37 °C. The *inset* shows a double reciprocal plot of the data.

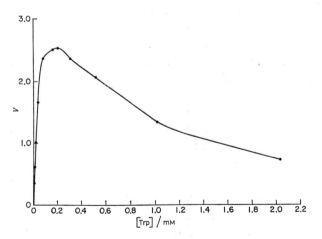

FIG. 7. Effect of tryptophan concentration on the rate of formation of 5-hydroxytryptophan
with 83μM-tetrahydrobiopterin as cofactor. Velocity (V) is expressed as nmol 5-hydroxy-
tryptophan formed in 25 min at 37 °C.

tion was described by a hyperbola, the data show that high concentrations of tryptophan inhibit the accumulation (Grahame-Smith 1971).

Unlike the other two pterin-dependent hydroxylases, partially purified tryptophan hydroxylase from brain does not appear to be inhibited by an excess of oxygen in the presence of either tetrahydrobiopterin or $DMPH_4$ (Friedman *et al.* 1972). Therefore, from the properties of the isolated enzyme, it is not possible to account for the observed diminution of the amount of 5-hydroxytryptamine in the brain by hyperbaric oxygen (Faiman *et al.* 1969). In view of the observation that some preparations of brain tryptophan hydroxylase which have lost part of their activity do show inhibition by an excess of oxygen (Friedman *et al.* 1972), however, it would be wise to regard this point as unsettled. Possibly, the sensitivity of the enzyme to inhibition by high oxygen concentrations depends on factors that have not yet been elucidated.

Does 5-hydroxytryptamine (or its precursor, 5-hydroxytryptophan) inhibit tryptophan hydroxylase, thereby inhibiting its own synthesis? Jequier *et al.* (1969) have shown that concentrations of 5-hydroxy-L-tryptophan and of the amine as high as 0.1 mM do not inhibit the enzyme from rat brain stem. Significant inhibition was observed with 1mM-5-hydroxytryptophan, but since such a high concentration was required, the authors discounted the physiological importance of end-product inhibition of the hydroxylase. Contradictory evidence was furnished by experiments with both whole animals and rat brain (striatum) slices by Macon *et al.* (1971) and Hamon *et al.* (1972) who increased the brain concentration of 5-hydroxytryptamine either with monoamine oxidase inhibitors or by exposing brain slices to large amounts of the amine. When the concentration of the amine was increased 2.5-fold, the accumulation of 5-hydroxy[^{3}H]tryptamine (and its oxidation product, 5-hydroxy[^{3}H]indoleacetic acid) from L-[G-^{3}H]tryptophan was reduced by 37%. Since the specific activity of the tryptophan pool was not significantly reduced during the experiment, the authors concluded that the end-product must have inhibited the tryptophan hydroxylase.

It is difficult to reconcile this conclusion with the relative insensitivity of partially purified tryptophan hydroxylase to inhibition by 5-hydroxytryptamine. The concentration of the amine in the slices at which the apparent inhibition was observed was only 7–20 μM (assuming uniform tissue distribution), values far below the 1 mM required for inhibition with the isolated enzyme. If the conclusion of Hamon *et al.* and Macon *et al.* is valid, one must assume that either the concentration of 5-hydroxytryptamine at the site of the hydroxylase is almost 100 times higher than that calculated on the assumption of uniform distribution or that the intracellular environment of the hydroxylase somehow confers on the enzyme high sensitivity to inhibition by 5-hydroxytryptamine.

An obvious difference between the intracellular and 'test tube' environment is that, within the cell, the enzyme probably functions with a cofactor that is closer in structure to tetrahydrobiopterin than to $DMPH_4$. Since the sensitivity toward product inhibition has been examined only with $DMPH_4$ (Jequier *et al.* 1969) and since it has been shown that substrate inhibition is seen with tetrahydrobiopterin and not with $DMPH_4$ (Friedman *et al.* 1972), we have looked at the inhibition by 5-hydroxytryptamine and by 5-hydroxytryptophan of rabbit brain tryptophan hydroxylase in the presence of tetrahydrobiopterin. In the presence of tetrahydrobiopterin, the enzyme is significantly inhibited by 0.1mM-5-hydroxytryptophan but not by the same concentration of the amine (Table 6; Tong & Kaufman, unpublished results).

If, in the experiments of the French workers, the various manipulations that were used to increase the concentration of 5-hydroxytryptamine also increased the concentration of 5-hydroxytryptophan, the data in Table 6 could help resolve the question of whether the activity of tryptophan hydroxylase is regulated by product inhibition.

I hope I have emphasized at least one major point about the pterin-dependent aromatic amino acid hydroxylases with sufficient force that it will no longer be ignored by workers in the field, namely, that the properties of this family of enzymes vary dramatically with the pterin cofactor with which they are assayed.

TABLE 6

Inhibition of tryptophan hydroxylase from rat brain by 5-hydroxytryptamine and by 5-hydroxytryptophan

Pterin	Inhibition (%)			
	1mM-5HT	*1mM-5HTP*	*0.1mM-5HT*	*0.1mM-5HTP*
Tetrahydrobiopterin(mM)				
0.57	11	96	2	71
0.28	10	94	4	52
0.14	15	92	4	46
0.07	10	90	0	27
$DMPH_4$(mM)				
1.0	19	58	8	30
0.50	28	55	7	16
0.20	28	30	8	9
0.10	22	28	0	6

5HT, 5-hydroxytryptamine; 5HTP, 5-hydroxytryptophan. The ten-fold purified rabbit brain hydroxylase (Friedman *et al.* 1972) was used.

106   S. KAUFMAN

References

BRENNEMAN, A. R. & KAUFMAN, S. (1964) The role of tetrahydropteridines in the enzymatic conversion of tyrosine to 3,4-dihydroxyphenylalanine. *Biochem. Biophys. Res. Commun.* 17, 177–183

BUCKINGHAM, S., SOMER, S. C. & MCNARY, W. F. (1966) Sympathetic activation and serotonin release as factors in pulmonary edema after hyperbaric oxygen. *Fed. Proc.* 25, 566

CARLSSON, A., CORRODI, H. & WALDECK, B. (1963) α-Substituierte Dopacetamide als Hemmer der Catechol O-Methyl-Transferase und der enzymatischen Hydroxylierung aromatischer Aminosäuren: in den Catecholamin Metabolism eingreifende Substanzen. *Helv. Chim. Acta* 46, 2271–2285

CLELAND, W. W. (1967) Enzyme kinetics. *Annu. Rev. Biochem.* 36, 77–112

CRAINE, J. E., HALL, E. S. & KAUFMAN, S. (1972) The isolation and characterization of dihydropteridine reductase from sheep liver. *J. Biol. Chem.* 247, 6082–6091

DAIRMAN, W. (1972) Catecholamine concentrations and the activity of tyrosine hydroxylase after an increase in the concentration of tyrosine in rat tissues. *Br. J. Pharmacol.* 44, 307–310

FAIMAN, M. D. & HEBLE, A. (1966) The effect of hyperbaric oxygen on cerebral amines. *Life Sci.* 5, 2225–2234

FAIMAN, M. D., HEBLE, A. & MEHL, R. G. (1966) Hyperbaric oxygenation and brain norepinephrine and 5-hydroxytryptamine: oxygen-pressure interactions. *Life Sci.* 8, 1163–1178

FERNSTROM, J. D. & WURTMAN, R. J. (1971) Brain serotonin content: physiological dependence on plasma tryptophan levels. *Science (Wash. D.C.)* 173, 149–151

FISHER, D. B. & KAUFMAN, S. (1972) The inhibition of phenylalanine and tyrosine hydroxylases by high oxygen levels. *J. Neurochem.* 19, 1359–1365

FRIEDMAN, P. A., KAPPELMAN, A. H. & KAUFMAN, S. (1972) Partial purification and characterization of tryptophan hydroxylase from rabbit hindbrain. *J. Biol. Chem.* 247, 4165–4173

GRAHAME-SMITH, D. G. (1964) Tryptophan hydroxylation in brain. *Biochem. Biophys. Res. Commun.* 16, 586–592

GRAHAME-SMITH, D. G. (1971) Studies *in vivo* on the relationship between brain tryptophan, brain 5-HT synthesis and hyperactivity in rats treated with a monoamine oxidase inhibitor and L-tryptophan. *J. Neurochem.* 18, 1053–1066

HÄGGENDAL, J. (1967) The effect of high pressure air or oxygen with and without carbon dioxide added on the catecholamine levels of rat brain. *Acta Physiol. Scand.* 69, 147–152

HALL, G., HILLARP, N. A. & THIEME, G. (1961) Phenylalanine and tyrosine in the adrenal medulla. *Acta Physiol. Scand.* 52, 49–52

HAMON, M., BOURGOIN, S., MOROT-GAUDRY, Y. & GLOWINSKI, J. (1972) End product inhibition of serotonin synthesis in the rat striatum. *Nat. New Biol.* 237, 184–187

HAUGAARD, N. (1968) Cellular mechanism of oxygen toxicity. *Physiol. Rev.* 48, 311–373

ICHIYAMA, A., NAKAMURA, S., MISHIZUKA, Y. & HAYAISHI, O. (1968) Tryptophan 5-hydroxylase in mammalian brain. *Adv. Pharmacol.* 64, 5–17

ICHIYAMA, A., NAKAMURA, S., NISHIZUKA, Y. & HAYAISHI, O. (1970) Enzymic studies on the biosynthesis of serotonin in mammalian brain. *J. Biol. Chem.* 245, 1699–1709

IKEDA, M., LEVITT, M. & UDENFRIEND, S. (1965) Hydroxylation of phenylalanine by purified preparations of adrenal and brain tyrosine hydroxylase. *Biochem. Biophys. Res. Commun.* 18, 482–488

IKEDA, M., FAHIEN, L. A. & UDENFRIEND, S. (1966) A kinetic study of bovine adrenal tyrosine hydroxylase. *J. Biol. Chem.* 241, 4452–4456

JAMIESON, D. & VAN DEN BRENK, H. A. J. (1965) Electrode size and tissue pO₂ measurement in rats exposed to air or high pressure oxygen. *J. Appl. Physiol.* 20, 514–518

JEQUIER, E., LOVENBERG, W. & SJOERDSMA, A. (1967) Tryptophan hydroxylase inhibition: the mechanism by which p-chlorophenylalanine depletes rat brain serotonin. *Mol. Pharmacol.* 3, 274–278

JEQUIER, E., ROBINSON, D. S., LOVENBERG, W. & SJOERDSMA, A. (1969) Further studies on trypto-phan hydroxylase in rat brainstem and beef pineal. *Biochem. Pharmacol.* **18**, 1071–1081

KAROBATH, M. & BALDESSARINI, R. J. (1972) Formation of catechol compounds from phenyl-alanine and tyrosine with isolated nerve endings. *Nat. New Biol.* **236**, 206–208

KAUFMAN, S. (1957) The enzymatic conversion of phenylalanine to tyrosine. *J. Biol. Chem.* **226**, 511–524

KAUFMAN, S. (1958) A new cofactor required for the enzymatic conversion of phenylalanine to tyrosine. *J. Biol. Chem.* **230**, 931–939

KAUFMAN, S. (1959) Studies on the mechanism of the enzymatic conversion of phenylalanine to tyrosine. *J. Biol. Chem.* **234**, 2677–2682

KAUFMAN, S. (1961) The enzymatic conversion of 4-fluorophenylalanine to tyrosine. *Biochem. Biophys. Acta* **61**, 619–621

KAUFMAN, S. (1963) The structure of phenylalanine hydroxylation cofactor. *Proc. Natl. Acad. Sci. U.S.A.* **50**, 1085–1093

KAUFMAN, S. (1964a) Further studies on the structure of the primary oxidation product formed from tetrahydropteridines during phenylalanine hydroxylation. *J. Biol. Chem.* **239**, 332–338

KAUFMAN, S. (1964b) The role of pteridines in the enzymatic conversion of phenylalanine to tyrosine. *Trans. N.Y. Acad. Sci. (Ser. II)* **26**, 977–983

KAUFMAN, S. (1970) A protein that stimulates rat liver phenylalanine hydroxylase. *J. Biol. Chem.* **245**, 4751–4758

KAUFMAN, S. (1971) The phenylalanine hydroxylating system from mammalian liver. *Adv. Enzymol.* **35**, 245–319

KAUFMAN, S. & LEVENBERG, B. (1959) Further studies on the phenylalanine hydroxylation cofactor. *J. Biol. Chem.* **234**, 2683–2688

LLOYD, T. & KAUFMAN, S. (1973) Production of antibodies to bovine adrenal tyrosine hydroxyl-ase; cross-reactivity studies with other pterin-dependent hydroxylases. *Mol. Pharmacol.* **9**, 438–444

LLOYD, T. & WEINER, N. (1971) Isolation and characterization of a tyrosine hydroxylase cofactor from bovine adrenal medulla. *Mol. Pharmacol.* **7**, 569–580

LOVENBERG, W., JEQUIER, E. & SJOERDSMA, A. (1967) Tryptophan hydroxylation: measurement in pineal gland, brainstem, and carcinoid tumor. *Science (Wash. D.C.)* **155**, 217–219

LOVENBERG, W., JEQUIER, E. & SJOERDSMA, A. (1968) Tryptophan hydroxylation in mammalian systems. *Adv. Pharmacol.* **6A**, 21–36

MACON, J. B., SOKOLOFF, L. & GLOWINSKI, J. (1971) Feedback control of rat brain 5-hydroxy-tryptamine synthesis. *J. Neurochem.* **18**, 323–331

MUSACCHIO, J. M., WURTZBURGER, R. J. & D'ANGELO, G. L. (1971) Different molecular forms of bovine adrenal tyrosine hydroxylase. *Mol. Pharmacol.* **7**, 136–146

NAGATSU, T., LEVITT, M. & UDENFRIEND, S. (1964) Tyrosine hydroxylase: the initial step in norepinephrine biosynthesis. *J. Biol. Chem.* **239**, 2910–2917

NIELSEN, K. M., SIMONSEN, V. & LIND, K. E. (1969) Dihydropteridine reductase: a method for the measurement of activity, and investigations of the specificity for NADH and NADPH. *Eur. J. Biochem.* **9**, 497–502

OLTEN, R. R. & PUTMAN, P. A. (1966) Plasma amino acids and nitrogen retention by steers fed purified diets containing urea or isolated soy protein. *J. Nutr.* **89**, 385–391

PETRACK, B., SHEPPY, F. & FETZER, V. (1968) Studies on tyrosine hydroxylase bovine adrenal medulla. *J. Biol. Chem.* **243**, 743–748

REMBOLD, H. (1964) in *Pteridine Chemistry* (Pfleiderer, W. & Taylor, E. C., eds.), pp. 465–484, Pergamon, New York

SCRIMGEOUR, K. G. & CHEEMA, S. (1971) Discussion paper: quinonoid dihydropterin reduct-ase. *Ann. N.Y. Acad. Sci.* **186**, 115–118

SHIMAN, R. & KAUFMAN, S. (1970) Tyrosine hydroxylase (bovine adrenal glands). *Methods Enzymol.* **17A**, 609–615

SHIMAN, R., AKINO, M. & KAUFMAN, S. (1971) Stimulation and partial purification of tyrosine
   hydroxylase from bovine adrenal medulla. *J. Biol. Chem.* **246**, 1330–1340
SPECTOR, S., GORDON, R., SJOERDSMA, A. & UDENFRIEND, S. (1967) End product inhibition of
   tyrosine hydroxylase as a possible mechanism for regulation of norepinephrine synthesis.
   Mol. Pharmacol. **3**, 549–555
STORM, C. B. & KAUFMAN, S. (1968) The effect of variation of cofactor and substrate structure
   on the action of phenylalanine hydroxylase. *Biochem. Biophys. Res. Commun.* **32**, 788–
   793
UDENFRIEND, S., ZALTZMAN-NIRENBERG, P. & NAGATSU, T. (1965) Inhibitors of purified beef
   adrenal tyrosine hydroxylase. *Biochem. Pharmacol.* **14**, 837–845
ZANNONI, V. G., BROWN, N. C. & LADU, B. N. (1963) Role of reduced folic acid and derivatives in tyrosine metabolism. *Fed. Proc.* **22**, 232

Discussion

Grahame-Smith: You refer to the naturally occurring cofactor with caution because that is the cofactor you isolated from the liver. Does that mean that there is not another one in the brain?

Kaufman: Biopterin is the only naturally occurring cofactor that is currently known. Whether biopterin is the natural cofactor in brain is still an open question: the situation is confused. As I mentioned, the K_m for tetrahydrobiopterin assayed with a partially purified preparation of tyrosine hydroxylase from brain is about 0.25 mM (Lloyd & Kaufman, unpublished results). Rembold (1964) estimated that the concentration of biopterin in rat brain is about 1 μM. This is a huge discrepancy. However, the brain concentration was calculated on the assumption that biopterin is uniformly distributed within the brain. This assumption is undoubtedly wrong. The discrepancy indicates that biopterin is considerably concentrated wherever it is needed.

Alternatively, the cofactor in brain is not biopterin but it almost certainly is a closely related compound. I cannot believe that all these subtle differences in regulatory properties of tyrosine hydroxylase in the presence of a compound like biopterin and, say, 6,7-dimethyltetrahydropterin, are just accidental. It could be a derivative which resembled biopterin or a stereoisomer. But until someone identifies the natural cofactor in brain, biopterin is the only naturally occurring cofactor in mammalian tissues.

Wurtman: If catecholamines interfere with the interaction between the pteridine and the enzyme by virtue of the catechol group, there doesn't appear to be much basis for expecting a monohydroxylated compound like 5-hydroxytryptamine to affect the hydroxylation of tryptophan, at least by competitive inhibition with a cofactor.

Weiner: Do catechols inhibit tryptophan hydroxylase?

Kaufman: We observed inhibition only with high concentrations even in the presence of biopterin.

Weiner: Why are the kinetics of inhibition by dopa (Table 1), which do not look competitive with the pterin cofactor, strikingly different from everything that has been done with 6,7-dimethyltetrahydropterin (Ikeda *et al.* 1966)? We also have made Lineweaver–Burk plots of the inhibition by catecholamines of several enzyme preparations of tyrosine hydroxylase and have always found the inhibition competitive with the pterin cofactor. We have not examined the kinetics of the inhibition by dopa.

Kaufman: Compare Figs. 4 and 5. 6,7-Dimethyltetrahydropterin was much more competitive than tetrahydrobiopterin. With the dimethylpterin, the intercept on the ordinate was only slightly displaced from the origin. But it was displaced. We rigorously checked this. It is interesting that the inhibition is much more severe with biopterin than with the dimethyl compound.

Weiner: Using intact tissues and a rather indirect kinetic analysis of tyrosine hydroxylase, we have estimated the concentration of pterin cofactor in the adrenergic neuron of the vas deferens (see later, pp. 135–147) to be equivalent to about 0.15 mmol/1 of 6,7-dimethyltetrahydropterin. Since the synthetic cofactor is about five times less potent than biopterin, as you find, we would expect the concentration of biopterin in the adrenergic nerve to be about 0.03 mmol/1.

Kaufman: To obtain a better estimate of the concentration of biopterin in brain and neural tissue, we intend to determine its concentration in synaptosomes.

Gessa: Was there any change in the pteridine cofactor after specific lesions?

Weiner: Musacchio & Wurzburger (1973) have not looked at pterins but at dihydropterin reductase and they find a high concentration of this enzyme throughout the brain.

Kaufman: We found the same thing. Musacchio *et al.* (1971) made the unlikely suggestion that dihydropteridine reductase was a rate-limiting enzyme in this system, although it was orders of magnitude more active than the known hydroxylases in brain. There must be other pterin-dependent reactions in the brain still to be discovered, which are probably quantitatively much more important.

Fernstrom: Dairman (1972) injected tyrosine (i.p. into rats) and found no depletion of catecholamine concentrations in the brain. This evidence does not convincingly establish inhibition of the hydroxylase. Is there any kinetic evidence that suggests this relationship?

Kaufman: Dr Weiner will discuss this (see pp. 135–147). Lloyd and I have some unpublished preliminary evidence which indicates that tyrosine hydroxyl-ase, isolated from brain without proteolysis and therefore presumably a native enzyme, is certainly much less sensitive (by a factor of 5–10) to inhibition by an excess of tyrosine in the presence of tetrahydrobiopterin than is the adrenal enzyme. In Dairman's experiment, one would not have expected any inhibition. Also, his assay for measuring the rate of reaction is not sufficiently sensitive. I doubt whether he could have detected inhibition if there was any.

I should emphasize that some of the properties reported in the paper do apply only to the adrenal enzyme.

Wurtman: If the K_m values of tyrosine hydroxylase and tyrosine amino-transferase are about 100 µM (Kaufman, p. 100) and 7.5 mM (Mandel, p. 74), respectively, and the concentration of tyrosine in the brain about 10 µM (about 14 µg/g), both enzymes would normally be unsaturated with their amino acid substrate. In those rare circumstances when the concentration of tyrosine within the brain was somehow raised beyond the point of saturating tyrosine hydroxyl-ase—that is higher than 100–150 µM—the consequent suppression of catechol-amine synthesis might be counteracted by the ability of tyrosine aminotrans-ferase to destroy the excess of amino acid. Perhaps this is the physiological function of brain tyrosine aminotransferase.

Kaufman: This is eminently possible in the adrenal gland, but I am very reluctant to believe it also happens in the brain.

Wurtman: It is true, isn't it, Dr Munro, that tyrosine is one of the most toxic amino acids?

Munro: Yes. The levels which cause toxicity are low, because the toxicity is due to specific products and to vascular changes.

Wurtman: Fig. 7 shows the inhibition of tryptophan hydroxylase by tryp-tophan. Surely the concentration of substrate in the animal's brain would never be sufficiently high to suppress the hydroxylation?

Kaufman: No. Grahame-Smith (1971), in the experiment I referred to (p. 102), achieved a concentration of tryptophan in the rat brain sufficient to cause 20% inhibition.

Wurtman: That would occur only in animals receiving large, pharmacological doses of tryptophan. In the normal dynamic range, the concentration of brain tryptophan would never be nearly so high.

Fernstrom: Dr Gál, repeating Dr Glowinski's experiments (Macon *et al.* 1971) with pargyline, found no end-product inhibition of tryptophan hydroxylase *in vivo* (Millard *et al.* 1972).

Gál: Our results diverge from those of Dr Kaufman with respect to the amount of 5-hydroxytryptamine necessary to inhibit tryptophan 5-hydroxylase (p. 104). Everybody agrees about the feedback inhibition of tyrosine hydroxylase by noradrenaline, but considerable disagreements arise over whether 5-hydroxytryptamine causes feedback inhibition of tryptophan 5-hydroxylase. If 5-hydroxytryptamine has any inhibitory effect *in vivo*, a concentration of 0.1 mmol/l is required. Clearly, our kinetic considerations are different.

Kaufman: Did you use partially purified tryptophan hydroxylase or is this an *in vivo* experiment?

Gál: These were *in vivo* experiments. For instance, both 5-hydroxy-α-methyltryptophan (Upjohn) *in vitro* and α-methyltryptophan *in vivo* partially inhibit the synthesis of noradrenaline. I cannot support results related to 'feedback inhibition' by 5-hydroxytryptamine.

Kaufman: Tong's data are derived from experiments on tryptophan hydroxylase from rabbit hindbrain. There may be a species difference. Tong's experiment is simple to reproduce. With a partially purified preparation of tryptophan hydroxylase from rabbit hindbrain, and [^{14}C]tryptophan as the substrate, 5-hydroxy[^{14}C]tryptophan was separated from the tryptophan by paper chromatography. We are therefore certain that the product is 5-hydroxytryptophan. In the presence of the product we have to use an assay based on chromatographic separation rather than one where the fluorescence of the 5-hydroxytryptophan is measured.

Gál: We agree that 5-hydroxytryptamine does not affect the enzyme *in vitro*. From physicochemical and enzymological considerations alone, a product which inhibits by a feedback mechanism should work *in vivo* as well as *in vitro*. I hope we can all agree about that. It is true for noradrenaline but not for 5-hydroxytryptamine.

Glowinski: End-product regulation of 5-hydroxytryptamine synthesis in central 5-hydroxytryptaminergic neurons has already been the subject of many discussions. Some authors first failed to detect this mechanism after inhibition of monoamine oxidase (Lin *et al.* 1969; Millard & Gál 1971); however, we showed that increased intraneuronal concentrations of the amine, induced by inhibitors of monoamine oxidase or uptake of exogenous amine, could inhibit synthesis of the amine from tryptophan both *in vivo* (Macon *et al.* 1971) and in brain slices (Hamon *et al.* 1972, 1973). These results have been confirmed by other workers (Carlsson *et al.* 1973; Karobath 1972). So, there is little doubt that synthesis of 5-hydroxytryptamine is reduced when concentrations of the amine are increased by inhibition of monoamine oxidase. Moreover, we noticed that this feedback regulation accompanies the first step of the amine's synthesis. Dr Kaufman's observation that tryptophan hydroxylase is inhibited by 5-hydroxytryptophan and not by 5-hydroxytryptamine is most interesting but does not yet explain our results.

Kaufman: We thought about the converse. Is it possible that in your experiments raising 5-hydroxytryptamine concentrations could have raised the concentration of 5-hydroxytryptophan? Is the decarboxylase inhibited by its own product? If 5-hydroxytryptamine inhibited the decarboxylation of 5-hydroxytryptophan, the amount of the latter would increase.

Table 6 showed the strange *in vitro* dependence of the extent of inhibition on the concentration of the tetrahydrobiopterin. I am sure that extrapolation of brain concentrations of biopterin *in vivo* would probably indicate zero inhibition so that on the basis of our *in vitro* results we cannot make a strong case for this type of inhibition occurring *in vivo*.

Gál: The metabolism of tryptophan to 5-hydroxytryptamine is not linear. The pathway branches at the transamination of 5-hydroxytryptophan even though only 10–15% of 5-hydroxytryptophan is going to 5-hydroxyindole-pyruvic acid. If the amount of 5-hydroxytryptamine is increased by administration of, for instance, pargyline or any other monoamine oxidase inhibitor, the transamination is suppressed. Thus, the transamination is inhibited and more 5-hydroxytryptophan is thereby conserved for decarboxylation.

Wurtman: Dr Carlsson, when you suppressed the decarboxylation of 5-hydroxytryptophan, did the accumulation of that compound in brain inhibit the hydroxylation of tryptophan?

Carlsson: The concentration of 5-hydroxytryptophan increases linearly during the first 30 min after giving the decarboxylase inhibitor. This suggests a lack of product inhibition. These concentrations are still low compared with those at which Dr Kaufman observed inhibition. I doubt if there is any physiological situation in which sufficient amounts accumulate to suppress tryptophan hydroxylase.

Gessa: Since 5-hydroxytryptophan is being produced in 5-hydroxytryptaminergic nerve endings, its concentration might well reach 0.1 mmol/l.

Curzon: Has anyone demonstrated end-product inhibition of the synthesis of 5-hydroxytryptamine within its physiological range of concentration? In Dr Glowinski's experiments with pargyline the rate of increase of 5-hydroxytryptamine did not decline until its concentration was about two and a half times the physiological one.

Glowinski: That is perfectly correct. This is a great difference between 5-hydroxytryptaminergic neurons and catecholaminergic neurons. However, we never concluded that the negative feedback regulation of the amine's synthesis took place in normal physiological states. This has still to be demonstrated. Surprisingly, in experiments on the daily variations in the synthesis of the amine, we observed that synthesis was maximal during the light period when cerebral amounts of the amine are increased (Hery *et al.* 1972).

Moir: Over several years we have systematically investigated the indolylalkylamines in brain under various conditions by chromatographic methods which detected variations in 5-hydroxytryptophan, 5-hydroxytryptamine and 5-hydroxyindoleacetic acid (Ashcroft *et al.* 1965). This has been most unfruitful, for only in extreme conditions (such as treatment with α-methyldopa) did we ever detect 5-hydroxytryptophan. Even pretreatment with monoamine oxidase inhibitors and subsequent loading with pharmacological amounts of tryptophan does not give rise to 5-hydroxytryptophan in detectable quantities (Moir & Yates 1972). It is difficult to see from this how 5-hydroxytryptophan concentrations could have an *in vivo* implication for feedback.

Lajtha: Dr Kaufman, do you think that proteolytic enzymes can change the inactive form of the hydroxylase into a more active form or change hydroxylase activity by altering other properties of the enzyme?

Kaufman: Fisher & I (1973) recently found a 20- to 50-fold activation of soluble phenylalanine hydroxylase on treating the native enzyme with chymotrypsin. We now suspect that what everybody, including us, has been working on before is very close to a zymogen. The situation is a little different from that of the classical pancreatic zymogen in that the undigested hydroxylase has 2–5% of the normal activity. This activation is only seen when biopterin is used as the cofactor. We are still investigating whether this activation phenomenon is physiologically significant.

The V_{max} of the reaction catalysed by phenylalanine hydroxylase with the natural cofactor is not high, but after proteolytic activation there is 20- to 50-fold increase in V_{max}, and so biopterin becomes a much better cofactor than any of the model pterin cofactors. We don't know whether this applies *in vivo*, although it is known that the stability of a lysosomal membrane—which may determine the extent of intracellular proteolysis—can be varied by the amount of vitamin A, selenium and vitamin E in the diet.

Carlsson: Recently, Nagatsu *et al.* (1972) tried the natural cofactor and several synthetic cofactors and found about fivefold differences not only in the affinities but also in V_{max}.

Eccleston: It is necessary to postulate rapid changes in the rate of tryptophan hydroxylation in brain, because if one stimulates raphe cells electrically, the rate of synthesis increases within 30 min. When this stimulation is stopped, the enzyme seems to 'switch off' within 30 min. If one stops the cells from firing by using small doses of LSD, the synthesis switches off rapidly (Shields & Eccleston 1973). So, our theories must incorporate a rapid mechanism for 'switching on' and 'switching off' the hydroxylase.

Gál: The factor that Dr Kaufman mentioned has recently been described for tyrosine hydroxylase as well. Tryptic digestion causes a change in all these enzymes with respect to activation by cofactors.

Kaufman: The activity of tyrosine hydroxylase was almost doubled with the dimethylpterin as a cofactor. The effect on phenylalanine hydroxylase shows the specificity which one hopes might mean that it is physiologically significant. The great activation of phenylalanine by phospholipids or by limited proteolysis is only seen when the enzyme is assayed with tetrahydrobiopterin; when assayed with the dimethyl compound, there is very little activation.

Gál: Did you find any change in molecular weight of the enzyme?

Kaufman: The proteolytic treatment splits off about 30% of the native molecule and we find one subunit.

References

ASHCROFT, G. W., ECCLESTON, D. & CRAWFORD, T. B. B. (1965) 5-Hydroxyindole metabolism in rat brain. A study of intermediate metabolism using the technique of tryptophan loading I. *J. Neurochem.* **12**, 483–492

CARLSSON, A., BÉDARD, P., LINDQUIST, M. & MAGNUSSON, T. (1973) The influence of nerve-impulse flow on the synthesis and metabolism of 5-hydroxytryptamine in the central nervous system. *Biochem. Soc. Symp.* **86**, 17–32

DAIRMAN, W. (1972) Catecholamine concentrations and the activity of tyrosine hydroxylase after an increase in the concentration of tyrosine in rat tissues. *Br. J. Pharmacol.* **44**, 307–310

FISHER, D. B. & KAUFMAN, S. (1973) The stimulation of rat liver phenylalanine hydroxylase by lysolecithin and α-chymotrypsin. *J. Biol. Chem.* **248**, 4345–4353

GRAHAME-SMITH, D. G. (1971) Studies *in vivo* on the relationship between brain tryptophan, brain 5-HT synthesis and hyperactivity in rats treated with a monoamine oxidase inhibitor and L-tryptophan. *J. Neurochem.* **18**, 1053–1066

HAMON, M., BOURGOIN, S., MOROT-GAUDRY, Y. & GLOWINSKI, J. (1972) End-product inhibition of serotonin synthesis in the rat striatum. *Nat. New Biol.* **237**, 184–187

HAMON, M., BOURGOIN, S. & GLOWINSKI, J. (1973) Feed-back regulation of serotonin synthesis in rat striatal slices. *J. Neurochem.* **20**, 1727–1745

HERY, F., ROUER, E. & GLOWINSKI, J. (1972) Daily variations of serotonin metabolism in the rat brain. *Brain Res.* **43**, 445–465

IKEDA, M., FAHIEN, L. A. & UDENFRIEND, S. (1966) A kinetic study of bovine adrenal tyrosine hydroxylase. *J. Biol. Chem.* **241**, 4452–4456

KAROBATH, M. (1972) Serotonin synthesis with rat brain synaptosomes effects of serotonin and monoamine oxidase inhibitors. *Biochem. Pharmacol.* **21**, 1253–1263

LIN, R. C., NEFF, N. H., NGAI, S. H. & COSTA, E. (1969) Turnover rates of serotonin and nor-epinephrine in brain of normal and pargyline treated rats. *Life Sci.* **8**, 1077–1084

MACON, J. B., SOKOLOFF, L. & GLOWINSKI, J. (1971) Feed-back control of rat brain 5-hydroxy-tryptamine synthesis. *J. Neurochem.* **18**, 323–331

MANDEL, P. & AUNIS, D. (1974) Tyrosine aminotransferase in the brain, *This Volume*, pp. 67–79

MILLARD, S. A. & GÁL, E. M. (1971) The contribution of 5-hydroxyindolepyruvic acid to cerebral 5-hydroxyindole metabolism. *Int. J. Neurosci.* **1**, 211–218

MILLARD, S. A., COSTA, E. & GÁL, E. M. (1972) On the control of brain serotonin turnover rate by end-product inhibition. *Brain Res.* **40**, 545–551

Moir, A. T. B. & Yates, C. M. (1972) Interaction in the cerebral metabolism of the biogenic amines. Effect of phenelzine on this interaction. *Br. J. Pharmacol.* **45**, 265–274

Musacchio, J. M. & Wurzburger, R. J. (1973) Catecholamine biosynthesis. Regional distribution of dihydropteridine reductase in the bovine brain. *Fed. Proc.* **32**, 707 Abs

Musacchio, J. M., D'Angelo, G. L. & McQueen, C. A. (1971) Dihydropteridine reductase: implication on the regulation of catecholamine biosynthesis. *Proc. Natl. Acad. Sci. U.S.A.* **68**, 2087–2091

Nagatsu, T., Mizutain, K., Nagatsu, J., Matsuura, S. & Sugimoto, T. (1972) Pteridines as cofactor or inhibitor of tyrosine hydroxylase. *Biochem. Pharmacol.* **21**, 1945–1953

Rembold, H. (1964) in *Pteridine Chemistry* (Pfleiderer, W. & Taylor, E. C., eds.), pp. 465–484, Pergamon, New York

Shields, P. J. & Eccleston, D. (1973) Evidence for synthesis and storage of 5-hydroxytryptamine in two separate pools in the brain. *J. Neurochem.* **20**, 881–888

Weiner, N., Lee, F.-L., Waymire, J. C. & Posiviata, M. (1974) The regulation of tyrosine hydroxylase activity in adrenergic nervous tissue, *This Volume*, pp. 135–147

The *in vivo* estimation of rates of tryptophan and tyrosine hydroxylation: effects of alterations in enzyme environment and neuronal activity

A. CARLSSON

Department of Pharmacology, University of Göteborg

Abstract The products of hydroxylation of tryptophan and tyrosine by the appropriate hydroxylases, namely 5-hydroxytryptophan and 3,4-dihydroxy-phenylalanine (dopa) respectively, have not yet been detected with certainty in normal mammalian brain since the concentrations are less than 10 ng/g. After inhibition of the aromatic amino acid decarboxylase (e.g. with 3-hydroxybenzyl-hydrazine), however, both these substrates accumulate in rat or mouse brain initially at a constant rate. This accumulation is completely blocked by inhibitors of the respective aromatic hydroxylases. If such inhibitors are given some time after 3-hydroxybenzylhydrazine, the amounts of accumulated 5-hydroxy-tryptophan and dopa remain almost constant; efflux is slow. These and other observations, as well as comparisons with other measurements of monoamine synthesis, suggest that the initial accumulation of the two intermediates after treatment with 3-hydroxybenzylhydrazine is a useful indicator of the rate of hydroxylation of tryptophan and of tyrosine.

With this procedure, we have shown that tyrosine (but not tryptophan) hydroxylase of rat or mouse brain is saturated with its physiological amino acid substrate; the Michaelis constant, K_m, for tryptophan hydroxylase, estimated *in vivo*, was 60 μM, in agreement with published *in vitro* data with the natural cofactor tetrahydrobiopterin. Even moderate hypoxia retarded the hydroxyla-tion of both amino acids, as predicted from *in vitro* observations. Data indicating a complex interaction between neuronal activity and tryptophan and tyrosine hydroxylation rates are reported.

Dr Kaufman (see this volume, pp. 85–108) has just surveyed some basic properties of the pterin-dependent aromatic amino acid hydroxylases as they appear in partially purified *in vitro* systems. I shall report the behaviour of these enzymes in the brains of intact organisms.

METHOD

Several methods are available for investigating the synthesis of monoamines in the intact animal, but only one method directly measures the first step in their synthesis, the hydroxylation of tryptophan or tyrosine (Carlsson et al. 1972a). In principle, this method measures the amounts of 5-hydroxytryptophan and dopa accumulating after inhibition of the aromatic amino acid decarboxylase. Normally, the concentrations of these intermediates are too low—less than 10 ng/g—to permit detection by available analytical techniques. After administration of an inhibitor of the decarboxylase, for example 3-hydroxybenzyl-hydrazine (NSD 1015; 100 mg/kg intraperitoneally), there is an initially linear accumulation of both these intermediates in rat or mouse brain. Regional distribution studies reveal that both intermediates are formed locally in the brain; uptake from the blood stream is negligible. Histochemical observations shows that 5-hydroxytryptophan and dopa accumulate intraneuronally, in cell bodies and fibres normally storing 5-hydroxytryptamine and catecholamines, respectively (Bédard et al. 1971). Moreover, when these fibres degenerate after axotomy, the accumulations decline markedly (Carlsson et al. 1973a and unpublished results).

The efflux of 5-hydroxytryptophan and dopa can be studied by administration of an inhibitor of the decarboxylase followed, after a suitable time interval, by an inhibitor of tryptophan or tyrosine hydroxylase. Efflux of both compounds is slow (Carlsson et al. 1972a; Carlsson & Lindqvist 1973). The initial accumulations of the two intermediates are thus essentially determined by their rates of formation and should thus serve as indicators of the *in vivo* activities of the two hydroxylases.

The rate constants for amine synthesis, calculated from data obtained with NSD 1015, were 0.21, 0.39 and 0.54 h^{-1} for rat striatal dopamine, hemisphere noradrenaline and whole brain 5-hydroxytryptamine, respectively. These values are similar to those obtained by other techniques (see Carlsson et al. 1972a).

At the doses used, which incidentally are not critical, NSD 1015 markedly inhibited monoamine oxidase as well. The depletion of monoamines resulting from the decarboxylase inhibition is thus much diminished, thereby reducing the risk of feedback-induced changes in hydroxylase activities.

SUBSTRATE CONCENTRATION

Confirming earlier suggestions, we found that the tryptophan hydroxylase of normal mouse or rat brain was slightly less than half-saturated with its amino acid substrate. The tryptophan content of the brain had to be increased several times by injection of the amino acid in order to observe the maximum rate of hydroxylation. The K_m we calculated assuming that the concentration of tryptophan in the fluid surrounding the hydroxylase was equal to that of whole brain was 60 µM (Carlsson & Lindqvist 1972), which is in excellent agreement with the value reported by Dr Kaufman. We did not find the inhibition by high concentrations of the amino acid observed by Kaufman in our *in vivo* studies, possibly because inhibitory concentrations were not reached even after large doses of the amino acid.

The rate of tyrosine hydroxylation in the rat brain was unaffected by exogenous tyrosine, even in large excess, indicating that tyrosine hydroxylase is normally saturated with its amino acid substrate (Carlsson *et al.* 1972*a*). This might be expected from the K_m (20 µM; Shiman & Kaufman 1970) and the normal tyrosine concentration in brain (80 µM).

HYDROXYLASE INHIBITORS

Tyrosine hydroxylation is inhibited by α-methyltyrosine, a well known competitive inhibitor of tyrosine hydroxylase; the inhibition depends on the dose. As might be expected, this inhibition can be overcome by the administration of large doses of tyrosine (Carlsson & Lindqvist, unpublished results).

The hydroxylation of tryptophan is retarded by known inhibitors of tryptophan hydroxylase, such as *p*-chlorophenylalanine and 2-(3,4-dihydroxyphenyl)-pentanamide, as well as by large doses of tyrosine or phenylalanine (Carlsson & Lindqvist 1972).

pO_2

The *in vitro* data reported by Dr Kaufman indicate a low affinity of the aromatic amino acid hydroxylases for oxygen. The possibility that the amount of oxygen in a tissue, pO_2, is rate-limiting for hydroxylation *in vivo* thus can not be excluded. Indeed, even moderate hypoxia, which does not alter cerebral carbohydrate metabolism and energy production, markedly inhibited tyrosine and tryptophan hydroxylase (Davis & Carlsson 1973 and unpublished results); total amine concentrations are but slightly influenced, presumably because oxidative deamination is also inhibited. This interference might be functionally relevant, because the behavioural disturbance caused by hypoxia can be significantly alleviated by L-dopa (Brown *et al.* 1973).

TABLE 1

Synthesis and turnover of 5-hydroxytryptamine in the brains of rats exposed to room air or 5.6% oxygen (Davis & Carlsson, unpublished data).

Inhibitor	Room air		5.6% O_2	
	Accumulation (nmol h^{-1})	Synthesis (k/h^{-1})	Accumulation (nmol h^{-1})	Synthesis (k/h^{-1})
NSD 1015	1.03^a	0.73	$0.56^a(P<0.025)$	0.40
Pargyline	1.26^b	0.89	$0.48^b(P<0.005)$	0.34
Probenecid	0.83^c	0.58	$0.06^c(P<0.01)$	0.04

The rats were exposed to low concentrations of oxygen for 60 min in all experiments. The animals were killed 30 min after administration of NSD 1015 (100 mg/kg) and 60 min after pargyline (75 mg/kg) and probenecid (200 mg/kg) (all injected intraperitoneally).

The rate constants for synthesis (k) were obtained by dividing the rates of accumulation by the normal concentration of 5-hydroxytryptamine in whole brain (1.42 nmol/h), disregarding the insignificant decrease during hypoxia.

Agreement between the three methods was reasonable except for probenecid in hypoxia where oxidative deamination might be inhibited.

[a] Rate of accumulation of 5-hydroxytryptophan.
[b] Rate of accumulation of 5-hydroxytryptamine.
[c] Rate of accumulation of 5-hydroxyindoleacetic acid.

In our hypoxia studies, we compared several methods for measuring mono-amine synthesis and turnover in the brain. Qualitatively, the same result was obtained from the three methods: measuring the accumulation of (i) 5-hydroxy-tryptophan or dopa after giving NSD 1015, (ii) monoamines after inhibition of monoamine oxidase by pargyline and (iii) 5-hydroxyindoleacetic acid or homovanillic acid after blocking the efflux by probenecid (see Tables 1–3). Accumulation of [^{3}H]dopamine or [^{3}H]noradrenaline 15 min after intravenous injection of [^{3}H]tyrosine was not significantly influenced by hypoxia. The net yield of the tritiated amines from [^{3}H]tyrosine may depend not only on syn-

TABLE 2

Synthesis of noradrenaline in the brains of rats exposed to room air or 5.6% oxygen (Davis & Carlsson, unpublished data).

Inhibitor	Room air		5.6% O_2	
	Accumulation (nmol h^{-1})^a	Synthesis (k/h^{-1})^a	Accumulation (nmol h^{-1})^a	Synthesis (k/h^{-1})^a
NSD 1015 (dopa in hemispheres)	0.38	0.30	0.24^b	0.20
Pargyline (whole brain noradrenaline)	0.66	0.36	0.36^b	0.20

[a] See notes to Table 1.
[b] $P<0.001$.

TABLE 3

Synthesis and turnover of dopamine in the brains of rats exposed to room air or 5.6% oxygen (Davis & Carlsson, unpublished data).

Inhibitor	Room air		5.6% O_2	
	Accumulation (nmol h^{-1})[a]	Synthesis (k/h^{-1})[a]	Accumulation (nmol h^{-1})[a]	Synthesis (k/h^{-1})[a]
NSD 1015 (dopa in striatum)	4.13	0.22	2.38[b]	0.13
Pargyline (whole brain dopamine)	1.48	0.34	0.38[b]	0.09
Probenecid (whole brain homovanillic acid)	0.31	0.07	0.03[b]	0.01

[a] See notes to Table 1.
[b] $P < 0.001$.
The low values after probenecid treatment suggest that the efflux of homovanillic acid is incompletely blocked. In addition, oxidative deamination may be inhibited by hypoxia (cf. results in Table 1).

thesis but also on the disposition of different transmitter pools and therefore we should exercise caution in interpreting such data.

The turnover of dopamine, measured after inhibition of synthesis, was moderately retarded by hypoxia while in contrast, the turnover of noradrenaline was accelerated. Otherwise the results with noradrenaline were similar to those for dopamine.

HYDROXYLATION RATES AFTER INTERRUPTION OF THE NERVOUS IMPULSE FLOW

Disregarding any possible excitatory axono-axonal synapses, axotomy should immediately deprive the terminal fibre system of the nerve impulses. Accordingly, acute axotomy has generally been found to cause a marked retardation of the turnover of monoamine transmitter stores as revealed, for example, by the delayed disappearance of monoamine after inhibition of synthesis (Andén *et al.* 1966). Therefore, we investigated the influence of acute axotomy on the rates of tryptophan and tyrosine hydroxylation.

Surprisingly, the response to axotomy was by no means uniform in the different terminal systems investigated. We encountered all possible responses; an increase, a decrease and no change in the rate of hydroxylation. The most striking effect was found in the nigrostriatal dopamine system, where axotomy (a transverse cerebral hemisection) initially caused a considerable rise in the rate of tyrosine hydroxylation (Kehr *et al.* 1972 and unpublished results). In the limbic dopamine systems, the effect was similar though less marked.

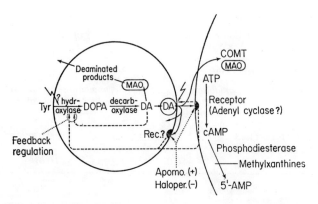

FIG. 1. Schematic illustration of the hypothetical mechanisms physiologically controlling the activity of tyrosine hydroxylase with a nerve terminal (left) and a postsynaptic neuron (right).

The release of transmitter (here dopamine, DA) causes a decrease in transmitter concentration intraneuronally, thereby reducing the hypothetical end-product inhibition of tyrosine hydroxylase. It is doubtful if the normal concentration of transmitter in free form is sufficient to cause appreciable end-product inhibition. Inhibitory transmitter levels may, however, be reached, for example, after inhibition of monoamine oxidase or interruption of nerve impulses by axotomy of the nigrostriatal dopamine tract.

The nerve impulses might influence tyrosine hydroxylase activity, for instance by influencing the availability of cofactor or hypothetical allosteric modulator.

Tyrosine hydroxylase activity might also be controlled via postsynaptic or hypothetical presynaptic monoaminergic receptors, independently of nerve impulses. Thus, tyrosine hydroxylase activity of the rat forebrain *in vivo* can be manipulated by a dopaminergic agonist (apomorphine; apomo.) and antagonist (haloperidol; haloper.), even after axotomy.

Changes in cerebral noradrenaline and 5-hydroxytryptamine systems were slight, whereas a 50% decrease in the rate of tryptophan hydroxylation was observed in the descendent spinal 5-hydroxytryptamine system. The spinal system we examined is thus most similar to the peripheral sympatho-adrenal system in this respect (cf. Weiner *et al.* 1972; Snider & Carlsson 1972; Snider *et al.* 1973).

These observations suggest that the nerve impulses influence the activities of aromatic amino acid hydroxylases by the following possible mechanisms: (1) receptor-mediated feedback control of aromatic amino acid hydroxylase activity; (2) end-product inhibition; or (3) a stimulating action of nerve impulses on the hydroxylase activity (see Fig. 1).

RECEPTOR-MEDIATED CONTROL OF TYROSINE AND TRYPTOPHAN HYDROXYLASE

The paradoxical enhancement of tyrosine hydroxylation in the striatal dopamine system after axotomy is probably due to the existence of a receptor-mediated feedback control mechanism (cf. Carlsson & Lindqvist 1963; Andén *et al.* 1969). In support of this hypothesis, tyrosine hydroxylation is stimulated by haloperidol, a blocking agent for dopamine receptors, and inhibited by apomorphine, a stimulating agent for dopamine receptors (Kehr *et al.* 1972). The stimulating action of haloperidol appeared to require the presence of an agonist such as dopamine or apomorphine and was thus not seen after axotomy unless apomorphine had been given. It should be emphasized that axotomy did not seem to uncouple this apparently receptor-mediated control mechanism, which thus seems to function in the absence of an impulse flow. Accordingly, we must postulate the existence of a messenger released by a change in the activity of postsynaptic (or hypothetical presynaptic) dopamine receptors and capable of influencing the activity of presynaptic tyrosine hydroxylase. Evidence in support of this mechanism has recently been obtained in an *in vitro* system of striatal slices (Goldstein *et al.* 1973).

Similarly, tryptophan hydroxylase activity appears to be controlled through 5-hydroxytryptamine receptors. The hallucinogenic agents *NN*-dimethyltryptamine and LSD-25, which appear to stimulate 5-hydroxytryptamine receptors and to slow its turnover (Andén *et al.* 1971), cause a moderate though significant retardation of tryptophan hydroxylation in mouse brain (Carlsson & Lindqvist 1972). Whether this effect persists after axotomy remains to be investigated.

END-PRODUCT INHIBITION OF AROMATIC AMINO ACID HYDROXYLASES

We suggested (Carlsson *et al.* 1960) that synthesis of catecholamines (and 5-hydroxytryptamine) is somehow controlled by amine concentrations in the store. The evidence for this hypothesis was essentially that within a few hours after inhibition of monoamine oxidase the initially rapid accumulation of amines (including basic metabolites) was brought to a standstill. We concluded that synthesis had been inhibited; rapid efflux of amines out of the brain seemed a less likely alternative.

Since then, numerous observations in support of this hypothesis have been published and it has been proposed that the mechanism is one of end-product inhibition (see Weiner *et al.* 1972 and this volume, pp. 135–147). This mechanism might act for catecholamines, but I should emphasize that a similar control appears to exist in 5-hydroxytryptamine neurons (see Carlsson & Lindqvist 1973) where direct product inhibition of tryptophan hydroxylase could not be demonstrated (Jequier *et al.* 1969).

STIMULATION OF HYDROXYLASE ACTIVITY BY NERVE IMPULSES

When the spinal, descendent fibre system carrying 5-hydroxytryptamine was acutely deprived of its impulse flow by a spinal transection, tryptophan hydroxylation was retarded by about 50% (see before). Likewise, stimulation of 5-hydroxytryptamine fibre systems appears to stimulate synthesis of the amine (Andén et al. 1964; Sheard & Aghajanian 1968). In 5-hydroxytryptamine neurons, direct end-product inhibition of tryptophan hydroxylase probably does not occur (see before). Thus, we have to consider the possibility that nerve impulses are capable of modulating tryptophan hydroxylase, and then perhaps also tyrosine hydroxylase by some other mechanism. Perhaps nerve impulses influence the availability of a factor controlling aromatic amino acid hydroxylases.

I should emphasize that the changes in aromatic amino acid hydroxylase activities discussed in this paper occurred mainly in short-term experiments where the total number of enzyme molecules may be regarded as constant (for further discussion of this and related problems, see Carlsson et al. 1972b,c, 1973b).

ACKNOWLEDGEMENTS

This study has been supported by a grant from the Swedish Medical Research Council.

References

ANDÉN, N.-E., CARLSSON, A., HILLARP, N.-Å. & MAGNUSSON, T. (1964) 5-Hydroxytryptamine release by nerve stimulation of the spinal cord. Life Sci. 3, 473–478
ANDÉN, N.-E., FUXE, K. & HÖKFELT, T. (1966) The importance of nervous impulse flow for the depletion of the monoamines from central neurons by some drugs. J. Pharm. Pharmacol. 18, 630–632
ANDÉN, N.-E., CARLSSON, A. & HÄGGENDAL, J. (1969) Adrenergic mechanisms. Annu. Rev. Pharmacol. 9, 119–134
ANDÉN, N.-E., CORRODI, H. & FUXE, K. (1971) Hallucinogenic drugs of the indolealkylamine type and central monoamine neurons. J. Pharmacol. Exp. Ther. 179, 236–249
BÉDARD, P., CARLSSON, A., FUXE, K. & LINDQVIST, M. (1971) Origin of 5-hydroxytryptophan and DOPA accumulating in brain following decarboxylase inhibition. Naunyn-Schmiedeberg's Arch. Pharmakol. 269, 1–6
BROWN, R., DAVIS, J. N. & CARLSSON, A. (1973) DOPA reversal of hypoxia-induced disruption of the conditioned avoidance response. J. Pharm. Pharmacol. 25, 412–414
CARLSSON, A. & LINDQVIST, M. (1963) Effect of chlorpromazine or haloperidol on formation of 3-methoxytyramine and normetanephrine in mouse brain. Acta Pharmacol. Toxicol. 20, 140–144

CARLSSON, A. & LINDQVIST, M. (1972) The effect of L-tryptophan and some psychotropic drugs on the formation of 5-hydroxytryptophan in the mouse brain *in vivo*. *J. Neural Transm.* **33**, 23–43

CARLSSON, A. & LINDQVIST, M. (1973) *In-vivo* measurements of tryptophan and tyrosine hydroxylase activities in mouse brain. *J. Neurol Transm.* **34**, 79–91

CARLSSON, A., LINDQVIST, M. & MAGNUSSON, T. (1960) On the biochemistry and possible functions of dopamine and noradrenaline in brain in *Adrenergic Mechanisms (Ciba Found. Symp.)* (Vane, J. R., Wolstenholme, G. E. W. & O'Connor, M., eds.), pp. 432–439, Churchill, London

CARLSSON, A., DAVIS, J. N., KEHR, W., LINDQVIST, M. & ATACK, C. V. (1972a) Simultaneous measurement of tyrosine and tryptophan hydroxylase activities in brain *in vivo* using an inhibitor of the aromatic amino acid decarboxylase. *Naunyn-Schmiedeberg's Arch. Pharmakol.* **275**, 153–168

CARLSSON, A., KEHR, W., LINDQVIST, M., MAGNUSSON, T. & ATACK, C. V. (1972b) Regulation of monoamine metabolism in the central nervous system. *Pharmacol. Rev.* **24**, 371–384

CARLSSON, A., BÉDARD, P., LINDQVIST, M. & MAGNUSSON, T. (1972c) The influence of nerve-impulse flow on the synthesis and metabolism of 5-hydroxytryptamine in the central nervous system. *Biochem. Soc. Symp.* **36**, 17–32

CARLSSON, A., LINDQVIST, M., MAGNUSSON, T. & ATACK, C. V. (1973a) Effect of acute trans-ection on the synthesis and turnover of 5-HT in the rat spinal cord. *Naunyn-Schmiedeberg's Arch. Pharmakol.* **277**, 1–12

CARLSSON, A., BÉDARD, P., DAVIS, J. N., KEHR, W., LINDQVIST, M. & MAGNUSSON, T. (1973b) Physiological control of 5-HT synthesis and turnover in the brain. *Proc. Vth Int. Congr. Pharmacol.* **4**, 286–298

DAVIS, J. N. & CARLSSON, A. (1973) Effect of hypoxia on tyrosine and tryptophan hydroxyla-tion in unanaesthetized rat brain. *J. Neurochem.* **20**, 913–915

GOLDSTEIN, M., ANAGNOSTE, B., & SHIRRON, C. (1973) The effect of trivastal, haloperidol and dibutyryl cyclic AMP on [¹⁴C]dopamine synthesis in rat striatum. *J. Pharm. Pharmacol.* **25**, 348–351

JEQUIER, E., ROBINSON, D. S., LOVENBERG, W. & SJOERDSMA, A. (1969) Further studies on tryptophan hydroxylase in rat brain stem and beef pineal. *Biochem. Pharmacol.* **18**, 1071–1081

KAUFMAN, S. (1974) Properties of pterin-dependent aromatic amino acid hydroxylases, *This Volume*, pp. 85–108

KEHR, W., CARLSSON, A., LINDQVIST, M., MAGNUSSON, T. & ATACK, C. V. (1972) Evidence for a receptor-mediated feedback control of striatal tyrosine hydroxylase activity. *J. Pharm. Pharmacol.* **24**, 744–747

SHEARD, M. H. & AGHAJANIAN, G. K. (1968) Stimulation of the midbrain raphe. Effect on serotonin metabolism. *J. Pharmacol. Exp. Ther.* **163**, 425–430

SHIMAN, R. & KAUFMAN, S. (1970) Tyrosine hydroxylase (bovine adrenal glands). *Methods Enzymol.* **17**, 609–615

SNIDER, S. R. & CARLSSON, A. (1972) The adrenal dopamine as an indicator of adreno-medullary hormone biosynthesis. *Naunyn-Schmiedeberg's Arch. Pharmakol.* **275**, 347–358

SNIDER, S. R., ALMGREN, O. & CARLSSON, A. (1973) The occurrence and functional significance of dopamine in some peripheral adrenergic nerves of the rat. *Naunyn-Schmiedeberg's Arch. Pharmakol.*, in press

WEINER, N., CLOUTIER, G., BJUR, R. & PFEFFER, R. J. (1972) Modification of norepinephrine synthesis in intact tissue by drugs and during short-term adrenergic nerve stimulation. *Pharmakol. Rev.* **24**, 203–222

WEINER, N., LEE, F.-L., WAYMIRE, T. C. & POSIVIATA, M. (1974) The regulation of tyrosine hydroxylase activity in adrenergic nervous tissue, *This Volume*, pp. 135–147

Discussion

Glowinski: We have recently developed a method which enables us to measure *in vivo* the rate of the first, limiting step of catecholamine synthesis in a discrete area of the brain. Using a microinjection procedure, we infuse L-[3,5-^{3}H$_2$]tyrosine locally in the striatum of the rat and measure the accumulation of ^{3}H$_2$O in tissue shortly after (three minutes) the end of the infusion of the labelled amino acid (Javoy *et al.* 1973, 1974). The ^{3}H$_2$O is formed during the conversion of L-[3,5-^{3}H$_2$]tyrosine into [^{3}H]dopa and thus gives a good index of the rate of the first step of catecholamine synthesis. With this new method which avoids the use of drugs we have obtained results similar to those presented by Dr Carlsson. For instance, we also observed an increase in the rate of conversion of tyrosine into dopa in the striatum shortly after interruption of nerve impulses in the nigro neostriatal dopaminergic pathway. This interruption was induced by an electrocoagulation of the substantia nigra or a microinjection of 6-hydroxydopamine in this area (Javoy *et al.* 1973). We observed a similar effect with 4-hydroxybutyric acid. This increased synthesis of dopamine which paralleled the enhanced concentrations of dopamine in dopaminergic terminals might be related to the interruption of dopamine release.

Carlsson: We have similar data for 4-hydroxybutyric acid (A. Carlsson & M. Lindqvist, unpublished findings).

Glowinski: The limitation of both our technique and that used by Dr Carlsson is that we are simultaneously measuring events in dopaminergic and noradrenergic terminals, if such terminals coexist in the brain structure analysed. In the striatum, the situation is particularly favourable, since this structure contains mainly dopaminergic terminals. This is not so, for example, in the cerebral cortex of the rat which contains similar proportions of dopaminergic and noradrenergic terminals (Thierry *et al.* 1973).

Carlsson: Fortunately, there are few areas where the two systems are equally mixed. One predominates in most areas, especially in discrete areas.

Gál: It augurs well that Professor Carlsson, Dr Glowinski and I agree. If we transect the medial forebrain bundle, we find absolutely no change in the activity of septal tryptophan hydroxylase.

Wurtman: Are you suggesting that the transection does not cut the axons of the ascending 5-hydroxytryptaminergic fibres?

Gál: No, it does. Otherwise one has to adopt other explanations, such as perikaryal synthesis of the enzyme or, heretical as this may sound, synthesis of hydroxylase by glia.

Carlsson: I don't like the idea of a glial localization for tyrosine hydroxylase or tryptophan hydroxylase. The virtually complete loss of enzyme activity in the brain after chronic axotomy is good support for a neuronal localization (Kuhar *et al.* 1973; McGeer *et al.* 1973). Moreover, our fluorescence histochemical studies show that dopa or 5-hydroxytryptophan accumulates in the neurons (Bédard *et al.* 1971). There is no support for any speculations about glial hydroxylases.

Munro: Does the build-up of dopamine in the receptor before feedback occurs represent the synthesis of an inhibitory compound, perhaps even a protein-dependent product? Is the time lag more important than the concentration?

Carlsson: Some evidence challenges this, arguing that the storage capacity is the prime factor. The lag appears to be the time necessary for filling the store to its maximum capacity, after which it overflows. In an animal treated with reserpine, which empties the store, we observe no build up of amine when we give a monoamine oxidase inhibitor, but the synthetic enzymes are immediately inhibited (Carlsson *et al.* 1960).

Wurtman: Presumably, the presynaptic receptor is different from the presynaptic uptake site.

Carlsson: Yes, they are entirely different.

Sandler: I always feel faintly uneasy about data of this type that rely on the use of drugs which are employed for a specific action on a particular enzyme system. We all know that there are really no such compounds; inhibitors, for instance, almost invariably act on a variety of different enzymes.

Let me cite an example of what I mean. You found that the amount of dopa generated increased when you inhibited decarboxylase with 3-hydroxy-benzylhydrazine but, as I mentioned before (p. 80), dopa might conceivably be formed by another route involving transamination. If you were to point out to me that the build-up of dopa could be countered with the tyrosine hydroxylase inhibitor, α-methyltyrosine, I would draw your attention to the work of Spector *et al.* (1965) who showed that α-methyltyrosine also inhibits tyrosine aminotransferase. I know that your approach is probably one of the best we have at the present time but, for the reasons I have stated, we must continue to be cautious.

Carlsson: We use diverse tools with entirely different chemical structures. If we obtain similar results, we start to feel confident about our data. We inhibit tyrosine hydroxylase not only by α-methyltyrosine but also by α-propyldopacetamide [2-(3,4-dihydroxybenzyl)pentanamide], and again completely prevent the accumulation of dopa. We have other data on the effect of 3-hydroxybenzylhydrazine which encourage us; for instance, the dose relationship shows a plateau of dopa accumulation between 50 and 200 mg/kg. Obviously other effects of this drug will be manifest in the higher dose range but they apparently do not influence the results.

Wurtman: The drugs used in this field have been differentiated into various classes on the basis of their 'dirtiness' or non-specificity. Probably, the least specific are those such as reserpine which release a great variety of chemically-heterogeneous compounds stored within subcellular organelles. Next are the synthetic aromatic amino acids such as *p*-chlorophenylalanine and 5-hydroxy-tryptophan; these compounds are taken up into virtually all cells by the same transport systems as those mediating the uptake of the related circulating amino acids (i.e., phenylalanine and tryptophan). Subsequently they are also decarboxylated within many cells. The third class consists of drugs like 3-hydroxybenzylhydrazine that inhibit enzymes metabolizing amino acids (e.g., by binding pyridoxine) even though they are not themselves amino acids. The fourth class includes drugs that inhibited monoamine uptake, such as the imipramines; these show somewhat more selectivity; the fifth and probably most specific class includes the receptor stimulants and antagonists, such as apomorphine.

Gessa: Goldstein *et al.* (1970) have shown that apomorphine effectively inhibits the biosynthesis of [^{14}C]dopamine from [^{14}C]tyrosine in slices of rat striatum even at 0.1 and 1.0 μM.

Carlsson: We have only used apomorphine in fairly high doses, 15 mg/kg. It might partly inhibit the enzyme, but I don't think this matters here. What is important is that haloperidol, which did not work in the absence of an agonist, starts to increase the rate of tyrosine hydroxylation in the presence of the agonist. Apomorphine is an agonist and acts by the receptor mechanism, an opinion supported by the *in vitro* data of Goldstein *et al.* (1973). Since the feedback regulation works after axotomy, when there is no more supply of impulses, it should work *in vitro* as well. Goldstein *et al.* tested this and found that this was so. Using another agonist, ET 495 [trivastal; 4-piperonyl-1-(2-pyrimidyl)piperazine], which is not a catechol and which did not inhibit tyrosine hydroxylase *in vitro*, they obtained the same results.

Gessa: What kind of transmitter operates between the receptor and tyrosine hydroxylase in the feed-back loop?

Carlsson: One compound that immediately comes to mind is cyclic AMP. Addition of the dibutyryl ester caused an increase in the activity of tyrosine hydroxylase (Goldstein *et al.* 1973), but the interaction between this and the receptor agonists is complicated. Goldstein's data have not proved satisfactorily that cyclic AMP is the messenger.

Weiner: Some of our unpublished results leave me sceptical about the quantitative aspects of this amino acid accumulation. We compared the rate of synthesis of [^{3}H]dopa from [3,5-^{3}H$_2$]tyrosine (which we measured after addition of either 4-bromo-3-hydroxybenzyloxyamine phosphate [NSD 1055] or decaborane) with the rate of formation of tritiated water and with the formation of the [³H]catecholamines, dopamine and noradrenaline. In both adrenal slices and the intact vas deferens, there was a good correlation between the production of ³H₂O, [³H]dopa and the [³H]catecholamines, indicating that the pathway from tyrosine through dopa to the catecholamines proceeded without significant side reactions in the presence of a monoamine oxidase inhibitor. However, in brain slices, we found that, in the presence of NSD 1055 or decaborane, tritiated water and [³H]dopa accumulated at almost twice the rate at which the [³H]-catecholamines accumulated.

Carlsson: 3-*O*-Methylation is another possible pathway. The discrepancy of results obtained by different methods does not surprise me. Brain slices bear no comparison with the intact brain where everything is enclosed by the blood–brain barrier. This barrier is absent from the slices.

Weiner: 3-Hydroxybenzylhydrazine (NSD 1015) is a pyridoxine antagonist. It therefore will block both dopa decarboxylase and dopa aminotransferase so that dopa will accumulate to a maximal concentration. What is measured is the total formation of dopa, yet possibly half the dopa is being transaminated rather than being converted into catecholamines. You are measuring dopa formation and not catecholamine formation.

Carlsson: Your results are open to an alternative explanation. Dopa can leak out of the nerve terminals in slices much more easily than it can in the brain, which is a closed system. The nerve terminals keep dopa inside because of the concentrating mechanism within the intact brain, and only the enzymes present there or in the close vicinity may act on dopa. *In vitro*, when the amino acid leaks out into the medium, it is exposed to all kinds of enzymes. This is the difference between the two systems. *In vivo*, dopa and catecholamines appear to be formed at similar rates.

Weiner: Possibly.

Sharman: In view of the evidence of a receptor-modulated feedback system, would you comment on the use of tyrosine hydroxylase inhibition to measure the turnover of catecholamines in the brain? As the concentration of dopamine falls and the rate of release of the amine is reduced, presumably the feedback system will have to compensate for this. Therefore, you will always get an overestimate of turnover rate from the fractional rate constant for the decrease in the concentration of dopamine. If the feedback mechanism is involved after inhibition of tyrosine hydroxylase, then the rate of decrease of the concentration of dopamine will increase until it reaches a maximum value. It is the time taken to reach this maximum value that will be related to the original turnover rate of dopamine in the brain.

Carlsson: In the hypoxia study, we administered α-methyltyrosine to rats and measured the rate of disappearance of dopamine. Hypoxia retarded this but conversely accelerated the rate of disappearance of noradrenaline. These findings ran counter to all the other results.

In the other experiment, we followed the accumulation of labelled dopamine and noradrenaline for 15 min after an intravenous injection of [^{3}H]tyrosine. We found no difference between hypoxic and control animals. I interpret this to mean that, in a more or less open system, we do not measure the rate of synthesis but rather the net amount of amine that collects in a particular fraction, that is in the store, which is not the same as the total amount.

Sharman: B. Werdinius & I (unpublished observations) measured the rate of formation of homovanillic acid and 3,4-dihydroxyphenylacetic acid together in the mouse brain and then estimated the turnover rate of dopamine by measuring the decrease in the concentration of this amine in the brain after inhibition of tyrosine hydroxylase with α-methyltyrosine. The estimated turnover rate for dopamine was about twice the combined rates of formation of the two acid metabolites.

Carlsson: You are right in your suspicions: we have good reasons for suspecting that blocking tyrosine hydroxylase activates feedback systems which could speed up the disappearance of dopamine particularly.

Sourkes: The evidence with apomorphine and haloperidol clearly suggests retrograde trans-synaptic regulation by dopamine (or apomorphine) on nerve endings, or by one of the other systems proposed, such as a short internuncial fibre from the postsynaptic site to the nerve ending. You have cut the nigrostriatal fibres without cutting the striato-nigral fibres. Should you not also check by specifically cutting the striato-nigral fibres in order to eliminate the possibility of a neuronal feedback? Quite rightly, you suggest that several types of modulatory or regulatory effect may be operating.

Carlsson: We hemisectioned the brain. Thus all nigro-striatal and striato-nigral connections were interrupted. But I agree we should eliminate the possibility.

Wurtman: Chlorimipramine acts on the presynaptic terminals of 5-hydroxytryptamine neurons to suppress the reuptake of this monoamine. In consequence, the molecules of the amine that are released into the synaptic cleft are inactivated at a slower rate and remain in the region of the postsynaptic receptor for a longer time. Because the physiological effects of the released transmitter are thereby potentiated, the neurons somehow 'learn' that less 5-hydroxytryptamine need be synthesized and released in unit time in order to transmit the desired number of 'messages' across the synapse. (This learning apparently involves a multineuronal reflex arc similar to the one described by Dr Carlsson and his associates which mediates the increased activity of dopaminergic neurons in animals treated with dopamine-receptor blocking agents.) After some time,

TABLE 1 (Wurtman)

Effects of chlorimipramine, tryptophan and diet on cerebral concentrations of 5-hydroxy-tryptamine

Treatment	Tryptophan ($\mu g/g$)	5-Hydroxy-tryptamine (ng/g)	5-Hydroxy-indoleacetic acid (ng/g)
Saline + saline	3.02 ± 0.43	550 ± 12	391 ± 21
Chlorimipramine + saline	3.02 ± 0.19	586 ± 25	309 ± 15^a
Saline + tryptophan	20.18 ± 2.29^{b,c}	815 ± 32^{b,c}	724 ± 37^{b,c}
Chlorimipramine + tryptophan	22.56 ± 1.53	837 ± 20^{b,c}	583 ± 21^{b,c}
Carbohydrate + saline	9.49 ± 0.56^{b,c}	761 ± 40^{b,c}	685 ± 28^{b,c}
Carbohydrate + chlorimipramine	8.19 ± 0.76^{b,c}	796 ± 16^{b,c}	475 ± 23^{b,c}

Groups of rats fasted overnight received chlorimipramine (25 mg/kg i.p.) or saline and, after 15 min, L-tryptophan (50 mg/kg i.p.) or saline. They were killed 60 min after the second injection, and brains were assayed for tryptophan, 5-hydroxytryptamine and 5-hydroxyindoleacetic acid. Other groups of fasted animals were given a carbohydrate diet (Fernstrom & Wurtman 1971) or no food, and, after 45 min, chlorimipramine or saline. They were killed 75 min after the injections. Data are given as mean ± standard error of the mean, and compared using the Student's *t* test. Chlorimipramine administration significantly depressed brain 5-hydroxyindoleacetic acid. This effect could be reversed (and brain 5-hydroxytryptamine elevated) by administering tryptophan or allowing rats to consume a carbohydrate meal that elevates brain tryptophan.

[a] $P < 0.01$ differs from saline–saline control.
[b] $P < 0.001$ differs from saline–saline control.
[c] $P < 0.001$ differs from chlorimipramine–saline group.

this decrease in the synthesis and release of the amine is reflected in the diminished concentration of its chief metabolite, 5-hydroxyindoleacetic acid within the brain.

J. Jacoby, J. Colmenares & I have recently found (Table 1; unpublished observations) that the decrease in the indoleacetic acid in brain observed in rats treated with chlorimipramine can be nullified by giving the animals either tryptophan or a carbohydrate diet that increases the amount of tryptophan in the brain. So, even though the synthesis of 5-hydroxytryptamine within brain neurons can be suppressed by an *external* feedback inhibition (i.e., feedback along multineuronal pathways and a decrease in the presynaptic input to the 5-hydroxytryptamine neurons), there is apparently no *internal* feedback control of the biosynthetic mechanisms operating within the cell itself. This control system probably differs considerably from the one operating within catecholaminergic neurons: I know of no evidence that increasing the amount of brain tyrosine can promote catecholamine synthesis, either in control animals or in rats given drugs that inhibit catecholamine uptake.

Moir: We chronically pretreated dogs over a period of days with the monoamine oxidase inhibitor, phenelzine, so that the biogenic amines reached a new steady state, as indicated by the constant amounts of metabolites in the cerebrospinal fluid. In the brain, the concentration of 5-hydroxytryptamine increases, that of tryptophan is normal and that of 5-hydroxyindoleacetic acid is reduced. If the animals are subsequently given tryptophan, the concentration of tryptophan in brain rises as expected but the concentration of 5-hydroxytryptamine is no more than was expected as the result of treatment with the monoamine oxidase inhibitor alone. Moreover, there is no increase in the amount of the indole acid in either brain or cerebrospinal fluid. This can be interpreted, as Dr Carlsson did, as signifying a 5-hydroxytryptamine feedback or one can take Dr Sandler's view that there is direct interaction on tryptophan hydroxylase. As 5-hydroxytryptophan does not become detectable in brain the indications are that phenelzine inhibits tryptophan 5-hydroxylase either directly or indirectly by feedback due to the elevation of the 5-hydroxytryptamine concentration caused by the monoamine oxidase inhibitor.

Wurtman: Could alternative metabolic pathways of 5-hydroxytryptamine account for this? For example, when monoamine oxidase is chronically blocked, might the amine instead be acetylated?

Grahame-Smith: While in Dr Glowinski's laboratory, Youdim (Agid *et al.* 1973) tested the effect of monoamine oxidase inhibitors on partially purified tryptophan hydroxylase (see Fig. 1). Pheniprazine inhibits most effectively and phenelzine less so.

Moir: This reinforces Dr Sandler's view that a drug does not have a simple and specific effect.

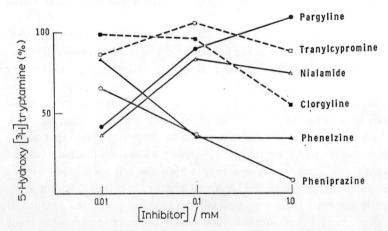

FIG. 1 (Grahame-Smith). The effect of monoamine oxidase inhibitors on the formation of 5-hydroxytryptamine. Partially purified tryptophan hydroxylase from pig brain stem was incubated with the inhibitor for 10 min, in the presence of the cofactor, $DMPH_4$, before the introduction of the substrate (Agid *et al.* 1973).

Kaufman: Dr Carlsson, from those data on hypoxia did you estimate a value for the K_m for oxygen for the hydroxylase or even a limiting value for the *in vivo* K_m?

Carlsson: No; that would be interesting but hard to estimate. It is difficult to measure the partial pressure of oxygen, pO_2, in the tissue, and of course, even more difficult to measure the intracellular pO_2 in the brain.

Kaufman: One estimate for the amount of oxygen in the brains of rats breathing normal air (20% oxygen) was 5% (Jamieson & Van den Brenk 1965).

Carlsson: Possibly, the amounts of intracellular oxygen in the brain are as little as 1–2 mmHg (0.2%); at such levels, mitochondria are still capable of maximum respiration (cf. Lubbers & Kessler 1968). So, an estimate of 1% wouldn't appear unreasonably low.

Kaufman: That corresponds roughly to the K_m values for oxygen that we determine with cerebral tyrosine hydroxylase and hepatic phenylalanine hydroxylase in the presence of tetrahydrobiopterin.

Weiner: The activity of tyrosine hydroxylase in the intact mouse vas deferens in 100% oxygen is twice what it is in room air (20% oxygen)—at 1750 m above sea level (Denver, Colorado) where the atmospheric pressure is about 630 mmHg. Of course, we don't know what the pO_2 is at the nerve endings.

References

AGID, Y., JAVOY, F. & YOUDIM, M. B. H. (1973) Monoamine oxidase and aldehyde dehydrogenase activity in striatum of rats after 6-hydroxytryptamine lesion of nigrostriatal pathway. *Br. J. Pharmacol.* **48**, 175

BÉDARD, P., CARLSSON, A., FUXE, K. & LINDQVIST, M. (1971) Origin of 5-hydroxytryptophan and DOPA accumulating in brain following decarboxylase inhibition. *Naunyn-Schmeideberg's Arch. Pharmakol.* **269**, 1–6

CARLSSON, A., LINDQVIST, M. & MAGNUSSON, T. (1960) in *Adrenergic Mechanisms (Ciba Found. Symp.)* (Vane, J. R., Wolstenholme, G. E. W. & O'Connor, M., eds.), pp. 432–439, Churchill, London

FERNSTROM, J. D. & WURTMAN, R. J. (1971) Brain serotonin content: increase following ingestion of carbohydrate diet. *Science (Wash. D.C.)* **174**, 1023–1025

GOLDSTEIN, M., FREEDMAN, L. S. & BACKSTROM, T. (1970) The inhibition of catecholamine biosynthesis by apomorphine. *J. Pharm. Pharmacol.* **22**, 715–716

GOLDSTEIN, M., ANAGNOSTE, B. & SHIRRON, C. (1973) The effect of trivastal, haloperidol and dibutyryl cyclic AMP on [^{14}C]dopamine synthesis in rat striatum. *J. Pharm. Pharmacol.* **25**, 348–351

JAMIESON, D. & VAN DEN BRENK, H. A. J. (1965) Electrode size and tissue pO_2 measurement in rats exposed to air or high pressure oxygen. *J. Appl. Physiol.* **20**, 514–518

JAVOY, F., AGID, Y., GLOWINSKI, J. & SOTELO, C. (1973) Biochemical and morphological changes in the dopamine neurons after interruption and degeneration of the nigro neostriatal pathway in *Dynamics of Degeneration and Growth in Neurons (Symp. 16–18 May)*, Stockholm

JAVOY, F., AGID, Y., BOUVET, D. & GLOWINSKI, J. (1974) *In vivo* estimation of the first step of dopamine synthesis in the dopaminergic terminals of the rat neostriatum. *J. Pharm. Pharmacol.*, in press

KUHAR, M. J., ROTH, R. H. & AGHAJANIAN, G. K. (1971) Selective reduction of tryptophan hydroxylase activity in rat forebrain after midbrain raphe lesions. *Brain Res.* **35**, 167–176

LUBBERS, D. & KESSLER, M. (1968) in *Oxygen Transport in Blood and Tissue* (Lubbers, D., Luft, U., Thews, O. & Witzleb, W., eds.), p. 90, Georg Thieme Verlag, Stuttgart

MCGEER, E. G., FIBIGER, H. C., MCGEER, P. L. & BROOKE, S. (1973) Temporal changes in amine synthesizing enzymes of rat extrapyramidal structures after hemitransections or 6-hydroxydopamine adminstration. *Brain Res.* **52**, 289–300

SPECTOR, S., SJOERDSMA, A. & UDENFRIEND, S. (1965) Blockade of endogenous norepinephrine synthesis by α-methyl-tyrosine, an inhibitor of tyrosine hydroxylase. *J. Pharmacol. Exp. Ther.* **147**, 86–95

THIERRY, A. M., STINUS, L., BLANC, G. & GLOWINSKI, J. (1973) Some evidence for the existence of dopaminergic neurons in the cortex. *Brain Res.* **50**, 230–234

The regulation of tyrosine hydroxylase activity in adrenergic nervous tissue

N. WEINER, F.-L. LEE, J. C. WAYMIRE* and M. POSIVIATA

Department of Pharmacology, University of Colorado School of Medicine, Denver, Colorado

Abstract Tyrosine hydroxylase catalyses the aromatic hydroxylation of both
L-phenylalanine and L-tyrosine to 3,4-dihydroxy-L-phenylalanine (dopa). This
mixed-function oxidase requires molecular oxygen, iron(II) and a reduced pterin
cofactor for activity. The catalytic behaviour of the isolated enzyme may vary
considerably, and in an incompletely understood manner, with the conditions
of the assay, the source and manner of preparation of the enzyme, and the concentrations and nature of the substrate and cofactor used in the assay. The
enzyme activity is regulated by end-product feedback inhibition, most notably
by noradrenaline although all catechols are inhibitory. This inhibition is competitively antagonized by reduced pterin cofactors including the putative natural
cofactor tetrahydrobiopterin and the synthetic cofactor, 6,7-dimethyltetra-
hydropterin. Noradrenaline and drugs which release noradrenaline intraneuronally (thereby increasing the amount of free intraneuronal amine) inhibit
enzyme activity in the intact tissue. These inhibitory effects are reversed by
addition of 6,7-dimethyltetrahydropterin to the medium. In the intact, adrener-
gically innervated, vas deferens of the guinea pig, tyrosine is the preferred substrate and tyrosine markedly inhibits the conversion of phenylalanine into dopa,
whereas phenylalanine has little effect on tyrosine hydroxylation *in situ*. The
conversion of both amino acids into the catecholamine is enhanced by adrenergic nerve stimulation. This effect, although inhibited by noradrenaline, does
not appear to result from reduced end-product feedback inhibition due to
neurotransmitter release, since it is not abolished by exogenous pterin cofactor.

Synthesis of noradrenaline, primarily from the aromatic amino acid tyrosine, is
known to take place in adrenal medulla cells, in peripheral post-ganglionic
neurons and in central adrenergic neurons. In all these cells it is generally
assumed that the enzymes which catalyse the biosynthesis of the catecholamines
and regulatory mechanisms for the synthesis and activity of these enzymes are
similar. Thus, analogues of tyrosine such as 3-iodotyrosine and α-methyltyrosine inhibit tyrosine hydroxylase in all mammalian tissues and repeated

* *Present address*: Department of Psychobiology, University of California, Irvine, California

administration of sufficient amounts of these compounds leads to the depletion of catecholamines throughout the body (Spector *et al.* 1965; Costa & Neff 1966). In all tissues so far examined, tyrosine hydroxylase requires a pterin cofactor for activity (Nagatsu *et al.* 1964; Coyle 1972) and it appears to be inhibited by several catechols including 3,4-dihydroxy-L-phenylalanine (dopa) and its metabolites, dopamine and noradrenaline (Nagatsu *et al.* 1964; Weiner 1970). Similarly, pyridoxine antagonists, α-methyldopa and related substances inhibit aromatic L-amino acid decarboxylase in brain, in peripheral adrenergic tissues and in the adrenal medulla (Sourkes 1966). Dopamine hydroxylase (EC 1.14.2.1) is a copper-dependent enzyme which is inhibited by a variety of thiols including disulfiram (Friedman & Kaufman 1965; Nagatsu *et al.* 1967). On administration of disulfiram, noradrenaline stores are depleted in tissues in which this catecholamine is ordinarily synthesized (Goldstein *et al.* 1964).

Of the enzymes involved in the biosynthesis of catecholamines, tyrosine hydroxylase has received the most attention in recent years. The enzyme is a mixed-function oxidase which requires molecular oxygen, a reduced pterin cofactor and either iron or catalase for optimal activity (Nagatsu *et al.* 1964). There is considerable controversy about whether iron(II) is an essential component of the enzyme or whether it merely serves to catalyse the destruction of the hydrogen peroxide produced during the enzymic reaction; hydrogen peroxide appears to be extremely deleterious to the enzyme (Petrack *et al.* 1968; Shiman & Kaufman 1970; Shiman *et al.* 1971).

An increasing body of evidence indicates that the activity of tyrosine hydroxylase may be regulated by a variety of factors, many of which may allosterically influence the enzyme system. The existence of such allosteric factors which regulate the activity of the key enzyme in a biosynthetic pathway is well known and is perhaps best exemplified by studies on aspartate carbamoyltransferase (EC 2.1.3.2) and phosphofructokinase (EC 2.7.1.11) ((Monod *et al.* 1963; Stadtman 1966; Koshland & Neet 1968). Our knowledge of allosteric influences and the effects of different concentrations of substrate and cofactor on tyrosine hydroxylase activity is still incomplete and the published reports frequently yield apparently conflicting results. The mechanism of the catalysis remains in dispute (Ikeda *et al.* 1966; Joh *et al.* 1969; Shiman *et al.* 1971). Much of the confusion probably arises primarily because various investigators have used different preparations of the enzyme which display different properties. It is also becoming recognized that different cofactors exhibit different kinetic properties (Shiman *et al.* 1971; Shiman & Kaufman 1970). The kinetics appear to depend on substrate concentration (Shiman *et al.* 1971), cofactor concentration (Ikeda *et al.* 1966), partial pressure of oxygen (Shiman *et al.* 1971) and very likely a host of allosteric effectors which have yet to be identified.

Because of its ready availability, the most commonly used source of the enzyme is the bovine adrenal medulla gland. The general applicability of the information obtained from this enzyme may, however, be questioned in view of some unique properties of the enzyme. Tyrosine hydroxylase from adrenal glands of non-ruminant animals is isolated from the cytosol as a soluble enzyme (Waymire *et al.* 1972). Likewise, the enzyme from a variety of peripheral mammalian tissues appears to be soluble (Weiner *et al.* 1972). Freshly prepared enzyme from bovine adrenal medulla also is recovered in the high-speed supernatant fraction but the enzyme shows a marked propensity either to self-aggregate or to adsorb onto other macromolecules or membranes in the homogenate (Musacchio *et al.* 1971). This particulate form of the enzyme may be solubilized by proteolytic enzymes, notably chymotrypsin and trypsin (Petrack *et al.* 1968; Shiman *et al.* 1971). The molecular weight of the partially digested enzyme is about a quarter that of the native enzyme and it seems reasonable to suppose that this enzyme does not exhibit the same kinetic properties or allosteric sensitivity of the native form (Musacchio *et al.* 1971; Joh *et al.* 1969).

Tyrosine hydroxylase can hydroxylate both phenylalanine and tyrosine (Ikeda *et al.* 1965, 1967). Ikeda *et al.* (1967) proposed that phenylalanine is hydroxylated to tyrosine which dissociates from the oxidized form of the enzyme. This oxidized form of the enzyme is reduced by a tetrahydropterin cofactor and is then capable of further hydroxylation. Using 6,7-dimethyl-5,6,7,8-tetrahydropterin (DMPH$_4$) as cofactor, Ikeda and his co-workers (1965) obtained a value for the K_m for tyrosine with bovine adrenal enzyme of 0.05 mM and a K_m of 0.3 mM for phenylalanine. Shiman & Kaufman (1970) obtained a tyrosine K_m value for the particulate enzyme of bovine adrenal medulla of 83 μM with DMPH$_4$ and 7 μM with tetrahydrobiopterin. These latter workers showed that, with tetrahydrobiopterin as cofactor, tyrosine inhibited the reaction at concentrations 2–3 times above its K_m value In contrast, Ikeda *et al.* (1966), using DMPH$_4$, were unable to demonstrate any inhibition of the enzyme by substrate when considerably higher concentrations of tyrosine were added to the medium.

We have found that an excess of the substrate (0.1 mM-tyrosine in the medium) inhibits tyrosine hydroxylase in homogenates of guinea-pig vasa deferentia with either tetrahydrobiopterin or DMPH$_4$ as cofactor. Shiman & Kaufman (1970) also examined the solubilized bovine adrenal enzyme (prepared by digestion of a particulate fraction of this tissue with chymotrypsin). The K_m values they found for both tetrahydrobiopterin and DMPH$_4$ were about three times greater than the values they obtained with the particulate enzyme. Using tetrahydrobiopterin, they also demonstrated enzyme inhibition with concentrations of tyrosine several times in excess of the K_m for tyrosine.

Of particular interest was the observation (Shiman *et al.* 1971) that the rate of hydroxylation of phenylalanine was considerably enhanced with tetra-hydrobiopterin instead of $DMPH_4$ as cofactor. Since inhibition by an excess of substrate was not demonstrable with phenylalanine, Shiman and his co-workers concluded that phenylalanine might be quantitatively an extremely important substrate for tyrosine hydroxylase *in vivo*.

We have examined this possibility by investigating the rate of formation of catecholamines from [*carboxy*-^{14}C]tyrosine and from [*carboxy*-^{14}C]phenyl-alanine in isolated, intact vasa deferentia of the guinea pig (Weiner *et al.* 1972). Since tetrahydrobiopterin is presumably the natural pterin cofactor, we sup-posed that we would observe considerable production of catecholamines from phenylalanine and, as Ikeda and his coworkers (1965, 1967) demonstrated, reciprocal substrate inhibition with both phenylalanine and tyrosine. Contrary to our expectations, tyrosine (at both 0.05 mM and 0.15 mM) was converted into catecholamines at a rate about four times faster than that of phenylalanine. Furthermore, although tyrosine was a potent inhibitor of the formation of catecholamines from phenylalanine, phenylalanine had virtually no effect on the conversion of tyrosine into catecholamines, at least up to a concentration twice that of tyrosine (Table 1). Obviously, in this complex intact system, several factors might account for this. The marked inhibition by tyrosine might be due to dilution of the [^{14}C]tyrosine formed from [^{14}C]phenylalanine, since the inter-

TABLE 1

Effect of variation of substrate and nerve stimulation on the activity of tyrosine hydroxylase (guinea-pig vas deferens preparation)

Substrates added (No. of experiments)	Amount of $^{14}CO_2$ produced per organ in 40 min (pmol)	
	Control[a]	Stimulated[a]
50μM-[*carboxy*-^{14}C]Tyrosine (6)	36.4 ± 4.3	79.0 ± 10.8
+ 0.10mM-Phenylalanine (4)	38.0 ± 3.3	71.2 ± 6.6
+ 50μM-Phenylalanine (4)	40.0 ± 8.2	70.3 ± 11.0
+ 0.10mM-Tyrosine (2)	73.3 (70.0; 76.7)	131.7 (113.3; 150.0)
50μM-[*carboxy*-^{14}C]Phenylalanine (10)	12.4 ± 1.4	34.6 ± 6.0
+ 0.10mM-Tyrosine (5)	2.7 ± 0.7	5.9 ± 0.9
+ 50μM-Tyrosine (4)	3.1 ± 1.3	11.7 ± 4.1
+ 25μM-Tyrosine (2)	2.6 (3.8; 1.4)	7.1 (7.8; 6.4)
+ 0.10mM-Phenylalanine (2)	23.7 (23.7; 23.7)	64.7 (69.3; 60.3)

[a] Values ± S.E.

Tissues were incubated in closed vessels in Krebs–Ringer hydrogen carbonate buffer previously equilibrated with 95% O_2 – 5% CO_2. $^{14}CO_2$ produced was collected in plastic cups containing NCS (Nuclear Chicago) organic-base solubilizer and counted by liquid scintillation spectro-metry.

mediate product presumably dissociates from the enzyme (Ikeda *et al.* 1967) and then mixes with the much larger pool of tyrosine which may have entered the nerve terminal from the medium. The lower rate of conversion for phenylalanine and the relative inability of phenylalanine to inhibit the hydroxylation of tyrosine might be explained if phenylalanine were taken up into the nerve terminal relatively slowly. Nevertheless, it appears that the quantitatively important precursor of catecholamines in the intact adrenergic neuron is tyrosine and that, in the presence of 50 μM-tyrosine and 50 μM-phenylalanine in the plasma, over 90% of the catecholamines formed are derived from tyrosine rather than phenylalanine. Dairman (1972) also failed to demonstrate inhibition of brain and heart catecholamine synthesis *in vivo* by high concentrations of tyrosine in plasma and in brain.

We have also studied the effect of nerve stimulation on the biosynthesis of catecholamines. The enhancement of synthesis from both amino acid precursors was similar, amounting to about a 2–3 fold increase with intermittent nerve stimulation of the hypogastric nerve of this preparation. Thus, the mechanism by which tyrosine hydroxylase activity is increased during nerve stimulation appears to pertain equally to both substrates (Table 1).

The kinetics of the interaction of oxygen with tyrosine hydroxylase also depend on the cofactor. Ikeda *et al.* (1966) demonstrated that 74 μM (6% oxygen in the atmosphere) is the approximate value of the apparent K_m for oxygen and the soluble bovine adrenal enzyme. Fisher & Kaufman (1972), using either tyrosine hydroxylase from the rat-brain caudate or solubilized bovine adrenal enzyme, observed that the apparent K_m for oxygen was equivalent to a partial pressure of oxygen of about 7.6 mmHg with either tetrahydrobiopterin or DMPH$_4$. Fisher & Kaufman demonstrated that when the partial pressure of oxygen in the atmosphere exceeded about 300 mmHg, the activity of the adrenal tyrosine hydroxylase was significantly inhibited. Inhibition of the brain enzyme begins at an oxygen pressure of about 60 mmHg with tetrahydrobiopterin as cofactor. We noticed that in the intact mouse vas deferens preparation formation of catecholamines from tyrosine when the tissue is equilibrated in an atmosphere of pure oxygen (630 mmHg at an altitude of 1610 m) is twice what it is when the system is equilibrated with air (130 mmHg oxygen). Although the partial pressure of oxygen in the atmosphere may be high, the diffusion of oxygen to the adrenergic nerve terminals within the intact tissue might conceivably be limiting and the concentration of oxygen in the depths of the tissue might be somewhat lower. Alternatively, the uptake of tyrosine into tissue equilibrated in air might be so much lower than its uptake in an atmosphere of oxygen that this step in the biosynthetic pathway becomes limiting. Evidence suggests, however, that this possibility is not likely (Goldstein *et al.* 1970*b*).

Reported values of the K_m for the synthetic pterin cofactor, $DMPH_4$, range between 0.09 and 0.75 mM. The K_m for tetrahydrobiopterin is about one third to one quarter that of $DMPH_4$ and has been reported to range from 0.02 to 0.12 mM. Excessive concentrations of $DMPH_4$ inhibit the soluble enzyme from the bovine adrenal medulla (Ikeda et al. 1966) particularly with low concentrations of tyrosine. In contrast, Shiman et al. (1971) could not demonstrate inhibition with high concentrations of either cofactor and with either the solubilized or particulate enzyme preparation from the bovine adrenal medulla.

Several years ago, we showed that a consequence of intermittent stimulation of the hypogastric nerve–vas deferens preparation of the guinea pig for 40–60 min was an increase in the synthesis of noradrenaline from tyrosine. Nerve stimulation altered neither the synthesis of catecholamines from dopa significantly nor the amount of tyrosine hydroxylase in the tissue during this short period (Alousi & Weiner 1966; Weiner & Rabadjija 1968a; Thoa et al. 1972). The effect on the synthesis of noradrenaline was at least partially overcome by the addition of 10 µM-noradrenaline to the medium. Therefore, we proposed that the enhanced activity of tyrosine hydroxylase results from reduced end-product feedback inhibition since the stimulated release of the neurotransmitter should reduce the amount of free catecholamines within the neuron. Noradrenaline and other catechols compete with the pterin cofactor to relieve the inhibition (Nagatsu et al. 1964; Ikeda et al. 1966). Thus, if pterin cofactor were added to the medium and if it were taken up by the nerve terminal in sufficiently large quantities, it presumably could overcome any degree of feedback inhibition by the catecholamine. The enzyme in control tissue might be stimulated by pterin cofactor to a greater degree than the (less inhibited) enzyme in stimulated tissue. The difference in the rate of catecholamine synthesis between control and stimulated preparations would, therefore, be reduced or abolished by addition of pterin cofactor. We had already demonstrated that a variety of drugs (e.g. amphetamine, tranylcypromine and reserpine) can inhibit the enzyme's activity in intact tissue by releasing noradrenaline from storage sites, thereby increasing the pool of free amine. These effects can be overcome competitively by $DMPH_4$ (Weiner et al. 1972). Analogous studies on stimulated preparations disclosed that $DMPH_4$ increased the enzyme activity in stimulated preparations to a degree equal to or greater than that in control preparations. Thus the enhanced activity of tyrosine hydroxylase associated with nerve stimulation does not result from reduced end-product feedback inhibition (Cloutier & Weiner 1973). We are now investigating other possible explanations for the increase in synthesis of noradrenaline which nerve stimulation induces.

Several factors besides catecholamines apparently influence tyrosine hydroxylase activity and these may be relevant to the physiological regulation of the activity of this enzyme. Kuczenski & Mandell (1972*a*) established that tyrosine hydroxylase activity is directly proportional to the ionic strength of the medium in which the assay is conducted. Roughly a 30% increase in activity is observed on increasing the ionic strength from 0.1 to 0.4. If the inorganic salt used is a sulphate, the activity almost doubles. Inorganic sulphate raises the value of V_{max} of tyrosine hydroxylase but has no effect on the K_m for DMPH$_4$. Kuczenski & Mandell also observed that sulphated polysaccharides, most notably heparin, activate preparations of tyrosine hydroxylase from the rat-brain caudate nucleus. In this activation, the value of the K_m for DMPH$_4$ drops fivefold from 0.74 to 0.15 mM while the V_{max} roughly doubles.

The soluble enzyme also appeared to be more sensitive to inhibition by dopamine than did the particulate enzyme, although the addition of heparin to the medium seemed to desensitize the soluble enzyme to the inhibitory effects of dopamine. Kuczenski & Mandell (1972*a*,*b*) have proposed that physiological inputs somehow transform the tyrosine hydroxylase from a soluble to a particulate state, thereby considerably increasing the activity of the enzyme and its affinity for the pterin cofactor. Simultaneously, the transformation would reduce the sensitivity of the enzyme to inhibition by dopamine. These two factors would facilitate the synthesis of catecholamines from tyrosine in the

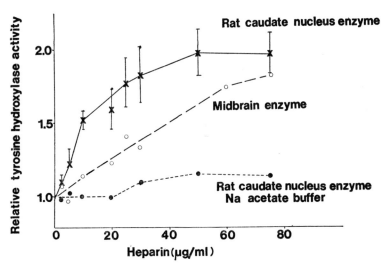

FIG. 1. Effect of heparin on the activity of soluble tyrosine hydroxylase from the rat brain caudate nucleus and rat midbrain. The enzyme preparations were prepared according to the procedure of Kuczenski & Mandell (1972*b*). Assay conditions were as described by Waymire *et al.* (1971), except that in two of the three series of experiments (upper two curves), Tris acetate (0.2 M, pH 6.1) was substituted for sodium acetate buffer (0.2 M, pH 6.1).

neuron. These workers further propose that soluble tyrosıne hydroxylase exists in a relatively inactive conformation and that binding to membrane modifies the conformation so that catecholamine synthesis is enhanced.

Our attempts to duplicate these results have met only limited success. The stimulation of soluble enzyme from human phaeochromocytoma tissue, from rat-brain caudate nucleus and from rat midbrain by heparin is related to the dose of heparin; the rat midbrain enzyme is less sensitive than is the hydroxylase from the caudate nucleus (Fig. 1). The K_m for DMPH$_4$ drops by about 65% in

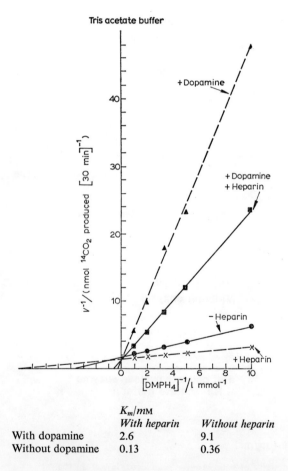

	K_m/mM	
	With heparin	*Without heparin*
With dopamine	2.6	9.1
Without dopamine	0.13	0.36

FIG. 2. Lineweaver–Burk plot of the effect of heparin and dopamine on the activity of soluble tyrosine hydroxylase (prepared from rat brain caudate nucleus) with K_m values set out below. The concentrations of heparin and dopamine were 28.6 µg/ml and 0.1 mM, respectively. The enzyme was prepared and assay was performed as described in Fig. 1 with Tris acetate buffer.

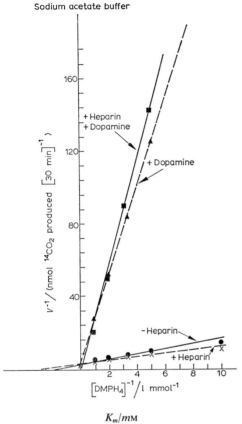

FIG. 3. Lineweaver–Burk plot of the effect of heparin and dopamine on the activity of soluble tyrosine hydroxylase (prepared from rat brain caudate nucleus) with K_m values set out below. The enzyme was prepared and assay was performed as described in Figs. 1 and 2 with sodium acetate buffer.

	K_m/mM	
	With heparin	*Without heparin*
With dopamine	>6.6	6.6
Without dopamine	0.61	0.67

the presence of heparin (Fig. 2), although the selection of buffer seems to be critical; the effect of heparin is much more modest in sodium acetate than in Tris acetate (cf. Figs. 2 and 3). The significance of this dependence on the buffer is uncertain.

Specific inorganic ions might regulate the activity of tyrosine hydroxylase. Gutman & Segal (1972) showed that an increased sodium concentration and decreased potassium concentration enhanced the activity of the soluble enzyme prepared from bovine adrenal medulla. This seems particularly intriguing since presumably sodium enters and potassium leaves the neuron during depolarization of the nerve terminal. According to Gutman & Segal, these ionic fluxes should enhance the activity of tyrosine hydroxylase. We have attempted to substantiate this possibility in the isolated hypogastric nerve–vas deferens preparation from the guinea pig. These preparations were stimulated in various concentrations of sodium ions and compared with unstimulated control preparations. The enzyme activity in the intact vas–deferens preparation did not change when the sodium concentration of the medium was lowered from 143 to 25 mM both with and without nerve stimulation (Table 2). However, in the complete absence of sodium ions the activity of tyrosine hydroxylase in the intact preparation increased significantly and the difference between control and stimulated preparations was eliminated. These results are difficult to interpret, however, since the tissue in the medium without sodium ions spontaneously contracted, albeit weakly, and responses to nerve stimulation were hard to provoke. Using homogenates of the vas deferens from guinea pig and optimal conditions for the tyrosine hydroxylase assay (Waymire *et al.* 1971) with 0.5 mM- or 0.2 mM-tetra-hydrobiopterin or 1.0 mM-DMPH$_4$ as cofactor, we found no significant effect of either NaCl, KCl or choline chloride (0–154 mM) on enzyme activity. Previously, Ikeda *et al.* (1966) had been unable to change the activity of the soluble enzyme (prepared from bovine adrenal medulla) with concentrations of sodium, potassium, calcium and chloride ions ranging from 0.1–10 mM. Although our homogenate studies make it seem unlikely that the enhanced activity of tyrosine

TABLE 2

Effect of diminution of sodium ion concentration on the activity of tyrosine hydroxylase during stimulation of a hypogastric nerve–vas deferens preparation from a guinea pig

| $[Na^+]/mM^a$ | Amount of $^{14}CO_2$ produced per organ in 40 min (pmol) | |
	Control[b]	Stimulated[b]
143 (11)	25.8 ± 2.3	55.8 ± 3.5[c]
104 (4)	23.2 ± 3.0	55.6 ± 7.0[c]
64 (4)	22.3 ± 1.7	48.7 ± 2.8[c]
25 (11)	29.7 ± 5.2	47.2 ± 6.6

[a] Isotonic saline in the Krebs–Ringer hydrogen carbonate solution was progressively replaced by isotonic sucrose.
[b] Values ± S.E.
[c] Significantly greater than corresponding control values ($P < 0.01$).
Tissues were incubated in closed vessels in Krebs–Ringer hydrogen carbonate buffer previously equilibrated with 95% O_2 – 5% CO_2. Medium contained 50μM-[*carboxy*-^{14}C]tyrosine (10 mCi/mmol). $^{14}CO_2$ produced was collected in plastic cups containing NCS (Nuclear Chicago) organic-base solubilizer and counted by liquid scintillation spectrometry.

hydroxylase during nerve stimulation is a consequence of the greater uptake of sodium ions by the nerve terminal and an activation of the enzyme by sodium ions, we cannot totally exclude this possibility. The state of the enzyme might differ sufficiently *in situ* and the ion sensitivity might be lost during (or subsequent to) homogenization.

The mechanism whereby nerve stimulation enhances synthesis of noradrenaline may be difficult to delineate, since many metabolic events are associated with nerve terminal depolarization and repolarization. For example, greater uptake of calcium ions should take place during nerve terminal depolarization and the uptake of calcium might facilitate the binding of tyrosine hydroxylase to membrane which could lead to an activation of the enzyme according to the mechanism proposed by Kuczenski & Mandell (1972*a,b*). Gutman & Segal (1972) demonstrated that 0.1mM-calcium slightly activated the bovine adrenal enzyme and that 0.2mM-calcium slightly inhibited the activity. Increasing calcium concentrations in the medium for rat-brain striatal slices inhibited synthesis of [^{14}C]dopamine from [^{14}C]tyrosine (Goldstein *et al.* 1970*a*). The uptake of calcium into the neuronal tissue under those conditions for incubation was not evaluated.

Although two mechanisms of regulation of tyrosine hydroxylase activity are well established (namely, end-product feedback inhibition and enzyme induction [see Weiner 1970]), it appears increasingly likely that the activity of this complex enzyme is regulated by other mechanisms, two of which relate to increased enzyme activity during (Alousi & Weiner 1966; Weiner & Rabadjija 1968*a*) and after (Weiner & Rabadjija 1968*b*) acute nerve stimulation. The mechanisms by which this regulation is achieved may involve ion fluxes associated with nerve terminal depolarization, but no clear evidence supports this or other mechanisms.

ACKNOWLEDGEMENTS

We thank Dr R. F. Long (Roche Ltd., Welwyn Garden City, England) for the generous supply of tetrahydrobiopterin used in these experiments. The work was supported by USPHS Grants (NSO7927 and NSO7642).

References

ALOUSI, A. & WEINER, N. (1966) The regulation of norepinephrine synthesis in sympathetic nerves: effect of nerve stimulation, cocaine, and catecholamine releasing agents. *Proc. Natl. Acad. Sci. U.S.A.* **56**, 1491–1496

CLOUTIER, G. & WEINER, N. (1973) Further studies on the increased synthesis of norepinephrine during nerve stimulation of the guinea-pig vas deferens preparation: effect of tyrosine and 6,7-dimethyltetrahydropterin. *J. Pharmacol. Exp. Ther.* **186**, 75–85

COSTA, E. & NEFF, N. H. (1966) Isotopic and non-isotopic measurements of the rate of catecholamine biosynthesis in *Biochemistry and Pharmacology of the Basal Ganglia* (Costa E., Côte, L. J. & Yahr, M. D., eds.), pp. 141–155, Raven Press, Hewlett, New York

COYLE, J. T. (1972) Tyrosine hydroxylase in rat brain—cofactor requirements, regional and subcellular distribution. *Biochem. Pharmacol.* **21**, 1935–1944

DAIRMAN, W. (1972) Catecholamine concentrations and the activity of tyrosine hydroxylase after an increase in the concentration of tyrosine in rat tissues. *Br. J. Pharmacol.* **44**, 307–310

FISHER, D. B. & KAUFMAN, S. (1972) The inhibition of phenylalanine and tyrosine hydroxylase by high oxygen levels. *J. Neurochem.* **19**, 1359–1365

FRIEDMAN, S. & KAUFMAN, S. (1965) 3,4-Dihydroxyphenylethylamine β-hydroxylase: physical properties, copper content, and role of copper in the catalytic activity. *J. Biol. Chem.* **240**, 4763–4773

GOLDSTEIN, M., ANAGNOSTE, B., LAUBER, E. & McKEREGHAN, M. R. (1964) Inhibition of dopamine-β-hydroxylase by disulfiram. *Life Sci.* 763–767

GOLDSTEIN, M., BACKSTROM, T., OHI, Y. & FRENKEL, R. (1970a) The effects of Ca^{++} ions on the C^{14} catecholamine biosynthesis from C^{14}-tyrosine in slices from the striatum of rats. *Life Sci.* **9**, 919–924

GOLDSTEIN, M., OHI, Y. & BACKSTROM, T. (1970b) The effect of ouabain on catecholamine biosynthesis in rat brain cortex slices. *J. Pharmacol. Exp. Ther.* **174**, 77–82

GUTMAN, Y. & SEGAL, J. (1972) Effect of calcium, sodium and potassium on adrenal tyrosine hydroxylase activity *in vitro*. *Biochem. Pharmacol.* **21**, 2664–2666

IKEDA, M., LEVITT, M. & UDENFRIEND, S. (1965) Hydroxylation of phenylalanine by purified preparations of adrenal and brain tyrosine hydroxylase. *Biochem. Biophys. Res. Commun.* **18**, 482–488

IKEDA, M., FAHIEN, L. A. & UDENFRIEND, S. (1966) A kinetic study of bovine adrenal tyrosine hydroxylase. *J. Biol. Chem.* **241**, 4452–4456

IKEDA, M., LEVITT, M. & UDENFRIEND, S. (1967) Phenylalanine as substrate and inhibitor of tyrosine hydroxylase. *Arch. Biochem. Biophys.* **120**, 420–427

JOH, T. H., KAPIT, R. & GOLDSTEIN, M. (1969) A kinetic study of particulate bovine adrenal tyrosine hydroxylase. *Biochim. Biophys. Acta* **171**, 378–380

KOSHLAND, D. E. & NEET, K. E. (1968) The catalytic and regulatory properties of enzymes. *Annu. Rev. Biochem.* **37**, 359–410

KUCZENSKI, R. T. & MANDELL, A. J. (1972a) Allosteric activation of hypothalamic tyrosine hydroxylase by ions and sulphated mucopolysaccharides. *J. Neurochem.* **19**, 131–137

KUCZENSKI, R. T. & MANDELL, A. J. (1972b) Regulatory properties of soluble and particulate rat brain tyrosine hydroxylase. *J. Biol. Chem.* **247**, 3114–3122

MONOD, J., CHANGEUX, J. P. & JACOB, F. (1963) Allosteric proteins and cellular control systems. *J. Mol. Biol.* **6**, 306–329

MUSACCHIO, J. M., WURZBURGER, R. J. & D'ANGELO, G. L. (1971) Different molecular forms of bovine adrenal tyrosine hydroxylase. *Mol. Pharmacol.* **7**, 136–146

NAGATSU, T., LEVITT, M. & UDENFRIEND, S. (1964) Tyrosine hydroxylase, the initial step in norepinephrine biosynthesis. *J. Biol. Chem.* **239**, 2910–2917

NAGATSU, T., KUZUYA, H. & HIDAKA, H. (1967) Inhibition of dopamine β-hydroxylase by sulfhydryl compounds and the nature of the natural inhibitors. *Biochem. Biophys. Acta* **139**, 319–327

PETRACK, B., SHEPPY, F. & FETZER, V. (1968) Studies on tyrosine hydroxylase from bovine adrenal medulla. *J. Biol. Chem.* **243**, 743–748

SHIMAN, R. & KAUFMAN, S. (1970) Tyrosine hydroxylase (bovine adrenal glands). *Methods Enzymol.* **17A**, 609–615

SHIMAN, R., AKINO, M. & KAUFMAN, S. (1971) Solubilization and partial purification of tyrosine hydroxylase from bovine adrenal medulla. *J. Biol. Chem.* **246**, 1130–1340

SOURKES, T. L. (1966) Dopa decarboxylase: substrates, coenzymes, inhibitors. *Pharmacol. Rev.* **18**, 53–60

SPECTOR, S. (1966) Inhibitors of endogenous catecholamine biosynthesis. *Pharmacol. Rev.* **18**, 599–609

SPECTOR, S., SJOERDSMA, A. & UDENFRIEND, S. (1965) Blockade of endogenous norepinephrine synthesis by α-methyltyrosine, an inhibitor of tyrosine hydroxylase. *J. Pharmacol. Exp. Ther.* **147**, 86–95

STADTMAN, E. R. (1966) Allosteric regulation of enzyme activity. *Adv. Enzymol.* **28**, 41–154

THOA, N. B., JOHNSON, D. G., KOPIN, I. J. & WEINER, N. (1971) Acceleration of catecholamine formation in the guinea-pig vas deferens after hypogastric nerve stimulation: roles of tyrosine hydroxylase and new protein synthesis. *J. Pharmacol. Exp. Ther.* **178**, 442–449

WAYMIRE, J. C., BJUR, R. & WEINER, N. (1971) Assay of tyrosine hydroxylase by coupled decarboxylation of dopa formed from [1-^{14}C]-L-tyrosine. *Anal. Biochem.* **43**, 588–600

WAYMIRE, J. C., WEINER, N., SCHNEIDER, F. H., GOLDSTEIN, M. & FREEDMAN, L. S. (1972) Tyrosine hydroxylase in human adrenal and pheochromocytoma: localization, kinetics, and catecholamine inhibition. *J. Clin. Invest.* **51**, 1798–1804

WEINER, N. (1970) Regulation of norepinephrine biosynthesis. *Annu. Rev. Pharmacol.* **10**, 273–290

WEINER, N. & RABADJIJA, M. (1968a) The effect of nerve stimulation on the synthesis and metabolism of norepinephrine in the isolated guinea-pig hypogastric nerve–vas deferens preparation. *J. Pharmacol. Exp. Ther.* **160**, 61–71

WEINER, N. & RABADJIJA, M. (1968b) The regulation of norepinephrine synthesis. Effect of puromycin on the accelerated synthesis of norepinephrine associated with nerve stimulation. *J. Pharmacol. Exp. Ther.* **164**, 103–114

WEINER, N., CLOUTIER, G., BJUR, R. & PFEFFER, R. I. (1972) Modification of norepinephrine synthesis in intact tissue by drugs and during short-term adrenergic nerve stimulation. *Pharmacol. Rev.* **24**, 203–221

Discussion

Fernstrom: With pargyline as the monoamine oxidase inhibitor, do you still see the increase in the rate of hydroxylation on addition of tyrosine?

Weiner: Yes. The activity of tyrosine hydroxylase still rises by about 70% during stimulation in the presence of pargyline and by about 100% in the absence of pargyline.

Carlsson: Did you obtain your results with or without pargyline?

Weiner: In the earlier stimulation experiments on the vas deferens preparations from the guinea pig, where we measured the production of [^{3}H]catecholamines from [^{3}H]tyrosine, pargyline was present. We have confirmed these results with pargyline by the CO_2 method. We omitted pargyline in the more recent experiments with the coupled decarboxylase assay, in the comparisons of the rate of conversion of phenylalanine and tyrosine into catecholamines and in the sodium deletion studies. We used no monoamine oxidase inhibitor in the studies with tranylcypromine and amphetamine; the only experiments with inhibitors were those on nerve stimulation of the guinea-pig vas deferens in the presence of different concentrations of tyrosine and 6,7-dimethyltetrahydropterin.

Wurtman: In Table 1 you showed that tyrosine hydroxylation continued to increase with concentrations of tyrosine up to 100 μM. At that concentration, Dr Kaufman found inhibition of the adrenal enzyme. Does the different source of enzyme explain the different response?

Kaufman: Until we know what the difference is, that seems likely.

Weiner: Tyrosine hydroxylase assayed in the supernatant fraction of the vas deferens homogenate of guinea pig was inhibited by 100μM-tyrosine but not by 50μM-tyrosine. The studies to which you refer (Table 1) were performed on intact vas deferens preparations and the substrate levels indicate the concentrations of tyrosine in the medium. In the guinea pig at any rate, the concentration of tyrosine is probably lower in the nerve ending than in the medium. We cannot compare our data on intact tissue with results obtained with homogenates or purified enzymes; we cannot yet measure accurately the concentration of tyrosine in nerve endings.

Wurtman: In view of Dr Lajtha's remarks on competition among acids for uptake into brain (see Lajtha, pp. 25–41), it is disturbing that so many researchers, ourselves included, tend to use media which contain only one amino acid. What would be the effect on tyrosine uptake and on catecholamine synthesis of the inclusion of leucine, isoleucine, valine, etc. in the medium? What happens to the *in vitro* biosynthesis of catecholamines in media designed to simulate the extracellular fluid?

Weiner: The presence of phenylalanine does not seem to make any difference to the hydroxylation of tyrosine in the intact tissue. This implies that uptake is not limiting or else is relatively little affected by the presence of what is ordinarily considered a competitive substrate.

Wurtman: That casts doubt on a widely accepted explanation of the neural damage caused by phenylketonuria, namely that the excess of phenylalanine inhibits tyrosine hydroxylation. Is catecholamine biosynthesis suppressed in phenylketonuria? Hsia's demonstration of small quantities of 4-hydroxy-3-methoxymandelic acid in urine (Nadler & Hsia 1961) was not very compelling.

Kaufman: Several years ago, Weil-Malherbe (1955) reported low levels of noradrenaline in phenylketonuric patients.

Weiner: He also found only small amounts of catecholamines in the plasma of mentally retarded children. The plasma concentrations of adrenaline in phenylketonuric children did not differ significantly from those in children who were mentally retarded from other causes.

Wurtman: Can the tyrosine hydroxylase present in nerve terminals or adrenal medulla cause tyrosine to be synthesized and liberated without further conversion into dopa? Could the neurons replace the liver in phenylketonuria?

Kaufman: With the enzyme isolated from brain, there is no doubt that tyrosine is liberated as an intermediate. Only negligible amounts of dopa are formed from phenylalanine without liberation of free tyrosine (Lloyd & Kaufman, unpublished findings, 1973). We also varied the size of the pool of unlabelled tyrosine and the results were consistent with the conclusion that the major pathway is through the free pool of tyrosine.

Weiner: Tyrosine rather than phenylalanine seems to be the quantitatively more important precursor of catecholamines (see pp. 138, 139). Possibly dilution explains the inhibition of the conversion of phenylalanine into catecholamines by tyrosine; [^{14}C]phenylalanine forms [^{14}C]tyrosine which is diluted by unlabelled tyrosine as the product (tyrosine) dissociates from the enzyme.

Wurtman: Three times as much catecholamine is synthesized with a given molarity of tyrosine than with the same molarity of phenylalanine (Table 1). Could this be explained by the release of the newly formed tyrosine before conversion into dopa or perhaps by its catabolism by tyrosine aminotransferase (cf. Mandel, pp. 67–79)?

Weiner: Yes, the conversion of phenylalanine into tyrosine might be much greater than we can measure in these experiments.

Kaufman: When I mentioned that in the presence of tetrahydrobiopterin the rate of hydroxylation of phenylalanine by tyrosine hydroxylase from the isolated bovine adrenal was equal to or greater than that of tyrosine (see Kaufman, pp. 91, 92), I was including all the hydroxylated products—tyrosine and catechols. I was not measuring the rate of hydroxylation of phenylalanine to tyrosine.

Our observations with the isolated brain enzyme confirm what you said about competition. The isolated brain enzyme hydroxylates phenylalanine and tyrosine at about the same rate, if we measure both tyrosine and dopa formed from phenylalanine. But if tyrosine and phenylalanine are presented to the enzyme together in equimolar amounts, tyrosine is the preferred substrate (by about four to one) as you found (Table 1), presumably because of its more favourable K_m, although we have not yet determined that.

Wurtman: These findings suggest that there should be no suppression of the first step, at least, in catecholamine synthesis in phenylketonuria.

Two assays for tyrosine hydroxylase are (i) treating dopa, the product of tyrosine hydroxylation, with aromatic L-amino acid decarboxylase and measuring the dopamine that is formed and (ii) measuring the tritium released from the *meta*-position of [^{3}H]tyrosine. What is your opinion, Dr Weiner, of the assay in which one measures the formation of all catechols, dopa, dopamine and noradrenaline, after their extraction on alumina columns?

Weiner: Although most assays are reliable when the tyrosine hydroxylase activity is high, when the activity of the enzyme is low, the assay using only adsorption chromatography on alumina yields an unacceptably large variation in the measurements of radioactivity obtained when the enzyme is absent or fully inhibited, owing to tyrosine adsorption to the column. We therefore use double-column chromatography. We adsorb the catecholamines and tyrosine remaining after chromatography on alumina onto a Dowex 50-H$^+$ column, elute the tyrosine selectively with sodium phosphate buffer (pH 6.5) and subsequently elute the catecholamines with HCl. For several years, we have been trying to isolate the neutral and acidic catechols from tyrosine but, when the c.p.m. of these acidic radioactive catechol products represent a minute fraction of the total, it is hard to separate them from contaminating precursor and other products, particularly when tyrosine may be converted to a considerable degree into acidic products, for instance, by transamination.

Sharman: The heparin results might be relevant to the physiological situation in the mast cells of the cow, sheep and goat. Could the large quantities of dopamine present in these cells be due to a small amount of tyrosine hydroxylase which is activated by the heparin?

Weiner: That is an interesting possibility.

Mandel: Did you control the ionic strength of the different buffers? We noticed that several buffers resulted in different activities of tyrosine aminotransferase but when we changed the buffers and kept the ionic strength constant, the activity did not change.

Weiner: No, we did not control it rigorously in the studies on the intact tissue. We did not calculate the ionic strength, but when we substituted equivalent amounts of potassium chloride or choline chloride for sodium chloride, we found no difference in activity. Kuczenski & Mandell (1972) noted that the activity of tyrosine hydroxylase is proportional to the ionic strength (see p. 141).

Mandel: The same happened with tyrosine aminotransferase.

Gál: I am interested in the lack of effect of phenylalanine on tyrosine hydroxylase shown in Table 1. We found about 30% inhibition of cerebral tyrosine hydroxylase with *p*-chlorophenylalanine in the first 48 h after its administration *in vitro* (Gál *et al.* 1970). Did you try *p*-chlorophenylalanine, chloroamphetamine or dihydroxyphenylserine? I am intrigued by the molecular interaction of the substrate and inhibitor with the enzyme and I suspect that the β-chloro and β-amino analogues of noradrenaline might be extremely active inhibitors.

Weiner: We have not tried any of those compounds.

Grahame-Smith: Dr Weiner, is there a large discrepancy between the amount of catechol synthesized and the amount released on nerve stimulation?

Weiner: We believe that the pool of newly synthesized noradrenaline is distinct from the pool of stored material in the vesicle. After labelling the cat spleen *in vivo* with [³H]noradrenaline four hours before removal of the organ, we found that the specific activity of the unchanged noradrenaline which is recovered in the perfusate from the isolated cat spleen in the intervals between stimulation is much less than the specific activity of the material coming out during stimulation (presumably by exocytosis). Furthermore, most of the [³H]noradrenaline which leaks out in the intervals between stimulation comes out of the spleen as deaminated metabolites. When the splenic nerve is stimulated, most of the radioactivity comes out of the spleen as [³H]noradrenaline. Obviously, the metabolic turnover is considerable even during the resting state, and the noradrenaline is handled differently during stimulation and when the nerve is at rest (Weiner *et al.* 1973).

These results suggest to us that newly synthesized material is more labile. It is not yet stored in the same manner, or to the same degree, as the bulk of the vesicle store and therefore has a much more rapid turnover. Since we stimulate the nerve at 20 Hz for 15 s every 15 min, we calculate that the total amount of [³H]noradrenaline and ³H-labelled metabolites coming out in the interval between stimulations is about twice as much as that which comes out during the brief period of stimulation.

Grahame-Smith: In that case, synthesis is a very wasteful process. It is surprising, therefore, that it should be carefully geared to the amount of noradrenaline in a pool.

Weiner: Even though it may be wasteful, it is a regulated process, since enhanced synthesis is geared to enhanced nervous activity.

Kaufman: Dr Weiner, in your coupled assay is dopa accumulated or continuously decarboxylated?

Weiner: It is continuously decarboxylated. The kidney aromatic amino acid decarboxylase is present in excess throughout the assay for tyrosine hydroxylase. We can add this decarboxylase preparation because tyrosine hydroxylase is absent from (or is present in extremely small amounts in) our kidney decarboxylase preparation. In the intact tissue, there appears to be no need to add exogenous aromatic amino acid decarboxylase. As far as we can tell, the conversion into dopamine is instantaneous. For example, we incubated paired mouse vas deferens preparations separately with [*carboxy*-¹⁴C]tyrosine. After 15 min, one incubation was stopped by addition of acid and the ¹⁴CO$_2$ was collected while to the second a large excess of 3-iodotyrosine, a tyrosine hydroxylase inhibitor, was added. The latter incubation was continued for another 30 min, after which the reaction was terminated with acid. The $^{14}CO_2$ was collected and counted by liquid scintillation spectrometry. The production of $^{14}CO_2$ was the same for both preparations, thereby indicating that residual dopa was not present after the first 15 min incubation, since no decarboxylation occurred in the next 30 min. Thus, there must be virtually instantaneous decarboxylation.

Kaufman: To establish that the inhibition by dopa is competitive in the kineticists' sense, one must extrapolate from the double reciprocal plot that is hyperbolic rather than linear. That requires an act of faith. It is difficult to distinguish a line that intercepts on the ordinate from one that only deviates from this ordinate intercept by about 10%, which is what we found with the dimethylpterin.

Weiner: I agree. However, the point I wish to emphasize is that our studies on the guinea-pig vas deferens reveal a marked difference between the apparent kinetics involving interactions between the pterin cofactor and noradrenaline and the pterin cofactor and nerve stimulation. To me, that implies that the mechanisms differ, whatever they are, and that we are not dealing with altered end-product feedback inhibition as a consequence of nerve stimulation. I am acutely aware that we are dealing with a multiphasic, complex system, and that other factors may complicate these studies leading us to possibly erroneous conclusions. However, at the moment, based on these studies this is our tentative conclusion.

References

GÁL, E. M., ROGGEVEEN, A. E. & MILLARD, S. A. (1970) DL-[2-^{14}C]p-Chlorophenylalanine as an inhibitor of tryptophan 5-hydroxylase. *J. Neurochem.* **17**, 1221–1235

KAUFMAN, S. (1974) Properties of pterin-dependent aromatic amino acid hydroxylases, *This Volume*, pp. 85–107

KUCZENSKI, R. T. & MANDELL, A. J. (1972) Allosteric activation of hypothalamic tyrosine hydroxylase by ions and sulphated mucopolysaccharides. *J. Neurochem.* **19**, 131–137

LAJTHA, A. (1974) Amino acid transport in the brain *in vivo* and *in vitro*, *This Volume*, pp. 24–41

MANDEL, P. & AUNIS, D. (1974) Tyrosine aminotransferase in the brain, *This Volume*, pp. 67–79

NADLER, H. L. & HSIA, D. Y. (1961) Epinephrine metabolism in phenylketonuria. *Proc. Soc. Exp. Biol. Med.* **107**, 721–723

WEIL-MALHERBE, H. (1955) The concentration of adrenaline in human plasma and its relation to mental activity. *J. Ment. Sci.* **101**, 733–755

WEINER, N., BOVE, F. C., BJUR, R., CLOUTIER, G. & LANGER, S. Z. (1973) in *New Concepts in Neurotransmitter Regulation* (Mandell, A. J., ed.), pp. 89–113, Plenum, New York

Nutritional control of the synthesis of 5-hydroxytryptamine in the brain

J. D. FERNSTROM, B. K. MADRAS, H. N. MUNRO and R. J. WURTMAN

Laboratories of Neuroendocrine Regulation and Physiological Chemistry, Department of Nutrition and Food Science, Massachusetts Institute of Technology

Abstract The concentrations of 5-hydroxytryptamine in the rat brain and of tryptophan in plasma and brain normally exhibit daily rhythms. The tryptophan rhythms are in phase; the daily increase in tryptophan content of the brain precedes that of 5-hydroxytryptamine by several hours. The injection of a small dose of tryptophan (e.g. 12.5 mg/kg), which increases plasma and brain tryptophan significantly but not beyond their normal daily ranges, causes brain 5-hydroxytryptamine to rise within an hour.

Insulin injection, or the optional consumption of a carbohydrate or carbohydrate–fat meal, *increases* plasma tryptophan concentrations in rats. The amounts of tryptophan and 5-hydroxytryptamine in the brain also rise. The plasma concentrations of most other amino acids (and of the portion of serum tryptophan that is not bound to albumin) fall. The consumption of a protein-containing diet induces even *larger* rises in plasma tryptophan but has no effect on the amount of brain tryptophan or 5-hydroxytryptamine. Both these rise if animals eat a synthetic amino acid mixture lacking the neutral amino acids that compete with tryptophan for entry into the brain (tyrosine, phenylalanine, leucine, isoleucine and valine). Injection of insulin or the consumption of carbohydrates depresses the amounts of serum free (non-albumin-bound) tryptophan and non-esterified fatty acids in rats, though increasing those of serum albumin-bound tryptophan and brain tryptophan. In humans, ingestion of glucose causes a decline in serum free tryptophan and non-esterified fatty acids but does not change the serum concentration of bound tryptophan. (The serum concentrations of other neutral amino acids, which compete with tryptophan for entry into the brain, fall markedly.) The concentration of 5-hydroxytryptamine in the rat brain is thus physiologically regulated by the diet through changes in the amount of brain tryptophan, which appears to reflect best the *ratio* of total plasma tryptophan to the sum of the competing neutral amino acids rather than the total tryptophan concentration of the serum or the smaller free pool. Neurons containing 5-hydroxytryptamine thus differ markedly from catecholaminergic neurons in which end-product inhibition rather than precursor availability seems to control the rate of neurotransmitter synthesis. If, after the consumption of diets that increase brain 5-hydroxytryptamine, more neurotransmitter is released into synapses, then the neurons which make 5-hydroxy-tryptamine may serve as amplifiers or sensors, providing the brain with information about peripheral metabolic state.

AMINO ACID RHYTHMS IN THE BODY

Our interest in the possibility that the nutritional state might control the synthesis of brain 5-hydroxytryptamine derived from the observation made several years ago that the plasma concentrations of tryptophan and most other amino acids exhibited characteristic fluctuations every day (Wurtman *et al.* 1968; Wurtman 1970). For instance, subjects on a standard mealtime schedule had the lowest plasma concentrations of tryptophan between 0200 and 0400. By late morning or early afternoon the concentrations levelled off 50–80% higher. The amplitude of the rhythm exhibited by any particular amino acid tended to vary inversely with its availability in the body; concentrations of the relatively scarce amino acids (e.g. methionine and cysteine) rose and fell as much as 100% of the nadir value during each 24 h period, while the more abundant amino acids (e.g. glycine and glutamate) rose only 10–30% above their daily nadir. Similar rhythms in the plasma concentrations of tryptophan (Fig. 1) and other amino acids were subsequently noted in rats and mice, but for humans the tryptophan peak, 35–150% above the daily nadir, came about 8–10 h later. This phenomenon was attributed to the rat's tendency to consume most of its food during the hours surrounding the onset of darkness. The plasma amino acid rhythms in humans, however, were not simply the result of the cyclic ingestion of dietary protein, since they persisted among volunteers who ate essentially no protein for two weeks (Wurtman *et al.* 1968). They disappeared, however, in subjects on a total fast (Marliss *et al.* 1970); this finding suggested that the rhythms were *not* truly circadian or of endogenous origin and that the most important factor in their genesis was nutritional, that is the release of insulin (and possibly other hormones) in response to dietary carbohydrate. (Insulin would be expected to raise or lower the plasma concentrations of amino acids by controlling their flux into muscle and other intracellular compartments.)

The mere existence of rhythms in the plasma concentrations of amino acids did not establish that such variations were of any consequence physiologically. To explore their possible significance, we set out to determine whether the metabolic fate of an amino acid could be influenced by the daily fluctuations in its plasma level. This task could be accomplished by showing that experimentally induced fluctuations of the same amplitude as those occurring spontaneously each day could cause parallel changes in the rate at which the amino acid

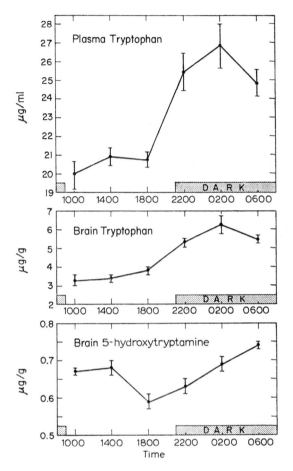

F_{IG}. 1. Daily rhythms in plasma tryptophan, brain tryptophan, and brain 5-hydroxy-tryptamine. Groups of ten rats kept in darkness from 2100 to 0900 were killed at intervals of four hours. Vertical bars indicate S.E.M.

was used for some purpose in the body (e.g. incorporation into proteins or conversion into a low molecular weight compound). The amino acid whose plasma concentration seemed most likely to influence its metabolic fate was tryptophan. Free and peptide-bound tryptophan was known to be the least abundant amino acid in the rat and in most foods (Wurtman & Fernstrom 1972); moreover, it was already known that daily rhythms in the ingestion of proteins containing tryptophan (and, presumably, in the concentration of tryptophan delivered to the liver through the portal venous circulation) generated parallel rhythms in the aggregation of hepatic polysomes (Fishman et al. 1969) and in the quantity of a particular protein, tyrosine aminotransferase, in the liver (Wurtman 1970).

As the dependent variable in our study, we chose to look for possible changes in the brain content of 5-hydroxytryptamine in rats exhibiting a spontaneous daily rhythm in plasma tryptophan or treated in such a way as to raise or lower plasma tryptophan. Three lines of evidence had suggested that the amount of tryptophan available to the brain might control synthesis of 5-hydroxytryptamine: (1) the existence of a daily rhythm in content of the amine in the brain of rats and mice (Albrecht *et al.* 1956); (2) the unusually high Michaelis constant, K_m, for tryptophan shown by tryptophan hydroxylase, the enzyme that catalyses the initial step of the biosynthesis (Lovenberg *et al.* 1968); and (3) the repeated demonstrations that high doses of tryptophan (50–1600 mg/kg) could greatly increase the brain concentrations of 5-hydroxytryptamine and its chief metabolite, 5-hydroxyindoleacetic acid (summarized by Wurtman & Fernstrom [1972]).

LOW DOSES OF TRYPTOPHAN AND BRAIN 5-HYDROXYTRYPTAMINE

Initial experiments were designed to determine whether the brain content of 5-hydroxytryptamine could be increased by giving rats low doses of tryptophan at a time of day when plasma and brain tryptophan and brain 5-hydroxytryptamine concentrations were known to be near their nadirs (Fig. 1). We searched for a dose of tryptophan that would raise the amount of 5-hydroxytryptamine in the brain but would be smaller than the amount of the amino acid normally consumed by rats each day and would not elevate plasma or brain tryptophan concentrations beyond their normal daily peaks. A dose of 12.5 mg/kg tryptophan injected (i.p.) into male rats weighing 150–200 g met these criteria, and constituted less than 5% of the amount of tryptophan that rats would be expected to ingest daily in 10–20 g of standard rat chow. It produced peak elevations in plasma and brain concentrations of tryptophan which were well within the nocturnal ranges of untreated animals and caused the amount of 5-hydroxytryptamine in brain to rise by 20–30% ($P < 0.01$) within one hour of treatment (Fernstrom & Wurtman 1971a). Doses of 25 mg/kg caused proportionately greater increases in both brain tryptophan and 5-hydroxytryptamine. Larger doses, which caused the brain tryptophan concentration to rise well above its physiological range, produced no further increments in brain 5-hydroxytryptamine (Fig. 2).

The observed rise in the amount of 5-hydroxytryptamine in the brains of rats receiving small doses of tryptophan was thought to be compatible with the

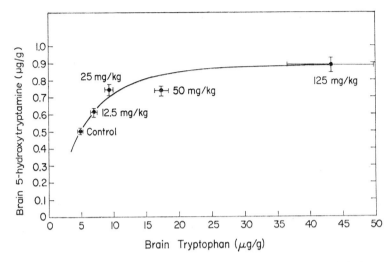

FIG. 2. Dose–response curve relating brain tryptophan and brain 5-hydroxytryptamine. Groups of ten rats received tryptophan intraperitoneally at noon, and were killed one hour later. All brain tryptophan concentrations were significantly higher than control levels ($P <$ 0.01). (Reproduced from Fernstrom & Wurtman 1971a).

hypothesis that the nocturnal rise in the brain amine in untreated rats was related to the daily rhythms in plasma and brain tryptophan (see Fig. 1). (However, this substrate-induced rhythm in the synthesis of 5-hydroxytryptamine is not necessarily the *only* factor responsible for the daily rhythm in its brain content; the rhythm might, for example, reflect changes in the rates at which the monoamine is released from neurons or metabolized intraneuronally.)

EFFECTS OF INSULIN OR DIETARY CARBOHYDRATE ON BRAIN TRYPTOPHAN AND 5-HYDROXYTRYPTAMINE

Now that small *increases* in plasma tryptophan had been shown to cause parallel changes in brain 5-hydroxytryptamine, we were interested to determine whether physiological *decreases* in the plasma amino acid could depress the amine content, and so we attempted to lower plasma tryptophan by giving insulin to the rats. It had not actually been shown that exogenous insulin depressed plasma tryptophan concentration in rats, probably because a simple assay for tryptophan only became available a few years ago. However, there was abundant evidence that insulin exerted this effect on almost all other amino acids, largely by enhancing their uptake into skeletal muscle (Wool 1965).

Rats similar to those used in the previous experiments received a dose of insulin (2 i.u./kg intraperitoneally) known to lower blood glucose concentrations. To our surprise, the hormone did not lower plasma tryptophan but instead increased its concentration by 30–40% (Fernstrom & Wurtman 1972*a*). This effect was independent of the route by which the insulin was administered. At the same time, the plasma glucose content fell 55% and there were major reductions in the plasma concentrations of most other amino acids, including the neutral amino acids generally believed to compete with tryptophan for uptake into the brain (Guroff & Udenfriend 1962; Blasberg & Lajtha 1965). Two hours after rats received the insulin, brain tryptophan levels were 36% higher ($P < 0.01$), and the amount of 5-hydroxytryptamine in the brain was up by 28% ($P < 0.01$) (Fernstrom & Wurtman 1971*b*).

This increase in brain 5-hydroxytryptamine might conceivably have been artifactual, resulting not from increased availability of substrate but from reflexes activated by the accompanying hypoglycaemia. To determine whether the physiological secretion of insulin in *normo*glycaemic rats also increased plasma and brain tryptophan concentrations and the content of 5-hydroxytryptamine in the brain, these indoles were measured in rats fasted for 15 h and then given access to a carbohydrate diet. In a typical experiment, the animals ate an average of 5 g/h during the first hour and 2 g/h during the second and third hours (Fernstrom & Wurtman 1971*b*; 1972*a*). The amount of plasma tryptophan was significantly raised one, two and three hours after food presentation but at the same times tyrosine concentrations were depressed.

On the basis of these observations, we proposed a mechanism by which dietary inputs affect 5-hydroxytryptamine in the brain. By eliciting insulin secretion, carbohydrate consumption should raise plasma tryptophan levels; this elevation should in turn cause a corresponding increase in brain tryptophan and thus in the amount of substrate presented to tryptophan hydroxylase, whereby the synthesis of 5-hydroxytryptamine, and so its brain concentrations, should increase. We predicted that the consumption of a diet containing both carbohydrates and protein should cause an even greater rise in brain 5-hydroxytryptamine: in addition to elevating plasma tryptophan by causing insulin secretion, the tryptophan molecules in the dietary proteins should also contribute directly to plasma tryptophan; brain tryptophan and the amine should presumably increase accordingly. When this model was tested by giving fasted rats access to diets containing either casein or a synthetic amino acid similar in composition to 18% casein, it was immediately apparent that our model needed major revision: as expected, protein consumption was followed by a major increase (about 60%, $P < 0.001$) in plasma tryptophan, but neither brain tryptophan nor brain 5-hydroxytryptamine was at all increased (Fernstrom & Wurtman 1972*b*).

EFFECTS OF DIETARY PROTEIN ON BRAIN TRYPTOPHAN AND 5-HYDROXY-TRYPTAMINE

Other investigators, using brain slices (Blasberg & Lajtha 1965) or animals treated with pharmacological doses of individual amino acids (Guroff & Udenfriend 1962), had shown that groups of amino acids (e.g. neutral, acidic and basic) are transported into brain by specific carrier systems, and that within a group the member amino acids compete with each other for common transport sites. Since protein ingestion introduces variable amounts of all the amino acids into the blood, it seemed possible that the amount of tryptophan in the brain failed to increase after protein ingestion because the plasma concentrations of other, competing, amino acids increased even more than that of tryptophan. To test this hypothesis, we allowed groups of animals to eat either a synthetic diet containing carbohydrates plus all the amino acids in the same proportions as are present in an 18% casein diet or this diet less five of the amino acids thought to share a common transport system with tryptophan (tyrosine, phenylalanine, leucine, isoleucine and valine). Both diets significantly increased plasma tryptophan concentrations above those found in the fasted control animals. However, only when the competing neutral amino acids had been removed from the diet were large increases in the brain concentrations of tryptophan, 5-hydroxytryptamine or 5-hydroxyindoleacetic acid observed (Fig. 3) (Fernstrom & Wurtman 1972b).

To rule out the possibility that this increase in brain 5-hydroxyindoles was simply a non-specific consequence of the omission of any group of amino acids from the diet, we repeated the above experiment omitting aspartate and glutamate instead of the five neutral amino acids. (These two amino acids comprise approximately the same percentage of the total α-amino nitrogen in casein as the five competing amino acids. Because they are charged at physiological pH, they are transported into the brain by a carrier system different from that transporting tryptophan [Blasberg & Lajtha 1965]. Thus, their absence would not be expected to alter the postprandial competition between tryptophan and other amino acids within its transport group for uptake into the brain.) One and two hours after presentation of this diet or the complete amino acid mixture, plasma tryptophan concentrations again increased 70–80% above those of fasted controls ($P < 0.001$). However, neither diet caused increases in the amount of tryptophan, 5-hydroxytryptamine or 5-hydroxyindoleacetic acid in the brain.

These results were interpreted to mean that brain tryptophan and 5-hydroxyindole concentrations do not simply reflect those of plasma tryptophan, but depend also upon the plasma concentrations of other neutral amino acids. This relationship was confirmed by a correlation analysis comparing the brain

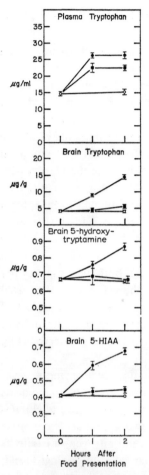

Fig. 3. Effect of ingestion of various diets containing amino acids on plasma and brain tryptophan and brain 5-hydroxyindole concentrations. Groups of eight rats were killed one or two hours after diet presentation (vertical bars represent s.e.m.): fasting controls, ○; complete amino acid mix diet, ■; mix diet less tyrosine, phenylalanine, leucine, isoleucine and valine, ●. The amount of plasma tryptophan after one and two hours was significantly greater in rats consuming both diets ($P < 0.001$) than in the fasting control animals. All the amounts of tryptophan, 5-hydroxytryptamine and 5-hydroxyindoleacetic acid in brain were significantly greater in rats consuming the diet lacking the five amino acids than in the fasting control animals ($P < 0.001$ for all but the value for 5-hydroxytryptamine after one hour, $P < 0.01$). (Reproduced from Fernstrom & Wurtman 1972b).

tryptophan level and the ratio of plasma tryptophan to the five competing amino acids among individual rats given various diets that contain different amounts of each amino acid. This analysis yielded a correlation coefficient of 0.95 ($P < 0.001$, $r = 0$), whereas the correlation between brain tryptophan and plasma tryptophan alone was less striking ($r = 0.66$; $P < 0.001, r = 0$). Similarly, the correlation coefficient for brain 5-hydroxytryptamine and 5-hydroxyindole-acetic acid against the plasma amino acid ratio was 0.89 ($P < 0.001$), whereas that of the two 5-hydroxyindoles against tryptophan alone was only 0.58 ($P < 0.001$) (Fig. 4). Thus, the brain concentrations of both tryptophan and the 5-hydroxy-

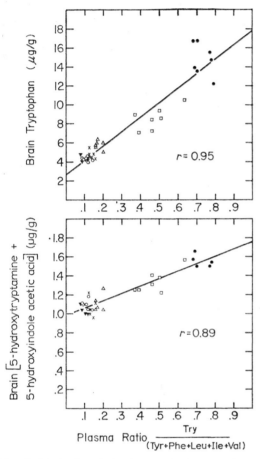

FIG. 4. (Top). Correlation between brain tryptophan concentration and the plasma ratio of tryptophan to the five competing amino acids in individual rats studied in the experiment described in Fig. 3; $r = 0.95$ ($P < 0.001$ that $r = 0$). (Below). Correlation between the sum of brain 5-hydroxytryptamine and 5-hydroxyindoleacetic acid, and the plasma ratio of tryptophan to the five competitor amino acids, in individual rats studied in the experiment described in Fig. 3, $r = 0.89$ ($P < 0.001$ that $r = 0$). One-hour control animals, ○; two-hour control animals, ▼; one hour complete amino acid mix diet, ×; two hour complete amino acid mix diet, △; one hour complete mix diet less five competing amino acids, □; two hour complete mix diet less five competing amino acids, ●. (From Fernstrom & Wurtman 1972b).

indoles more nearly reflect the ratio of plasma tryptophan to competing amino acids than the plasma tryptophan concentration alone. The reason that brain tryptophan and 5-hydroxytryptamine appeared, in our earlier formulation, to depend upon plasma tryptophan alone was that all the physiological manipulations tested at that time (tryptophan injections, insulin injections and carbohydrate consumption) raised the *numerator* in the ratio of plasma tryptophan to competitor while either lowering the denominator or leaving it unaltered. Only when rats consumed protein were both the numerator and the denominator elevated.

The effect of food consumption on 5-hydroxyindoles in rat brain may now be represented as in Fig. 5. Since carbohydrate ingestion elicits insulin secretion, it simultaneously raises plasma tryptophan and lowers the concentrations of the competing neutral amino acids in rats (Fernstrom & Wurtman 1972a);

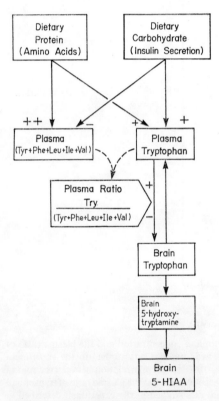

FIG. 5. Proposed sequence describing diet-induced changes in the concentration of 5-hydroxytryptamine in the rat brain. The ratio of tryptophan to the combined levels of tyrosine, phenylalanine, leucine, isoleucine, and valine in the plasma is thought to control the tryptophan level in the brain. (From Fernstrom & Wurtman 1972b).

hence the ratio of plasma tryptophan to competing amino acids increases, leading to elevations in brain tryptophan and 5-hydroxytryptamine. In contrast, consumption of protein provides the plasma with an exogenous source of all the amino acids; however, the ratio of tryptophan to its competitor amino acids is almost always lower in dietary proteins than it is in plasma. Probably for this reason, protein ingestion increases the plasma concentrations of tryptophan less than it does the concentrations of competing amino acids, and thereby decreases the ratio of tryptophan to competitor. The insulin secretion elicited by protein consumption will, by itself, produce an opposite change in this ratio. Thus, the amounts of tryptophan and 5-hydroxyindole in the brain can decrease, increase or remain unchanged after eating, according to the proportion of protein to carbohydrates in the diet and the amino acid composition of the particular proteins. Our most recent observations suggest that diets containing high concentrations of protein (such as 40% casein, or as high a proportion of protein as would be consumed in a steak) actually *decrease* the synthesis of 5-hydroxytryptamine in the brains of rats.

RELATIONSHIP BETWEEN BINDING OF SERUM TRYPTOPHAN TO ALBUMIN AND ITS AVAILABILITY TO THE BRAIN

Tryptophan in plasma is distributed between two pools: about 10–20% circulates as the free amino acid, while the remainder is bound to serum albumin (McMenamy *et al.* 1957). No other amino acid binds appreciably to plasma proteins. Because binding in general implies storage, several investigators have suggested that the plasma free-tryptophan pool is biologically important in determining the availability of circulating tryptophan to brain and other tissues (Knott & Curzon 1972).

A variety of other lipid-soluble compounds also bind to albumin in the blood (e.g., hormones, drugs and fatty acids), and increases in the binding of one compound may cause the displacement of another. Thus, for example, if the concentration of free fatty acids in serum *in vitro* increases, the amount of serum free tryptophan rises too but the albumin-bound tryptophan falls (McMenamy & Oncley 1958). We tested this relationship *in vivo*, by feeding rats diets that would be expected to alter the amount of free fatty acids in their serum. We then measured the new concentrations of free and bound tryptophan. Subsequently, we studied the response of brain tryptophan to diet-induced changes in serum free and bound tryptophan as an approach to determining whether the *physiological* control of brain tryptophan depends on the total or the free tryptophan concentration in the blood.

TABLE 1

Effect of glucose ingestion on serum and brain tryptophan

	Control	Glucose (1 h)	Glucose (2 h)
Serum total tryptophan (μg/ml)	16.2 ± 0.2	19.6 ± 0.6[b]	19.9 ± 0.4[c]
Serum free tryptophan (μg/ml)	5.5 ± 0.1	4.8 ± 0.3[a]	4.2 ± 0.2[c]
Free (% of total)	34	25	21
Serum bound tryptophan (μg/ml)	10.7 ± 0.3	14.8 ± 0.6[b]	15.7 ± 0.5[c]
Non-esterified fatty acid (mmol/l)	1.147 ± 0.034	0.648 ± 0.077[c]	0.604 ± 0.044[c]
Brain tryptophan (μg/g)	4.16 ± 0.42	6.42 ± 0.56[b]	5.93 ± 0.72[b]

D-Glucose (2 g/4 ml tap water) was administered to three groups of 35 rats (108–146 g) through a stomach tube. Control animals received tap water in the same way. Serum was pooled from seven samples; all values are given as mean ± s.e.m. (Table taken from Madras *et al.* 1973).
[a] $P < 0.05$, differs from controls.
[b] $P < 0.01$, differs from controls.
[c] $P < 0.001$, differs from controls.

Rats were killed one or two hours after receiving a glucose load (2 g in 4 ml water) per os. The glucose caused a major elevation in total serum tryptophan although the rise was restricted to the albumin-bound fraction; as expected, serum free tryptophan and serum non-esterified fatty acids decreased markedly (Table 1). Brain tryptophan concentrations rose significantly in these animals (Madras *et al.* 1973).

In other rats, the total amount of tryptophan in plasma and brain was increased by insulin injection (Fernstrom & Wurtman 1972*a*). This treatment altered the proportion of serum tryptophan that was free or bound to albumin in a manner similar to the result of glucose administration: *total* serum tryptophan increased, *albumin-bound* tryptophan increased but serum *free* tryptophan decreased (Madras *et al.* 1973).

In another experiment, groups of rats were fasted overnight and presented the next day with a diet containing either carbohydrates only or fats as well. Both diets increased the total amount of tryptophan in serum, albumin-bound tryptophan and brain tryptophan (Table 2) while they depressed serum free tryptophan. In both groups, the amounts of non-esterified fatty acids in the serum declined but to a much lesser extent in the group consuming fat. Serum free tryptophan also fell less in this group. Both dietary groups showed significant reductions in the blood concentration of tyrosine, one of the five neutral amino acids which have been shown (Fernstrom & Wurtman 1972*b*) to compete physiologically with tryptophan for entry into the brain (Madras *et al.* 1973).

The good correlation between the amounts of non-esterified fatty acids and free tryptophan in the serum suggests that carbohydrates (and insulin) increase the albumin-binding of serum tryptophan by decreasing the extent to which

TABLE 2

Effects of carbohydrate or carbohydrate–fat diets on serum and brain tryptophan

	Fasted controls	Diets	
		Carbohydrate + fat	Carbohydrate
Serum total tryptophan (µg/ml)	16.5 ± 0.3	18.4 ± 0.5^b	19.1 ± 0.4^c
Serum free tryptophan (µg/ml)	6.2 ± 0.1	5.7 ± 0.2^a	3.4 ± 0.2^c
Free (% of total)	37	33	18
Serum bound tryptophan (µg/ml)	10.3 ± 0.4	12.7 ± 0.7^a	15.7 ± 0.5^c
Serum tyrosine (µg/ml)	19.5 ± 0.7	11.7 ± 0.4	14.4 ± 0.5
Non-esterified fatty acid (mmol/l)	0.831 ± 0.021	0.615 ± 0.029^c	0.301 ± 0.024^c
Brain tryptophan (µg/g)	2.24 ± 0.11	3.07 ± 0.18^c	3.45 ± 0.19^c

Groups of 22 rats weighing 170–200 g were deprived of food but not water at 1400 and present-ed with one of the experimental diets at 1030 the next day. Two hours later, the animals were decapitated and serum and brains taken for assay. Control animals had free access to water and were killed throughout the experiment. Each serum value is obtained from two pooled samples. Each diet contained dextrose (270 g), sucrose (221 g), dextrin (270 g), Harper's salt mix (40 g), vitamin mix (10 g) and choline (2 g), to which was added agar (35 g) in water (1 l). The fat diet also contained Mazola oil (150 g). All values are given as mean ± S.E.M. (Table taken from Madras et al. 1973).
^{a} $P < 0.05$, differs from controls.
^{b} $P < 0.01$, differs from controls.
^{c} $P < 0.001$, differs from controls.

albumin is saturated with free fatty acids and thus enhancing its affinity for tryptophan. No correlation was observed in any of these experiments between physiological changes in serum *free* tryptophan and brain tryptophan concentrations.

We have observed similar changes in serum tryptophan and free fatty acid concentrations in humans after a glucose load (Lipsett et al. 1973). Although *total* serum tryptophan does not increase after consumption of carbohydrate, as it does in rats, it does not decrease. However, serum free tryptophan concentrations do decline, about 35% within 90 min, coincident with decreases in serum free fatty acid concentrations. The ratio of total blood tryptophan to other neutral amino acids increases markedly in humans after carbohydrate consumption, as it does in rats. There is at present no information available concerning the effects of dietary carbohydrate on brain tryptophan or 5-hydroxytryptamine in humans.

ACKNOWLEDGEMENTS

These studies were supported in part by grants from the John A. Hartford Foundation, NASA and the United States Public Health Service (NS-10459 and AM-142281).

References

ALBRECHT, P., VISSCHER, M. B., BITTNER, J. J. & HALBERG, F. (1956) Daily changes in 5-hydroxytryptamine concentration in mouse brain. *Proc. Soc. Exp. Biol. Med.* **92**, 702–706

BLASBERG, R. & LAJTHA, A. (1965) Substrate specificity of steady-state amino acid transport in mouse brain slices. *Arch. Biochem. Biophys.* **112**, 361–377

FERNSTROM, J. D. & WURTMAN, R. J. (1971a) Brain serotonin content: physiological dependence on plasma tryptophan levels. *Science (Wash. D.C.)* **173**, 149–152

FERNSTROM, J. D. & WURTMAN, R. J. (1971b) Brain serotonin content: increase following ingestion of carbohydrate diet. *Science (Wash. D.C.)* **174**, 1023–1025

FERNSTROM, J. D. & WURTMAN, R. J. (1972a) Elevation of plasma tryptophan by insulin in the rat. *Metabolism* **21**, 337–343

FERNSTROM, J. D. & WURTMAN, R. J. (1972b) Brain serotonin content: physiological regulation by plasma neutral amino acids. *Science (Wash. D.C.)* **178**, 414–416

FISHMAN, B., WURTMAN, R. J. & MUNRO, H. N. (1969) Daily rhythms in hepatic polysome profiles and tyrosine transaminase activity: role of dietary protein. *Proc. Natl. Acad. Sci. U.S.A.* **64**, 677–682

GUROFF, G. & UDENFRIEND, S. (1962) Studies on aromatic amino acid uptake by rat brain *in vivo*. *J. Biol. Chem.* **237**, 803–806

KNOTT, P. J. & CURZON, G. (1972) Free tryptophan in plasma and brain tryptophan metabolism. *Nature (Lond.)* **239**, 452–453

LIPSETT, D., MADRAS, B. K., WURTMAN, R. J. & MUNRO, H. N. (1973) Carbohydrate ingestion causes a selective decline in non-albumin-bound plasma tryptophan coincident with the reduction in plasma free fatty acids. *Life Sci.* **12**, 57–64

LOVENBERG, W., JEQUIER, E. & SJOERDSMA, A. (1968) A tryptophan hydroxylation in mammalian systems. *Adv. Pharmacol.* **6A**, 21–36

MADRAS, B. K., COHEN, E. L., FERNSTROM, J. D., LARIN, F., MUNRO, H. N. & WURTMAN, R. J. (1973) Dietary carbohydrate increases brain tryptophan and decreases free plasma tryptophan. *Nature (Lond.)* **244**, 34–35

MARLISS, E. B., AOKI, T. T., UNGER, R. H., SOELDNER, J. S. & CAHILL, G. F. (1970) Glucagon levels and metabolic effects in fasting man. *J. Clin. Invest.* **49**, 2256–2270

MCMENAMY, R. H. & ONCLEY, J. L. (1958) The specific binding of L-tryptophan to serum albumin. *J. Biol. Chem.* **233**, 1436–1440

MCMENAMY, R. H., LUND, C. C. & ONCLEY, J. L. (1957) Unbound amino acid concentrations in human blood plasma. *J. Clin. Invest.* **36**, 1672–1679

WOOL, I. G. (1965) Relation of effects of insulin on amino acid transport and on protein synthesis. *Fed. Proc.* **24**, 1060–1070

WURTMAN, R. J. (1970) in *Mammalian Protein Metabolism*, Vol. 4 (Munro, H. N., ed.), pp. 445–479, Academic Press, New York

WURTMAN, R. J. & FERNSTROM, J. D. (1972) in *Perspectives in Neuropharmacology* (Snyder, S. H., ed.), pp. 145–193, Oxford University Press, London

WURTMAN, R. J., ROSE, C. M., CHOU, C. & LARIN, F. (1968) Daily rhythms in the concentrations of various amino acids in human plasma. *N. Engl. J. Med.* **279**, 171–175

Discussion

Moir: When the rats were given glucose, did the 5-hydroxytryptamine concentration in the brain rise with that of tryptophan?

Fernstrom: Yes.

Gessa: Were the doses of insulin you used within the physiological range?

Fernstrom: The dose used (2 i.u./kg, see p. 158) is standard in experiments of this kind (e.g. Sanders & Riggs 1967). I cannot say whether it caused elevations in blood insulin that exceeded normal concentrations found after hormone secretion; however, the same effects were observed after animals ate carbohydrate, and thus presumably secreted insulin. It is hard to imagine a more physiological situation.

Gessa: Does insulin *in vitro* stimulate the uptake of tryptophan and other amino acids by brain slices?

Fernstrom: The influence of insulin on amino acid uptake into brain has only been studied for those amino acids that are gluconeogenic intermediates. After administration of insulin or carbohydrate, the concentrations of total tryptophan in plasma and in the brains of rats increase to about the same extent—about 40% over control in two hours. Naively, this suggests that the ratio of brain tryptophan to plasma tryptophan does not change and consequently that insulin affects brain tryptophan through plasma tryptophan rather than by direct control over the uptake of tryptophan into the brain.

Wurtman: Also insulin lowers the plasma concentrations of tyrosine, phenylalanine, isoleucine, leucine and valine, all of which compete with tryptophan for uptake into the brain.

Lajtha: Present evidence suggests that insulin has no direct effect on cerebral amino acid transport. Is there any evidence of compartmentation of either tryptophan or the newly formed 5-hydroxytryptamine?

Eccleston: Our evidence suggests two pools for synthesis of 5-hydroxytryptamine (Shields & Eccleston 1972, 1973); this implies two pools of tryptophan for such synthesis. We believe that the newly taken up tryptophan enters the pool for the synthesis of amine which is rapidly turning over at the nerve terminal.

Wurtman: The physiological increase in cerebral 5-hydroxytryptamine in response to consumption of carbohydrate is found both in the brain stem (the locus of the perikarya of neurons containing 5-hydroxytryptamine) and in the telencephalon (i.e. within nerve terminals).

Glowinski: It is clear that the amount of 5-hydroxytryptamine in the rat brain increases during the light period of the day. As I have already mentioned we also observed higher levels of the amine in different structures of the rat brain at various times of the light period compared with those seen at various times of the dark period. Moreover, isotopic techniques or measurement of the initial accumulation of the amine after monoamine oxidase inhibition shows that synthesis of the amine is maximal during the light period. During this period, the amount of tryptophan in plasma decreases (Hery *et al.* 1972). This points to a poor correlation between plasma tryptophan levels and the rate of synthesis of the amine in the brain. This does not exclude the possibility that the amount of free tryptophan in plasma parallels the daily rhythm of monoamine synthesis. However, since we can demonstrate *in vitro* (in slices) that the daily changes in the synthesis of 5-hydroxytryptamine were related to changes in tryptophan transport, we believe that the tryptophan uptake process may also contribute to the regulation of the synthesis of the amine. Changes in the affinity of tryptophan for its carrier are perhaps as important as those of free tryptophan in plasma.

Fernstrom: Certainly, mechanisms other than tryptophan availability may also influence synthesis of 5-hydroxytryptamine.

Curzon: You withdrew the mixed diet and in relatively short-term experiments put the rats on either a carbohydrate or a fat-rich diet. Did you keep the rats on these new diets? Did the changes in brain tryptophan metabolism persist or did adaptive differences oppose them?

Fernstrom: If rats consume a diet for several weeks that is poor in tryptophan, such as a diet containing corn as the only source of protein (Fernstrom & Wurtman 1972), blood and brain concentrations of tryptophan and brain 5-hydroxytryptamine all decline profoundly. So presumably adaptive differences do not oppose these changes. However, such animals are far from normal. In our acute experiments, we attempt to study rats that are as near to normal as possible. To this end, we are designing experiments in which food is withdrawn only during the latter portion of the light period, when rats normally eat little, and replaced two or three hours later just before the onset of darkness, the time when rodents usually begin to eat, instead of chronic exposure as you suggest. Measurements of the changes in tryptophan and 5-hydroxytryptamine elicited by food consumption should clarify the relationship between the cerebral concentration of 5-hydroxytryptamine and diet in these animals.

Curzon: But isn't it stressful suddenly to present a rat accustomed to one diet with a diet which, I presume, he notices is different? What would happen in the reverse experiment where you fed animals on a high carbohydrate diet, withdraw it, and then confront them with a mixed diet?

Fernstrom: A week or so on a carbohydrate diet might be a stress; the animal might begin to suffer from protein malnutrition.

Wurtman: Even though *mean* brain tryptophan and 5-hydroxytryptamine concentrations, averaged throughout each 24 h period, are depressed in animals on a diet containing corn as the sole source of protein for a long time, the pattern of daily rhythms for both these compounds in brain is preserved and is similar to that observed in rats eating normal diets (Fernstrom & Wurtman 1971).

No evidence suggests that brain tryptophan or 5-hydroxytryptamine undergo 'adaptive changes' in response to chronic dietary tryptophan deficiency.

Curzon: Suppose you used two diets, both nutritionally adequate but tasting different. The stress resulting from the change from one to the other might induce changes in the fatty acid concentrations and therefore changes in the amount of tryptophan in the brain.

Wurtman: These diets do not raise plasma glucocorticoid concentrations as we should expect if there were a stress. The pure carbohydrate and carbohydrate plus fat diets induce enormous differential changes in serum free and albumin-bound tryptophan: free tryptophan falls while bound tryptophan accumulates. There is no correlation between the changes in serum free tryptophan (which falls, along with non-esterified fatty acids) and those in brain tryptophan and 5-hydroxytryptamine (which rise) (Madras *et al.* 1973). Total serum tryptophan, brain tryptophan and brain 5-hydroxytryptamine also rise. Protein-containing diets also elevate serum total tryptophan but without increasing brain trypto-phan or 5-hydroxytryptamine (see p. 158 and Fernstrom & Wurtman 1972). As Dr Fernstrom indicated, this differential effect of protein-free and protein-containing diets probably results from the increase that the latter causes in blood concentrations of competing neutral amino acids.

Sandler: The unique properties of tryptophan seem to fit it for its role of essential amino acid in least abundance in nature. What happens to 5-hydroxy-indole synthesis in extreme scarcity situations such as carcinoid and pellagra?

Fernstrom: Several weeks of tryptophan deficiency reduce brain concentra-tions of the monoamine in experimental animals (Gál & Drewes 1962). We have confirmed these results using artificial amino acid diets and found a decline in brain and plasma tryptophan levels which parallels the decrease of brain 5-hydroxytryptamine. We have also demonstrated this effect in rats consuming natural proteins instead of amino acid mixtures. Wheat (*Sorghum vulgare*) contains an adequate amount of tryptophan and usable nicotinic acid but has a high leucine content. People who consume this grain as their main protein source suffer from a form of pellagra associated with low urinary concentrations of 5-hydroxyindoleacetic acid. Supplementing the diet with leucine further lowers the excretion of this metabolite in humans (Belavady *et al.* 1963). If rats are fed *ad libitum* a diet containing 10% *Sorghum vulgare*, brain levels of 5-hydroxytryptamine are considerably depressed (Ramanamurthy & Srikantia 1970). The mechanism of this effect is possibly to increase competi-tion of leucine with tryptophan for uptake into the brain.

Wurtman: Experiments in which rats acutely consume a normal diet tell us more about the physiology of 5-hydroxytryptamine neurons than those in which animals are chronically tryptophan-deficient. We can speculate that 5-

hydroxytryptamine neurons have the special function of continuously monitoring the amino acid pattern of the plasma. This pattern changes in characteristic ways in response to various diets, fasting and the secretion of various hormones. These neurons may provide the brain with information about systemic metabolism. The brain could then use this information to decide on strategies, for example, whether to sleep, and whether to eat. Preliminary results suggest that the quantity of food that the animal opts to consume can be modified by experimental alterations that raise brain 5-hydroxytryptamine.

Curzon: Commonly in life, an animal does not get enough food. Then we observe what could be a homeostatic mechanism; that is, the amount of fatty acids goes up, tryptophan is displaced from the albumin and, temporarily at least, more tryptophan may be made available for brain protein synthesis. I don't know how long this mechanism might continue but possibly it could be a protection against short periods of food deprivation, especially at vulnerable stages of development.

Roberts: The difficulties in determining the relationship between plasma tryptophan, brain tryptophan and the synthesis of 5-hydroxytryptamine may be partially resolved by introducing a dynamic factor, the turnover of brain tryptophan. When the concentrations of other amino acids in plasma and their transport into the brain are depressed, what happens to the turnover of tryptophan in the brain? In those circumstances where you observe an elevation of plasma tryptophan, is it possible that the source is the liver and that the elevation is due to the inhibition of other uses of tryptophan in this organ?

Fernstrom: Turnover measurements would be useful. We are just beginning to look at the effects of different diets on the brain concentrations of other neutral amino acids.

We have not yet identified the source of the tryptophan that causes the rise in plasma tryptophan after injection of insulin or consumption of carbohydrate. The amount of tryptophan in liver does not decrease and falls only slightly (about 10%) in muscle. However, since muscle represents 40% of total body mass, a 10% decline in muscle tryptophan could supply sufficient amino acid to the blood to raise the plasma concentration by 50%.

Gál: L-[^{14}C]Tryptophan (90 μCi/kg), incorporated into protein in rat brains, is distributed evenly across the brain (Harvey & Gál 1974), if you calculate the incorporation per g brain rather than per area. After three days, there is a noticeable loss of label from the cerebral proteins.

Oja: If tryptophan regulates protein synthesis in the brain, since its supply from plasma may be a limiting factor, am I correct in assuming that most of the increase in tryptophan in the brain is used for increased protein synthesis and that only a small portion is used in the synthesis of 5-hydroxytryptamine?

Gál: Yes; about 20% is used for the synthesis of the amine.

Wurtman: Of course, the brain is heterogeneous and only few of its cells are able to synthesize 5-hydroxyindoles. Ideally, one would like to know the proportions of tryptophan used *within these cells* for the synthesis of protein and of 5-hydroxyindoles.

Kaufman: Is it heresy to suggest that, perhaps, the amount of plasma-bound tryptophan determines the brain concentration of tryptophan?

Wurtman: Fig. 1 summarizes one possible explanation for the paradoxical finding that serum total tryptophan and albumin-bound tryptophan, but not serum free tryptophan, change in parallel with brain tryptophan in response to various diets. Let us suppose that serum free tryptophan is in equilibrium with circulating albumin-bound tryptophan, and with the tryptophan in such tissues as brain, heart and skeletal muscle. Let us further postulate that in order for the albumin-bound tryptophan to enter any of these tissues it must pass—even transiently—through a free tryptophan phase. Suppose that the affinity of albumin for tryptophan molecules is greater than that of muscle or heart but less than that of brain. The phenomenon of albumin binding will then favour tryptophan molecules entering brain at the expense of heart and skeletal muscle—even though the size of the total tryptophan pool within the latter tissues may be greater than in brain. These affinity constants can presumably be varied by other circulating factors, for example, insulin may affect tryptophan uptake into muscle, while competing neutral amino acids will affect its uptake into heart. In any case, all the tryptophan in the serum would be available to brain, even though only the free part actually entered brain tissue.

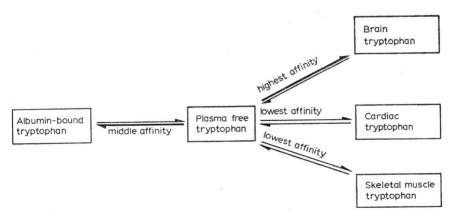

FIG. 1 (Wurtman). Possible equilibria between circulating tryptophan and tryptophan in tissues. Brain is assumed to have the highest affinity for tryptophan, heart and skeletal muscle the lowest, while the affinity of albumin lies in between.

One general consequence of the binding of circulating compounds to albumin might be to facilitate their uptake into certain organs (whose affinity constants exceed that of albumin) at the expense of others. This formulation assumes that the binding to the albumin molecule is not very 'tight' and that the bound molecule is easily stripped off the albumin to enter the tissue. Probably, we are all accustomed to thinking of pharmacological analogies in which lipid-soluble drugs that are bound to albumin are displaced by other drugs thereby causing toxic effects. These pharmacological analogies may not tell us much about the binding of tryptophan to albumin. The affinity constant of albumin for tryptophan is low, about 10 μmol/l (McMenamy & Oncley 1958); this is much lower than, for example, the affinities of circulating steroids to globulin-binding proteins. It is not hard to imagine that brain might be the winner in its competition with albumin for tryptophan.

Barondes: One factor determining synaptic efficacy is the amount of neurotransmitter released at a given synapse. For a 25% change in nerve ending 5-hydroxytryptamine to produce a physiologically significant change it should be reflected in the amount released each time the presynaptic terminal fired. Is there any evidence for this? Alternatively the altered amount of 5-hydroxytryptamine might provide a change in the size of the 'reservoir' in nerve terminals that are close to being depleted. It is not clear that the 5-hydroxytryptamine system is modulated in this way. Does it seem likely that the change which you observe is physiologically significant?

Fernstrom: Currently, we are unable to measure directly the amount of 5-hydroxytryptamine released when 5-hydroxytryptaminergic neurons depolarize. So, we cannot yet say whether more monoamine is released from nerve endings when brain concentrations of 5-hydroxytryptamine rise. One indirect approach might be to measure alterations in body functions that are thought to be influenced by central 5-hydroxytryptaminergic neurons (for example, sleeping, eating and temperature regulation) in response to a change in brain concentration of the monoamine.

Wurtman: This assumes that the release of 5-hydroxytryptamine is quantal.

References

BELAVADY, B., SRIKANTIA, S. G. & GOPALAN, C. (1963) The effect of the oral administration of leucine on the metabolism of tryptophan. *Biochem. J.* **87**, 652–655

FERNSTROM, J. D. & WURTMAN, R. J. (1971) Effect of chronic corn consumption on serotonin content of rat brain. *Nat. New Biol.* **234**, 62–64

FERNSTROM, J. D. & WURTMAN, R. J. (1972) Brain serotonin content: physiological regulation by plasma neutral amino acids. *Science (Wash. D.C.)* **178**, 414–416

GÁL, E. M. & DREWES, P. A. (1962) Studies on the metabolism of 5-hydroxytryptamine (serotonin) II: effect of tryptophan deficiency in rats. *Proc. Soc. Exp. Biol. Med.* **110**, 368-371

HARVEY, J. A. & GÁL, E. M. (1974) *Science (Wash. D.C.)*, in press

HERY, F., ROUER, E. & GLOWINSKI, J. (1972) Daily variations of serotonin metabolism in the rat brain. *Brain Res.* **43**, 445–465

MADRAS, B. K., COHEN, E. L., FERNSTROM, J. D., LARIN, F., MUNRO, H. N. & WURTMAN, R. J. (1973) Dietary carbohydrate raises brain tryptophan while lowering non-albumin-bound plasma tryptophan. *Nature (Lond.)* **244**, 34–35

MCMENAMY, R. H. & ONCLEY, J. L. (1958) The specific binding of L-tryptophan to serum albumin. *J. Biol. Chem.* **233**, 1436–1447

RAMANAMURTHY, P. S. V. & SRIKANTIA, S. G. (1970) Effects of leucine on brain serotonin. *J. Neurochem.* **17**, 27–32

SANDERS, R. B. & RIGGS, T. R. (1967) Modification by insulin of the distribution of two model amino acids in the rat. *Endocrinology* **80**, 28–37

SHIELDS, P. J. & ECCLESTON, D. (1972) Effects of electrical stimulation of rat midbrain on 5-hydroxytryptamine synthesis as determined by a sensitive radioisotope method. *J. Neurochem.* **19**, 265–272

SHIELDS, P. J. & ECCLESTON, D. (1973) Evidence for the synthesis and storage of 5-hydroxy-tryptamine in two separate pools in the brain. *J. Neurochem.* **20**, 881–888

The transfer of tryptophan across the synaptosome membrane

A. PARFITT* and D. G. GRAHAME-SMITH

MRC Unit and Department of Clinical Pharmacology, University of Oxford, Oxford and
*National Institute of Child Health and Human Development, National Naval Medical Center, Bethesda, Maryland

Abstract Synaptosomes contain 5-hydroxytryptamine and enzymes for its synthesis. Since the rate of synthesis of 5-hydroxytryptamine in the brain is dependent on the concentration of L-tryptophan at tryptophan hydroxylase, the synaptosome has been used to study *in vitro* how transport across the synaptosomal membrane determines intrasynaptosomal concentrations of tryptophan.

Tryptophan is taken up by two systems, one of high affinity (K_m 11.1 μM), the other of low affinity (K_m 0.32 mM). It is not yet known whether this indicates two systems in synaptosome membrane or heterogeneity of synaptosome function. Uptake by both systems is dependent on temperature and energy.

The low affinity system is insensitive to changes in external concentration of sodium. Uptake by the high affinity system initially overshoots the eventual steady state and the degree of overshoot is proportional to the incubation temperature. Uptake of tryptophan is inhibited by certain amino acids and for L-phenylalanine the K_i is 0.163 mM. In the low affinity uptake system, the ratio of internal to external tryptophan concentration may reach 4. This is a saturable carrier-mediated transport system, present in the rat at birth, and shows stereospecificity for L-isomers.

The efflux of tryptophan and phenylalanine from preloaded synaptosomes is stimulated markedly by certain other amino acids and their influx promoted by preloading with either amino acid. On the basis of this countertransport or exchange diffusion data, there appear to be several systems of overlapping specificity for transporting amino acids across synaptosomal membranes. Countertransport data also indicate that at least two functionally separate intrasynaptosomal compartments exist which have different propensities for participation in the exchange diffusion process. The integrity of these compartments appears to depend on temperature.

The disturbance of the steady-state distribution of L-tryptophan and L-phenylalanine by the introduction of a second amino acid promoting exchange diffusion has revealed the existence of an oscillatory response in the ensuing redistribution of the labelled amino acid.

On the data so far available it is *not* possible to say that a synaptosomal membrane transport system exists which is specifically geared to determining the concentration of tryptophan in a compartment subserving synthesis of 5-hydroxytryptamine.

At present controversy surrounds the value of the Michaelis constant, K_m, of L-tryptophan 5-hydroxylase in the brain, the enzyme responsible for the first and rate-limiting step in 5-hydroxytryptamine biosynthesis. Estimates of the K_m value for this enzyme prepared from rabbit and rat brain vary from 50 μM (Friedman et al. 1972) to 300 μM (Ichiyama et al. 1970; Jequier et al. 1969). Tryptophan concentrations in the brain are in the range of 20–50 μM (Grahame-Smith 1971; Fernstrom & Wurtman 1971). On the assumption of an even distribution of tryptophan at both anatomical and subcellular levels, even with the lower values of K_m reported, the enzyme should not be saturated physiologically with its substrate. In vivo the rate of synthesis of 5-hydroxytryptamine increases as the concentration of brain tryptophan increases until the brain tryptophan concentration reaches about 80 μg/g (400 μM; Grahame-Smith 1971). These experiments indicate that the enzyme is not saturated with its substrate in vivo and that changes in the concentration of tryptophan 5-hydroxylase alter the rate of synthesis of 5-hydroxytryptamine in the brain.

The factors regulating the intraneuronal concentration of tryptophan may therefore be important in the regulation of the rate of synthesis of 5-hydroxytryptamine in the brain. From first principles, we may assume these factors to be: (1) the concentration of free plasma tryptophan (Knott & Curzon 1972); (2) the passage of tryptophan across the walls of brain capillaries; (3) the concentration of tryptophan presented to the neuronal membrane; and (4) the mechanism of transport of tryptophan across the neuronal membrane. The work described here concerns this fourth factor.

Pinched-off nerve endings (synaptosomes) contain L-tryptophan 5-hydroxylase, aromatic amino acid decarboxylase and 5-hydroxytryptamine and they are able to synthesize the amine from L-tryptophan with 5-hydroxytryptophan as the intermediate (see Grahame-Smith 1967). Accordingly, the passage of tryptophan across the limiting membrane of synaptosomes has been studied.

We have already described in detail most of our methods (Grahame-Smith & Parfitt 1970) but some general points should be made. Crude mitochondrial (P_2) fractions were prepared from brains, from which the cerebella had been removed, by the methods described by Gray & Whittaker (1962). Initially, the brains were homogenized in 0.32M-sucrose and, after preparation, the P_2 fraction was resuspended in a Krebs–Ringer phosphate solution containing 40mM-D-glucose. This resuspended P_2 fraction was used in routine transport studies. The P_2 fraction was further fractionated on a discontinuous sucrose gradient (Gray & Whittaker 1962) to give fractions A (myelin), B (synaptosomes), and C (mitochondria). Each of these was centrifuged again and the resulting pellets were again suspended in Krebs–Ringer phosphate solution for transport studies. Carefully calculated iso-osmotic conditions were maintained for all incubations to prevent break-up of synaptosomes. At the end of incubation, samples were vacuum-filtered through a Millipore filter (25 mm diameter; 0.65 μm pore size), and the residue was briefly washed with 0.32M-sucrose at 2 °C. The radioactivity of the material (derived from the radioactive substrates used) retained by the filter was measured by scintillation counting. By this method, we can study transport processes over extremely short times. For tryptophan uptake, the average reproducibility was $\pm 1.5\%$ of the total activity (as c.p.m.) on the filter. This describes in essence the experimental approach and variations on this will be described as they arise.

Fig. 1 shows the progress of the uptake of 0.5mM-L-[3-^{14}C]tryptophan by a rat-brain P_2 fraction at 37 and 2 °C. Uptake was extremely rapid at 37 °C and reached equilibrium in 10–20 min. The uptake at 2 °C was slow and probably represents diffusion. The amount of tryptophan taken up increases with temperature up to 40 °C above which it apparently declines probably owing to

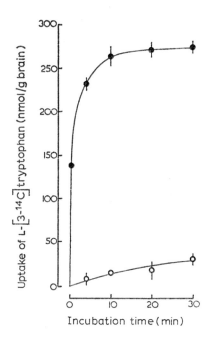

FIG. 1. Time course of uptake of 0.5mM-L-[3-^{14}C]tryptophan (specific activity 0.08 Ci/mol) by the P_2 fraction from rat brain at 2 °C (○) and 37 °C (●).

disruption of the synaptosomes. The radioactivity measured after such uptake experiments has been shown by paper chromatography and autoradiography to be associated only with tryptophan.

The criteria upon which we based our conclusion that the uptake by P_2 fractions was due to its synaptosome content are as follows:—

(1) Treatment of crude mitochondrial fractions in hypo-osmolar conditions, with detergents, by ultrasonication or by freezing and thawing decreased the ability of P_2 fractions to take up L-[3-^{14}C]tryptophan. Such procedures break up synaptosomes.

(2) Marchbanks (1967) found that synaptosomes are eluted intact with their contents from iso-osmotic Sephadex G50 columns but on hypo-osmotic columns they break up and release their contents, in which event small molecules such as tryptophan are retained by the gel. When P_2 fractions preloaded with L-[3-^{14}C]-tryptophan were run through iso-osmotic columns, the radioactivity was associated with the protein. When P_2 fractions, similarly preloaded, were placed on a hypo-osmotic Sephadex column, the protein was eluted devoid of radioactivity. The synaptosomes had lysed on the Sephadex column suspended in hypotonic solution, whereupon tryptophan was released and retained by the gel.

(3) When P_2 fractions were further fractionated on a discontinuous sucrose density gradient by centrifugation, fraction B (the synaptosomal fraction) had the greatest uptake activity (see Fig. 2). This experiment was repeated with

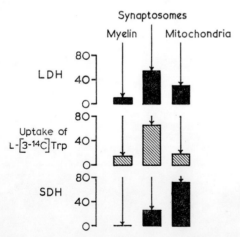

FIG. 2. Distribution of lactate dehydrogenase (LDH) activity, L-[3-^{14}C]tryptophan uptake activity and succinate dehydrogenase (SDH) activity in subfractions of rat-brain P_2 fraction: fraction A (myelin fraction); fraction B (synaptosomal fraction); fraction C (mitochondrial fraction).

TABLE 1

The percentage distribution of endogenous tryptophan, lactate dehydrogenase and succinate dehydrogenase in subfractions of rat-brain P_2 fraction.

% Distribution	Subfraction[a]		
	A	B	C
Lactate dehydrogenase activity	15	55	30
Succinate dehydrogenase activity	2	31	67
Endogenous tryptophan	18	56	26

Recoveries (%) from gradient: tryptophan 83; lactate dehydrogenase 77; succinate dehydrogenase 74; protein 87. [a] See Fig. 2.

brains of one-day-old rats, which contain only a little myelin, and the pattern indicating predominant synaptosomal uptake of tryptophan was even more marked. Similar experiments showed that endogenous tryptophan in the P_2 fraction had a similar synaptosomal disposition (Table 1). Experiments on the uptake of L-[U-^{14}C]phenylalanine showed this also to have a predominantly synaptosomal uptake.

These results provide reasonable evidence that indeed synaptosomal uptake is being measured. For routine work, we studied the uptake by P_2 fractions.

KINETICS OF TRYPTOPHAN UPTAKE

Initial investigations with external concentrations of L-[3-^{14}C]tryptophan of 0.167–0.837 mM showed that tryptophan uptake follows saturable Michaelis–Menten kinetics with a K_m of 1.0 ± 0.28 mM with a V_{max} of 225 ± 39, where V is the amount of L-tryptophan (in μmol) taken up in 5 min by the P_2 fraction prepared from 1 g of brain (Grahame-Smith & Parfitt 1970). These experiments have been repeated more recently with two ranges of tryptophan concentrations: 0.5–8.0 μM and 0.2–3.2 mM. The initial uptake of L-[2,3-^{3}H$_2$]tryptophan was studied over 20 s: Table 2 shows the values for the K_m and V_{max} obtained. Two uptake systems, one of low and the other of high affinity, are apparent for tryptophan. The functional importance and precise individual properties of these two systems are not yet known nor is it known whether the two systems are present in all synaptosomes or whether the results indicate a heterogeneity of synaptosomal function. If we assume that the concentration presented to the neuronal membrane is somewhere in the region of the brain tryptophan concentration (i.e. 25–50 μM), then probably the high affinity system, with a K_m of about 10 μM, is operating.

TABLE 2

Values of K_m and V_{max} for L-[2,3-^{3}H$_2$]tryptophan with rat-brain P$_2$ fraction.

Concentration of L-[2,3-^{3}H$_2$]tryptophan	K_m/M	V_{max}/nmol (g brain)$^{-1}$ min^{-1}
0.5–8.0 μM	11.1×10^{-6}	128
0.2–3.2 mM	3.2×10^{-4}	621

STEREOSPECIFICITY OF TRYPTOPHAN UPTAKE

Fig. 3 shows the effect of L-tryptophan and D-tryptophan on the uptake of L-[3-^{14}C]tryptophan. L-Tryptophan 'inhibits' the uptake of its radioactive isotope by dilution but D-tryptophan is a poor inhibitor of L-tryptophan uptake. Clearly, the transport process is stereospecific.

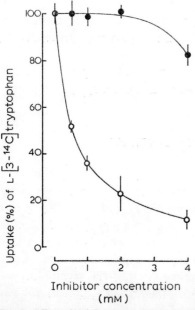

FIG. 3. Effect of D-(●) and L-tryptophan (○) on uptake of L-[3-^{14}C]tryptophan by rat-brain P$_2$ fraction. Incubations were for 2 min at 37 °C with 0.5mM-L-[3-^{14}C]tryptophan and either D- or L-tryptophan at the concentrations shown.

THE EFFECT OF OTHER AMINO ACIDS ON TRYPTOPHAN UPTAKE

The effect of L-methionine, L-tyrosine, L-leucine and DL-p-chlorophenylalanine on L-[3-^{14}C]tryptophan uptake is shown in Table 3. Phenylalanine appears to

TABLE 3

Effect of various amino acids on the uptake of L-[3-^{14}C]tryptophan by rat-brain P$_2$ fractions.

Amino acid	Inhibition of tryptophan uptake (%)
Glycine	16.0
L-Methionine	18.3
L-Tyrosine	27.4
L-Leucine	36.7
DL-p-Chlorophenylalanine	56.6

be a competitive inhibitor of tryptophan uptake with a K_i of between 0.09–0.2 μM (Grahame-Smith & Parfitt 1970). In addition, L-tryptophan inhibits the uptake of L-phenylalanine.

We have tried to measure the intrasynaptosomal volume of both P$_2$ fractions and purified synaptosomal fractions using the method described by Marchbanks (1967) but our attempts have not been wholly successful. Ratios of the internal to external concentration of between 2:1 and 4:1 were obtained after the incubation of P$_2$ fractions with 0.5 mM-L-[3-^{14}C]tryptophan. There did seem to be a concentrative uptake at these concentrations but not too much reliance should be placed upon these figures.

Changes in the sodium, calcium, magnesium and potassium concentrations of the external medium had little reproducible effect upon tryptophan uptake. Certainly no critical sodium dependence was demonstrable.

Acute replacement of oxygen by nitrogen over the incubation medium had no effect on tryptophan uptake. Likewise, the acute omission of D-glucose had no effect. Potassium cyanide and deoxyglucose alone and in combination did decrease uptake, particularly after preparations had been preincubated with these metabolic inhibitors (see Table 4). At present, we do not know whether this metabolic inhibition affects the high affinity or low affinity system or whether it is active influx, concentrative uptake or both which is primarily affected.

TABLE 4

The effect of potassium cyanide and 2-deoxy-D-glucose on uptake of 0.02mM-L-[3-^{14}C]-tryptophan by rat-brain P$_2$ fraction.

Additive	Concentration (mM)	Uptake (%) L-[3-^{14}C]tryptophan
None	—	100
Cyanide	5	40
2-Deoxy-D-glucose	40	74
Cyanide + 2-deoxy-D-glucose	5 + 40	48

In experiments with 2-deoxy-D-glucose, D-glucose was omitted from the incubation medium. Incubations lasted 20 min at 37 °C.

EXCHANGE DIFFUSION PHENOMENA

By exchange diffusion we mean the process by which the transport of a molecule in one direction is linked to the transport of a molecule in the other direction, that is, influx linked to efflux or vice versa. We observed that when a P_2 fraction was preloaded with unlabelled L-tryptophan and reprepared, the subsequent uptake of L-[3-^{14}C]tryptophan was greatly accelerated compared with a control suspension (see Fig. 4). When P_2 fractions were preloaded with radioactive tryptophan and then incubated with non-radioactive tryptophan the efflux of radioactive tryptophan was greatly increased (see Fig. 5). Because the stimulation of efflux of an internal radioactive amino acid by an external amino acid is technically easier to study than stimulated influx, we have determined the effect of several natural and unnatural amino acids on the rate of efflux of L-[3-^{14}C]tryptophan (see Table 5). Amino acids 1–13 effectively cause apparent efflux of tryptophan. Alanine, glycine and glutamic acid (14–16) form an intermediate group while the rest, 17–21, were not effective in accelerating the loss of tryptophan. However, we found no clear cut-off point. Clearly, a wide variety of amino acids can undergo exchange diffusion with L-tryptophan. The effects of L- and D-isomers of both phenylalanine and tyrosine are shown in

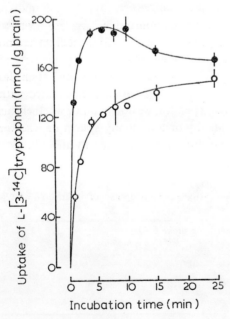

FIG. 4. Effect of preloading P_2 fraction with L-tryptophan on the subsequent uptake of L-[3-^{14}C]tryptophan: not preloaded (○); preloaded (●).

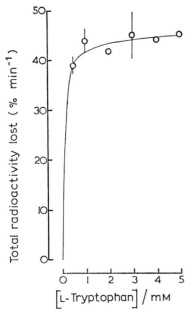

FIG. 5. Effect of L-tryptophan on the efflux of L-[3-^{14}C]tryptophan from a rat-brain P_2 fraction preloaded with L-[3-^{14}C]tryptophan. Results are expressed as the percentage of the total radioactivity accumulated by the P_2 fraction during the preloading incubation that was released in 1 min in the presence of the external amino acid.

Table 6. The L-isomers were clearly more effective at causing efflux though there is not absolute stereospecificity. 5-Hydroxytryptophan stimulates efflux of tryptophan, 5-hydroxyindoleacetic acid and 5-hydroxytryptamine do not, and L-kynurenine and L-3-hydroxykynurenine both modestly exchange.

The existence of such powerful exchange diffusion systems poses certain conceptual problems. The realization that exchange diffusion can occur to such a degree makes it very difficult to be certain that when *uptake* is studied, only one way influx is being measured. Since L-tryptophan can exchange with many amino acids which are found in high concentrations within the synaptosome (Mangan & Whittaker 1966; Bradford & Thomas 1969) and since many of these amino acids are qualitatively indistinguishable as far as transport is concerned, then in order to say whether a thermodynamically uphill concentrative uptake of L-tryptophan has taken place we should monitor the intracellular concentrations of all the molecular species capable of reaction with the L-tryptophan transport system. Similarly, we cannot yet say whether the metabolic inhibitors are affecting concentrative uptake or exchange diffusion in one direction or the other. Exchange diffusion poses many technical problems in the study of directional fluxes.

TABLE 5

The effect of 21 amino acids (at 1.0 mM and 5.0 mM in the medium) on the rate of loss of
L-[3-^{14}C]tryptophan from a preloaded P$_2$ fraction.

Amino acid	Percentage efflux	
	At 1.0 mM	At 5.0 mM
1 L-Isoleucine	46	56
2 L-Valine	40.5	53.5
3 L-Cysteine	29	49
4 L-Serine	35	47
5 L-Methionine	31.5	46
6 L-Histidine	39	44
7 L-Leucine	35	44
8 L-Glutamine	28	44
9 L-Threonine	32.5	43
10 L-Asparagine	29	42
11 L-Phenylalanine	31.5	41
12 L-Tryptophan	39	40
13 DL-p-Chlorophenylalanine	37	40
14 L-Alanine	18.5	35
15 Glycine	4	19
16 L-Glutamic acid	4	16
17 L-Hydroxyproline	10	11
18 L-Proline	5.5	9.5
19 L-Lysine	4.5	6
20 L-2,4-Diaminobutyric acid	4	6
21 L-Arginine	0	4.5

The crude mitochondrial fraction was preloaded by incubation with 3.0mM-L-[3-^{14}C]trypt-
ophan and then washed by centrifugation. Subsequent incubations were carried out for one
minute at 37 °C. The amount of L-[3-^{14}C]tryptophan released is expressed as a percentage
of the total intrasynaptosomal L-[3-^{14}C]tryptophan before exposure to the test amino acid,
each of which was present at a concentration of 1.0 mM and 5.0 mM.

TABLE 6

The effect of the D- and L-isomers of phenylalanine and tyrosine (at 0.1 mM and 5.0 mM) on the
rate of loss of L-[3-^{14}C]tryptophan from a preloaded crude mitochondrial fraction.

Amino acid	Percentage efflux	
	At 0.1 mM	At 0.5 mM
L-Phenylalanine	49	56
D-Phenylalanine	17	24
L-Tyrosine	34	47
D-Tyrosine	7	12

Since it has been suggested that L-glutamic acid might be a transmitter in the
mammalian central nervous system (see Krnjevic 1970), we were interested to
see whether this amino acid could participate in exchange diffusion. Bradford

(1970) has shown that L-glutamate is selectively released from synaptosomes by electrical stimulation and by potassium ions. Our preliminary results can be summarized as follows:—

(1) Crude mitochondrial fractions of rat brain rapidly take up L-[U-^{14}C]-glutamic acid. The amount retained by the crude mitochondrial fraction is dependent upon the concentration of glutamic acid in the external medium.

(2) After incubation of a crude mitochondrial fraction with L-[U-^{14}C]-glutamic acid for eight minutes, 72% of the radioactivity in the crude mitochondrial fraction was associated with the glutamate, 16% with aspartate and 12% with aminobutyric acid.

(3) When crude mitochondrial fractions were incubated with L-[U-^{14}C]-glutamate and at a point when uptake had almost reached a maximum, non-radioactive L-glutamate was added to the medium. There was a dramatic and swift fall in the amount of L-[U-^{14}C]glutamate retained by the crude mitochondrial fraction. This strongly suggested that efflux of glutamic acid had been promoted by the addition of an excessive concentration of external glutamic acid.

In concurrent experiments, we found that external non-radioactive glutamic acid was as effective as increasing the potassium ion concentration in the medium in causing the efflux of internal L-[U-^{14}C]glutamate and that the efflux follows a similar time course. Obviously, exchange diffusion processes must be considered when studying transport of excitatory amino acids and their inactivation by 'reuptake'. In addition, these results suggest two systems for the efflux of glutamic acid, one subserving a neurotransmitter function and one a 'metabolic' or 'nutritional' function.

THE EFFECT OF TEMPERATURE ON THE TRANSPORT OF L-TRYPTOPHAN AND L-PHENYLALANINE

When we studied the uptake of L-[3-^{14}C]tryptophan by crude mitochondrial fractions at various temperatures, we observed an odd phenomenon. Although the initial rate of apparent uptake increased as the temperature was raised to 37 °C, the higher the temperature the greater was an apparent overshoot over the equilibrium position some 40–80 min later. This is clear from Fig. 6. The apparent equilibrium point (expressed as the amount of radioactivity bound to an aliquot portion of a crude mitochondrial fraction) was highest at about 15 °C.

There are thus two effects of temperature, one upon the eventual equilibration of the amino acid and the other affecting transport across the membrane. Since temperature has a differential effect upon these processes, it would seem that

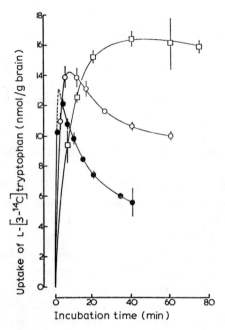

FIG. 6. The effect of temperature on the time course for the uptake of L-[3-^{14}C]tryptophan by a P$_2$ fraction.

Portions of a P$_2$ fraction were incubated at 12 °C (□), 25 °C (○) or 37 °C (●) with 5μM-L-[3-^{14}C]tryptophan.

they must be independently controlled but at the moment there is no good explanation of mechanisms involved.

These experiments led us to consider the possibility of compartmentation of the pools of amino acids present within the synaptosome and available for transport processes. Because phenylalanine showed the same differential temperature effects upon its steady-state accumulation and initial rate of uptake and also because phenylalanine is easily measured at low concentrations by a fluorimetric technique, we used phenylalanine in these studies. Our experiment was based on the following hypothesis. Suppose synaptosomes were preloaded with L-[^{14}C]phenylalanine and at a given temperature allowed to 'equilibrate' internally. If the phenylalanine was totally and diffusely equilibrated with its internal environment and then efflux was induced by placing outside the limiting membrane of the synaptosome an amino acid which would undergo exchange diffusion with the internal [^{14}C]phenylalanine, then, if there was no compartmentation, the specific activity of the effluxing phenylalanine should be equal to that of the equilibrated internal phenylalanine. If, however, there existed within the synaptosome two or more compartments that had not equilibrated with the

[^{14}C]phenylalanine initially taken up during the preloading procedure, then when an external amino acid exchanging with phenylalanine was presented to the synaptosomal membrane the effluxing phenylalanine should not have the same specific activity as the total phenylalanine within the synaptosome.

We preloaded P$_2$ fractions by incubation with 10μM-L-[U-^{14}C]phenylalanine for 40 min at 37 °C. The incubation was terminated by cooling and the resultant P$_2$ fraction was collected by centrifugation and washed. Aliquot portions of the washed fraction were then incubated for one minute at 37 °C with 0.5mM-L-methionine, which exchanges rapidly with L-phenylalanine. Incubations were terminated by Millipore filtration. We estimated the specific activity of the L-[U-^{14}C]phenylalanine in the preloaded P$_2$ fraction by separating the phenylalanine by paper chromatography and measuring its radioactivity and its amount by fluorimetry. The specific activity of the phenylalanine effluxing after incubation with the L-methionine was estimated in the same way. Table 7 shows that the specific activity of the effluxing phenylalanine was at least twice that of the phenylalanine within the synaptosomes.

Two interpretations can be put upon this finding. There might be a heterogenous population of synaptosomes, some containing phenylalanine of much higher specific activity than others and from which methionine was differentially causing efflux. The alternative, which we favour, is that there is compartmentation of amino acids within synaptosomes. Equilibration throughout the internal environment of the synaptosome is slow and perhaps temperature dependent and the pool partaking in exchange diffusion contains a higher concentration of 'newly taken up' phenylalanine. It has yet to be proved that this conclusion is correct but if it is, it has important implications in regard to precursor pools for neurotransmitter synthesis (e.g. tryptophan for 5-hydroxytryptamine and tyrosine for catecholamines), to pools of amino acids suitable for protein synthesis, and also to the pools of putative neurotransmitting amino acids.

TABLE 7

The specific activity of L-[U-^{14}C]phenylalanine present intrasynaptosomally after washing the preloaded fraction and present in the medium after displacement from the washed fraction by exchange with L-methionine.

Source of L-[U-^{14}C]phenylalanine	Specific activity (μCi/μmol)
Intrasynaptosomal after washing the preloaded fraction	0.89
Incubation medium after exposure to L-Met (i.e. effluxed)	2.08

The P$_2$ fraction was preloaded by incubation with 10μM-L-[U-^{14}C]phenylalanine for 40 min at 37 °C. The incubation was terminated by cooling and centrifugation and the resultant pellet was washed once. Aliquot portions of the washed fraction were then incubated for one minute at 37 °C with 0.5mM-L-methionine. Incubations were terminated by Millipore filtration and the filtrates collected.

HUNTING DURING PERTURBATION OF THE STEADY-STATE DISTRIBUTION OF
L-TRYPTOPHAN AND L-PHENYLALANINE WITHIN THE SYNAPTOSOME

We have also investigated the differential effects of temperature upon the internal
distribution of amino acids and transmembrane flux, with unexpected results.

We incubated crude mitochondrial fractions at either 37 or 10 °C for various
times (see Figs. 7 and 8) with either L-[3-^{14}C]tryptophan or L-[U-^{14}C]phenyl-
alanine. When we considered that equilibrium had been reached, we added
either methionine or phenylalanine to the incubation. The efflux of the radioactive amino acid was then followed at extremely short time intervals.

At 37 °C, L-phenylalanine caused a sharp drop in the radioactivity of L-
[U-^{14}C]phenylalanine associated with the P$_2$ fraction (Fig. 7), but this measured
radioactivity oscillated for about 15 min until a new equilibrium value was
reached. We did not observe such oscillations when the incubation temperature
was 10 °C (Fig. 8). Similar effects were seen when L-tryptophan was induced to
efflux by the addition of L-methionine to the medium. We have not yet determined the specific activity of internal and effluxing amino acids and thus we
cannot be precise about the mechanism underlying these phenomena. It seems,
however, as if the membrane is 'hunting' for a new equilibrium when the internal

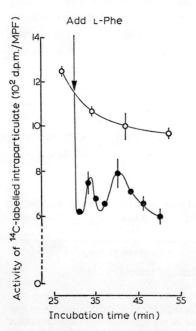

FIG. 7. The effect of introducing L-phenylalanine during the incubation at 37 °C of a P$_2$ fraction with L-[U-^{14}C]phenylalanine: control (○); after addition of L-phenylalanine (●): MPF,
Millipore filter.

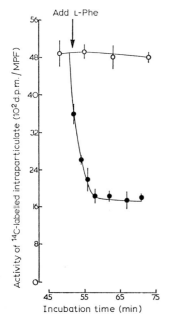

FIG. 8. The effect of introducing L-phenylalanine during the incubation at 10 °C of a P_2 fraction with L-[U-^{14}C]phenylalanine: control (○); after addition of L-phenylalanine (●).

and external environments are disturbed by introducing an amino acid which causes rapid and marked cross exchange diffusion. It also appears that this hunting is temperature dependent either because of the rapidity with which changes take place at 37 °C or because it is dependent upon a supply of metabolic energy perhaps absent at 10 °C.

CONCLUSIONS

These investigations were initiated by our interest in the function of tryptophan transport in the cerebral synthesis of 5-hydroxytryptamine and they have revealed a situation of great complexity. Many questions can be raised, some of which relate to the mechanisms of transport of amino acids across synaptosomal membrane and compartmentation of the amino acids with the nerve ending, others to the functional importance of these phenomena.

Undoubtedly, the synaptosomal membrane possesses carrier-mediated transport mechanisms for amino acids. Inward flux, outward flux, exchange diffusion and concentrative uptake can be demonstrated for tryptophan. It is, however, justifiable to ask how amino acids enter nerve endings *in vivo*. Do they arrive by axonal flow or do they pass through the neuronal membrane? How important is the neuronal membrane as a part of the blood–brain barrier? This barrier is presumably composed in part of brain capillary walls. But because capillaries are enveloped by glial end-feet, there is no direct contact between neurons and the vascular system; neurons and their processes are suspended within a sponge of neuroglial tissue (see Glees 1973). From this anatomical consideration, it is apparent that glial tissues as well as having a supportive function must also have a metabolic transport function and our results tell us nothing about this. The neuronal membrane, of which the limiting membrane of the synaptosome is an example, also forms part of the blood–brain barrier. However, it is worth remembering that part of the limiting membrane of the synaptosome has been sheared off and though it appears to reseal itself nothing is known of the function of this damaged membrane. In addition, one part of the surface of the synaptosome is that area contiguous to the synapse and this must be highly specialized. Beside these two surface areas there is the rest of the synaptosome, and at present it is impossible to say across which area the transport processes reported here occur.

What is the significance of at least two pools of phenylalanine within the synaptosome, one of which contains 'newly-taken-up' phenylalanine and which participates more readily in cross exchange diffusion? Is this an artifact of the experimental situation or does it exist *in vivo*? We should be able to design experiments to answer this question. When one considers the compartmentation of amino acids within the nerve ending, the very fast rates of flow and exchange, the delicate control mechanism of transport as exemplified by the hunting phenomenon, then perhaps the most intriguing question is this: given a stable internal concentration for the whole spectrum of amino acids, how is information about the environment inside the membrane relayed to the outside of the membrane, so that carrier-mediated influx and efflux are set at the right level? It is obvious that we must know much more about the actual mechanics by which amino acids are transported across membranes before we can begin to answer this question.

What conclusions can be drawn about the role of tryptophan transport in brain synthesis of 5-hydroxytryptamine? We have produced no evidence of a system specifically translocating tryptophan across the synaptosomal membrane. The systems we have studied appear to have a rather wide specificity and for this reason we think it unlikely that synthesis of 5-hydroxytryptamine is controlled by a step regulating the transport of tryptophan across the synaptosomal membrane and thus determining a crucial intracellular concentration of tryptophan in the vicinity of tryptophan 5-hydroxylase. This is not to say that the intracellular concentration of tryptophan is not important as a factor regulating the synthesis; obviously, it is. There is no doubt that tryptophan concentrations in the brain are vulnerable to variations in the plasma concentration of a variety of amino acids (McKean *et al*. 1968), though of course the fact that many amino acids share a common exchange diffusion mechanism means that a rise in the external concentration of any one amino acid might cause the efflux of several amino acids, potential changes in each being 'suffered' by changes in another.

As one of us has argued elsewhere (Grahame-Smith 1973), there is presumably a functional pool of 5-hydroxytryptamine within the nerve ending. One presumes that, in terms of physiological function, it is the size of this pool which is important in neurotransmission. How is the size of this pool controlled? Obviously there must be adequate synthesis, but there is no evidence available to judge how much of the 5-hydroxytryptamine which is totally synthesized actually enters this pool, how much is rendered functionally inactive by intraneuronal metabolism by monoamine oxidase and how much is bound to synaptic vesicles in 'reserve' or 'sequestered'. It may be that under physiological circumstances synthesis of the amine is in excess of neuronal needs. If so, control of synthesis as a factor regulating the size of a functional pool of 5-hydroxytryptamine might be of little importance compared with the mechanisms of binding and intraneuronal metabolism. In the context of the transport studies described here, one of the most interesting studies which could be undertaken would be to investigate whether 5-hydroxytryptamine within synaptosomes is synthesized from a pool of tryptophan separate from the general synaptosomal pool. Experiments designed to answer definitively this question will require some ingenuity.

ACKNOWLEDGEMENT

We thank the *Journal of Neurochemistry* for allowing reproductions of Figs. 1–5.

References

BRADFORD, H. F. & THOMAS, A. J. (1969) Metabolism of glucose and glutamate by synaptosomes from mammalian cerebral cortex. *J. Neurochem.* **16**, 1495–1504

BRADFORD, H. F. (1970) Metabolic response of synaptosomes to electrical stimulation: release of amino acids. *Brain Res.* **19**, 239–247

FERNSTROM, J. D. & WURTMAN, R. J. (1971) Brain serotonin content: physiological dependence on plasma tryptophan levels. *Science (Wash. D.C.)* **173**, 149–152

FRIEDMAN, P. A., KAPPELMAN, A. H. & KAUFMAN, S. (1972) Partial purification and characterisation of tryptophan hydroxylase from rabbit hindbrain. *J. Biol. Chem.* **247**, 4165–4173

GLEES, P. (1973) The neuroglial compartments at light microscopic and electron microscopic levels in *Metabolic Compartmentation in the Brain* (Balazs, R. & Cremer, J. E., eds.), pp. 209–231, Macmillan, London

GRAHAME-SMITH, D. G. (1967) The biosynthesis of 5-hydroxytryptamine in brain. *Biochem. J.* **105**, 351–360

GRAHAME-SMITH, D. G. (1971) Studies *in vivo* on the relationship between brain tryptophan, brain 5HT synthesis, and hyperactivity in rats treated with a monoamine oxidase inhibitor and L-tryptophan. *J. Neurochem.* **18**, 1053–1066

GRAHAME-SMITH, D. G. (1973) The metabolic compartmentation of brain monoamines in *Metabolic Compartmentation in the Brain* (Balazs, R. & Cremer, J. E., eds.), pp. 47–56, Macmillan, London

GRAHAME-SMITH, D. G. & PARFITT, A. (1970) Tryptophan transport across the synaptosmal membrane. *J. Neurochem.* **17**, 1339–1353

GRAY, E. G. & WHITTAKER, V. P. (1962) The isolation of nerve endings from brain: an electron-microscopic study of cell fragments derived by homogenisation and centrifugation. *J. Anat.* **96**, 79–88

ICHIYAMA, A., NAKAMURA, S., NISHIZUKA, Y. & HAYAISHI, O. (1970) Enzymic studies on the biosynthesis of serotonin in mammalian brain. *J. Biol. Chem.* **245**, 1699–1709

JEQUIER, E., ROBINSON, D. S., LOVENBERG, W. & SJOERDSMA, A. (1969) Further studies on tryptophan hydroxylase in rat brain stem and beef pineal. *Biochem. Pharmacol.* **18**, 1071–1081

KNOTT, P. J. & CURZON, G. (1972) Free tryptophan in plasma and brain tryptophan metabolism. *Nature (Lond.)* **239**, 452–453

KRNJEVIC, K. (1970) Glutamate and GABA in brain. *Nature (Lond.)* **228**, 119–124

MANGAN, J. L. & WHITTAKER, V. P. (1966) The distribution of free amino acids in subcellular functions of guinea pig brain. *Biochem. J.* **98**, 128–137

MARCHBANKS, R. M. (1967) The osmotically sensitive potassium and sodium compartments of synaptosomes. *Biochem. J.* **104**, 148–157

McKEAN, C. M., BOGGS, D. E. & PETERSON, N. A. (1968) The influence of high phenylalanine and tyrosine on the concentrations of essential amino acids in brain. *J. Neurochem.* **15**, 235–241

Discussion

Wurtman: It has been suggested that fractions of synaptosomes can be prepared which have a special affinity for one particular amino acid and that this property can be used as a criterion for identifying neurotransmitters. Is it possible to separate populations of synaptosomes that will selectively take up tryptophan? If so, does this mean that tryptophan could be a neurotransmitter?

Grahame-Smith: As far as I know, no one has prepared synaptosomes which selectively take up tryptophan nor do I think that this alone would mean tryptophan was a neurotransmitter.

Fernstrom: Since brain capillaries are surrounded by astrocytes, an amino acid entering the brain must presumably pass through them; it might even be transported by them. If so, it would not be surprising to find that glial cells which surround synapses (and the vesicles that are formed from them when they are disrupted) also transport amino acids.

Artifactual vesicles are invariably formed when membranes are disrupted by homogenization. Couldn't the uptake process you observe really describe transport into such vesicles rather than into synaptosomes?

Grahame-Smith: No. We have prepared a range of tissues, including heart, liver, muscle and gut mucosa as if we were preparing synaptosomes and only preparations from brain and spleen take up tryptophan. Being a soft organ, spleen is perhaps easily homogenized like brain, preserving some of the nerve endings. The uptake by spleen preparations was about 5% of that of brain.

Mandel: Is this behaviour specific for synaptosomes? What happens with microsomes involved in protein synthesis or with blood platelets?

Grahame-Smith: Blood platelets actively take up amino acids (Boullin & Green 1972). Microsomes, mitochondria and other particles do take up a little tryptophan and phenylalanine but only a fraction of the quantity that synaptosomes take up. We have tried to estimate the concentration inside the synaptosome, but to measure the volume of a synaptosome preparation is technically extremely difficult. Marchbanks (1967) tried by equilibrating the synaptosomes with potassium but he had to leave them in a cold room overnight in a medium unsuitable for our incubations. The estimated value, therefore, may not be very reliable. Using such measurements, however, we have achieved internal to external concentration gradients of 4:1 for tryptophan uptake by synaptosomes; there is concentrative uptake.

Wurtman: The transport system that takes up tryptophan into a synaptosome is inhibited by the presence of other neutral amino acids. Couldn't the *in vivo* structures that give rise to synaptosomes also take up leucine, valine, etc.? What might be the function of these amino acids in the synaptic region? Doesn't the synaptosomal uptake of the compounds—which do not give rise to known neurotransmitters—make one worry about the significance of tryptophan uptake?

Grahame-Smith: Presumably, in part the functions are the same as those of amino acids in every cell. The membrane has to monitor what is inside and outside to maintain the internal environment. This, I presume, is the basis of the particular transport system that we have been studying. I'm sure it is not specific to nerves.

Wurtman: But the subcellular structures that use other amino acids for protein synthesis are, of course, far away from the nerve terminals (see Barondes, pp. 265-275).

Sourkes: Have you measured the extrasynaptosomal space using inulin? If this volume is significant and you have not measured it, then the amount of uptake into the synaptosome that you calculate will be biased by the amount of amino acid outside the synaptosome in a compartment contiguous with the external medium.

Grahame-Smith: We have measured it, but we do not do it routinely, because our technique renders that unnecessary. The preparation (see p. 176) removes most of the non-particulate contaminants. I don't think the amounts of amino acids left in the extrasynaptosomal fluid are important.

Sourkes: The results of our inhibitor studies on slices of rat-brain cortex (Keily & Sourkes 1972) confirm yours, but we could find no evidence for a low affinity uptake system.

Fernstrom: The concentration of tryptophan in the extracellular fluid might be the same as that of free tryptophan (i.e. non-albumin-bound) in serum. This is about the same concentration as the K_m for the high affinity uptake system for tryptophan you described. Hence, the tryptophan transport system might not be saturated *in vivo*.

The lower concentrations of phenylalanine (10 μM; see p. 187) approach the concentrations in plasma and thus perhaps in the extracellular fluid (Bito *et al.* 1966), since phenylalanine does not bind to serum proteins. It would be interesting to test even lower concentrations of phenylalanine in this synaptosome system, especially with respect to competition with other neutral amino acids, to assess whether your findings extend into the physiological range.

Grahame-Smith: The drawback then is that one is studying a totally unphysiological situation.

Wurtman: What do these preparations tell us about the biology of the 5-hydroxytryptamine neuron?

Grahame-Smith: The transport phenomena we have found in synaptosomes do not appear very different from those observed in other cells (e.g. Ehrlich ascites cells) and we believe that there is little evidence that the tryptophan uptake process in synaptosomes is peculiar or specifically geared to the function of this amino acid as a precursor of 5-hydroxytryptamine in the synaptosome. I agree with Dr Fernstrom and Dr Lajtha that it is evident that other amino acids compete for the precursors of the neurotransmitters at the membrane. It seems unlikely that the availability of tryptophan, and consequently the rate of tryptophan hydroxylation, is acute and crucial in regulating the amount of 5-hydroxytryptamine release during neurotransmission.

Wurtman: How does that follow?

Grahame-Smith: This harks back to the loaded question that I asked Dr Weiner (p. 151) about how synthesis is geared to functional activity. If only

5% of the material synthesized is used at the nerve ending for neurotransmission, then control of synthesis in unstressed and physiological circumstances becomes relatively unimportant, because the mechanisms of intraneuronal binding and intraneuronal metabolism could control the size of the pool of neurotransmitter available for release much better than synthesis. That is a real possibility; acute control of synthesis may be relatively unimportant.

Wurtman: We need to know whether the quantity of neurotransmitter stored in the terminals of 5-hydroxytryptaminergic neurons is related to the quantity of neurotransmitter released into synapses.

Oja: Dr Grahame-Smith, considering unidirectional transport of tryptophan into synaptosomes, you assume two transport mechanisms, one of high affinity and the other of low affinity. A further assumption must be that some tryptophan diffuses passively into synaptosomes. Basing our calculation on the presence of two saturable systems and one non-saturable system, we can express the velocity of uptake, v, by the following equation:

$$v = V'S/(K_m' + S) + V''S/(K_m'' + S) + K_dS$$

where V' and V'' are the maximal velocities of the two saturable transport systems, K_m' and K_m'' are the transport constants equivalent to Michaelis constants, K_d is the diffusion constant and S is the concentration of tryptophan outside the synaptosomes. If two concentration ranges, one low and one high

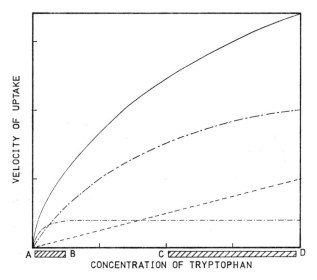

FIG. 1 (Oja). Theoretical curves for the uptake of tryptophan into synaptosomes. Unbroken line depicts the total uptake over the various concentrations of tryptophan. Broken lines denote the underlying one non-saturable (straight line) and two saturable components (hyperbolic curves) of the uptake.

(e.g. AB and CD in Fig. 1), are tested independently, it is usually easy to construct Lineweaver–Burk plots from the two sets of data to obtain arbitrary values for K_m', V', K_m'' and V''. Regrettably, these values, especially those for the high-affinity transport, may be severely in error if the total uptake is not properly resolved into its components before analyses. We have just faced a similar problem; we found two saturable systems and one non-saturable system for influx of taurine into brain slices (Lähdesmäki & Oja 1973). Lineweaver–Burk plots derived directly from the results within low and high ranges of concentration gave quite simply arbitrary values for K_m', V', K_m'' and V''. These preliminary estimates appeared later, however, to be completely erroneous when we resolved the total influx into three components which we analysed separately. Furthermore, even with computers it was extremely difficult to resolve the total uptake of tryptophan in Fig. 1 into its three components in view of experimental and biological variation. So I feel that the values of K_m and V for tryptophan you quote, derived from a graphical analysis of the total uptake, are open to criticism.

Grahame-Smith: I agree. My comments in regard to K_m should have been prefaced by the proviso that they represent my interpretation of my data. Obviously, if one studies only certain substrate concentrations, the conclusions are only applicable to those conditions, particularly if a Lineweaver–Burk plot is manually constructed.

References

BARONDES, S. (1974) Do tryptophan concentrations limit protein synthesis of specific in sites in the brain? *This Volume*, pp. 265–275

BITO, L., DAVSON, H., LEVIN, E., MURRAY, M. & SNIDER, N. (1966) The concentrations of amino acids and other electrolytes in cerebrospinal fluid, *in vivo* dialysate of brain, and blood plasma of dog. *J. Neurochem.* **13**, 1057–1067

BOULLIN, D. J. & GREEN, A. R. (1972) Mechanisms by which human blood platelets accumulate glycine, GABA, and amino acid precursors of putative neurotransmitters. *Br. J. Pharmacol.* **45**, 83–94

KEILY, M. & SOURKES, T. L. (1972) Transport of L-tryptophan into slices of rat cerebral cortex. *J. Neurochem.* **19**, 2863–2872

LÄHDESMÄKI, O. & OJA, S. S. (1973) On the mechanism of taurine transport at brain cell membranes. *J. Neurochem.* **20**, 1411–1417

MARCHBANKS, R. M. (1967) The osmotically sensitive potassium and sodium compartments of synaptosomes. *Biochem. J.* **104**, 148–157

Tryptophan concentration in the brain

A. T. B. MOIR*

MRC Brain Metabolism Unit, University Department of Pharmacology, Edinburgh

Abstract Before considering the changes induced by tryptophan administration, we must consider what we mean by brain tryptophan concentrations and how these concentrations are related to the concentrations of tryptophan in body fluids, to the concentrations of similar amino acids and to the subsequent metabolism of these amino acids.

Most methods of estimating the amount of tryptophan in the brain include a step in which protein is precipitated. By such methods, significant differences in the concentration of tryptophan in different regions of brain have been revealed. However, the pattern of these inter-regional differences varies with the method of extraction used; the pattern found in dog brain by Price & West (1960) differs from that which we found by precipitating the tissue with trichloroacetic acid (Eccleston *et al.* 1968). This precipitation is the method used by most workers to estimate the cerebral concentration of tryptophan. Although McMenamy and his co-workers have shown that this method allows the quantitative recovery of tryptophan from plasma, the work of Gál *et al.* (1964) suggests that this method does not release a small fraction of the tryptophan which remains with the precipitated brain protein, perhaps incorporated in the protein.

TYROSINE IN BRAIN AND IN BODY FLUIDS

In experiments with dogs, I found that, as in men (McMenamy & Oncley 1958), about 80% of the tryptophan in plasma is strongly bound to protein and only 20% is 'free' as measured either by ultrafiltration (Moir 1971*a*) or

---

* *Present address*: Scottish Home and Health Department, St Andrew's House, Edinburgh

dialysis (Geddes & Moir 1969; Geddes *et al.* 1973). This free tryptophan in plasma is obviously the only form that is capable of equilibrating with tryptophan in other body compartments. Table 1 compares the mean concentrations of tryptophan and tyrosine in canine body fluids with the range of the mean concentrations of these amino acids in different regions of brain. The gradient for both amino acids from plasma ultrafiltrate to cerebrospinal fluid is small, about 2:1, and for tryptophan, at least, this gradient is controlled by simple diffusion mechanisms (Geddes & Moir 1969; Geddes *et al.* 1973). The concentrations in plasma ultrafiltrate and in erythrocytes in the dog are extremely interesting: the gradient rise for both amino acids is considerable, 1:3.8 for tryptophan and 1:5.1 for tyrosine. The concentration of tryptophan in the erythrocytes appears to be due predominantly to binding mechanisms of high affinity while the high concentrations of tyrosine in canine erythrocytes have been shown to be the result of active transport mechanisms. These results for dogs contrast with those for humans (McMenamy *et al.* 1960) where the concentration ratio between plasma and erythrocytes was 1:0.83 for diffusable tryptophan and 1:0.97 for tyrosine (which was fully diffusable at both sites). Elwyn (1966) has suggested that the difference in the patterns of amino acid concentration between human and canine erythrocytes arises because human erythrocytes contain an active transport system with a preference for leucine while the system in canine erythrocytes prefers alanine. However, the behaviour of the concentrations of tryptophan and tyrosine after the administration of tryptophan to dogs suggests that such a mechanism alone is insufficient to explain the results for tryptophan for which high-affinity binding is also probably important. It has not been emphasized sufficiently that the active transport systems for amino acids in erythrocytes, which are usually cation linked, have to work in different ionic environments in man and dog (see Table 1).

The data in Table 1 suggest concentration gradients for tryptophan and tyrosine from plasma ultrafiltrate to brain of between 1:2.5 and 1:8 (depending

TABLE 1

Concentrations of tryptophan, tyrosine, sodium and potassium in canine body fluids and brain

	Tryptophan	Tyrosine	Na	K
Total plasma[a]	56	48	146	3.5
Plasma ultrafiltrate[a]	11	34	—	—
Erythrocytes[a]	43	174	106	8.4
Cisternal cerebrospinal fluid[a]	5.8	18	154	3.1
Brain regions[b]	25–42	120–280	—	—

[a] Mean concentrations in nmol/l
[b] Range of mean concentrations in nmol/kg

TABLE 2

Total and free concentrations of tryptophan in the rat

	Total concentration[a]	Concentration in ultrafiltrate (free)[a]
Plasma	78.4 ± 29.2(10) nmol/ml	9.4 ± 4.0(9) nmol/ml
Brain	37.6 ± 3.2 nmol/g wet tissue	29.1 ± 8.1(9) nmol/ml
Cisternal cerebrospinal fluid	1.0 ± 0.24(7) nmol/ml	

[a] Mean concentration ± S.D. (number of animals)

on the amino acid and the brain region) and of about twice these values from cerebrospinal fluid to brain. However, we should remember that here we are contrasting the free molal concentrations in plasma or cerebrospinal fluid with the total molar concentrations of brain and there is no reason to expect that the total concentration of tryptophan in brain is not related to the concentration of free tryptophan in the brain in a similar manner to the situation in plasma and, further, that the free concentration in brain and plasma might be simply related.

To examine this hypothesis, I prepared an ultrafiltrate of rat brain after homogenizing the tissue and disrupting it by ultrasonic vibrations in either the presence or the absence of a few crystals of sodium cyanide. The amount of tryptophan in this brain ultrafiltrate should be an index of the concentrations of free tryptophan in brain. In Table 2 the concentrations in brain ultrafiltrate are compared with the tryptophan concentrations in whole brain, plasma, plasma ultrafiltrate and cisternal cerebrospinal fluid from the same animals. The results show that the molal concentration of free brain tryptophan is almost equivalent to the molar concentration of brain total tryptophan in the control animal and that there are substantial gradients between free tryptophan in brain and in both plasma (3:1) and cerebrospinal fluid (25:1). This is in accord with the stereospecific uptake system defined *in vivo* for rat brain by Yuwiler (1973).

TRYPTOPHAN CONCENTRATIONS AFTER TRYPTOPHAN ADMINISTRATION

Administration of single doses of tryptophan to animals and to humans either orally or intravenously results in relatively transient increases in tryptophan concentrations accompanied by parallel transient alterations in the concentrations of its metabolites in the metabolic pathways. I tried to prolong these alterations in tryptophan metabolism so that they were less critically dependent on the point at which a rapid biphasic response was sampled by administering tryptophan to dogs in the form of a loading intravenous injection followed by a constant infusion (Moir 1971*b*). Despite the constant infusion of tryptophan, the concentration of tryptophan in plasma fell rapidly from a maximum value at one hour, when as much as 50% of the total plasma tryptophan was in the free form (as indicated by the concentrations in ultrafiltrate; see Fig. 1). The proportion of plasma tryptophan in the free form declined through the constant infusion of tryptophan and the concentrations of tryptophan in cisternal cerebrospinal

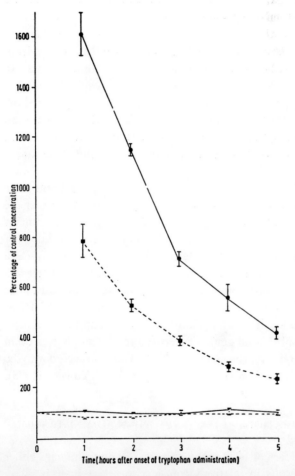

FIG. 1. Concentrations of tryptophan in plasma ultrafiltrate (●) and in whole plasma (■) and of tyrosine in plasma ultrafiltrate (lower solid line) and in whole plasma (lower broken line) after administration of L-tryptophan. Results are expressed as percentage of control values of each dog. Ultrafiltrate results are the means and s.e.m. from three experiments. Whole plasma results are means and s.e.m. from six experiments except the values at five hours which are the means and s.e.m. from three experiments (Moir 1971*b*; reproduced with permission from *Br. J. Pharmacol.*).

fluid (Fig. 2) also declined after reaching the peak concentration two hours after the start of tryptophan administration.

Two parameters of tryptophan metabolism were, however, maintained in an altered steady state after this form of tryptophan load. First, tryptophan concentrations in erythrocytes (Fig. 2) stayed at a concentration 350% of their control concentrations. This implied that the cells were saturated with tryptophan. Second, the concentration of 5-hydroxyindoleacetic acid in cisternal cerebrospinal fluid was maintained at steady values, 250% of control concentrations, during the 3–5 h after the start of the tryptophan infusion. This second observation suggested that the cerebral metabolism of 5-hydroxyindoles was likely to be in a new steady state during that period. I found total concentrations of

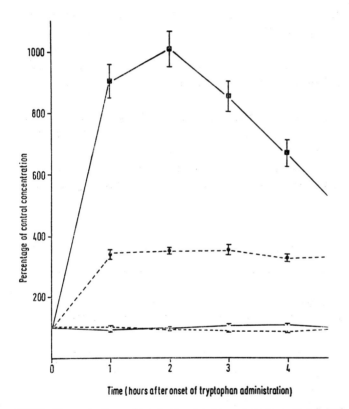

FIG. 2. Concentrations of tryptophan in cisternal cerebrospinal fluid (■) and in erythrocytes (●) and of tyrosine in cisternal cerebrospinal fluid (lower solid line) and in erythrocytes (lower broken line) after administration of L-tryptophan. Results expressed as percentage of control values of each dog. Results are the means and S.E.M. from six experiments except the values at five hours (not shown) which are the means and S.E.M. from three experiments (Moir 1971*b*; reproduced with permission from *Br. J. Pharmacol.*).

tryptophan of 200 nmol/g in all regions of brain in dogs killed at four hours and also that there was no longer a significant inter-regional variation in tryptophan concentrations. I concluded that, after the administration of a large amount of tryptophan, inter-regional variations, perhaps due to binding differences, tend to be obscured possibly owing to the relative increase in the proportion of free tryptophan in brain (cf. the similar occurrence in plasma).

Although I have not examined the free concentrations of tryptophan in brain after administration of tryptophan, I have examined the changes in these concentrations in rats induced by administration of amphetamine (Tagliamonte *et al.* 1971; Schubert & Sedvall 1972). Table 3 shows the concentrations of total and free tryptophan in plasma and in brain as well as the concentrations in cisternal cerebrospinal fluid before and 90 min after administration of D-amphetamine. The total plasma concentration of tryptophan fell slightly (to 95% of control) but insignificantly after administration of amphetamine while the concentration of free tryptophan in plasma increased to 138% of control values. This was accompanied by similar, significant, increases of a similar degree in the total and free concentrations of brain tryptophan (137% and 141%, respectively).

It is interesting that the cerebrospinal fluid showed the greatest increase in concentration following this treatment (160% control values). Perhaps the small size and relatively slow turnover of this cerebrospinal fluid pool allow it to be a sensitive index of transient alterations in brain tryptophan concentration. From their preliminary investigations, Coppen *et al.* (1971) have suggested that

TABLE 3

Concentrations of tryptophan in the rat[a] before and 90 min after administration of D-amphetamine (10 mg/kg intraperitoneally).[b]

	Plasma		Brain			
	Total (nmol/ml)	Ultrafiltrate (nmol/ml)	Total (nmol/g)	Ultrafiltrate (nmol/ml)	Cisternal cerebrospinal fluid (nmol/ml)	Temperature (°C)
Control	78.4 ± 29.2(10)	9.4 ± 4.0(9)	37.6 ± 3.2(10)	29.1 ± 8.1(9)	1.00 ± 0.24(7)	36.8
Amphetamine	74.3 ± 16.6(10)	13.0 ± 5.8(10)	51.5 ± 10.1(10)	41.0 ± 15.2(10)	1.86 ± 0.67(6)	39.7
Significance (*t* test)	N.S.[c]	$0.05 < P < 0.1$	$P < 0.001$	$0.01 < P < 0.05$	$0.005 < P < 0.01$	$P < 0.001$

[a] The rats were grouped three or four to a cage.
[b] Concentrations expressed as means ± S.D. (number of animals).
[c] Not significant.

patients suffering from depression have lower concentrations of tryptophan in lumbar cerebrospinal fluid than normal, as well as low concentrations of 5-hydroxyindoleacetic acid (Ashcroft *et al.* 1966). Further investigation of a small sample of depressed patients (Coppen *et al.* 1972) suggests that they have low levels of free tryptophan in the plasma. At the moment, little confidence can be placed in these initial results as there is insufficient evidence that the conditions of the group studied were adequately controlled. Strict control is essential (Moir *et al.* 1970) because many factors, such as small alterations in the metabolic load of tryptophan (Fernstrom & Wurtman 1971), food deprivation or 'stress' (Knott & Curzon 1972; Curzon *et al.* 1972), the concentration of free fatty acids (Curzon *et al.* 1973) or a wide variety of drugs (Gessa & Tagliamonte, this volume, pp. 207–216), may influence tryptophan and its cerebral metabolism.

TYROSINE CONCENTRATIONS AFTER TRYPTOPHAN ADMINISTRATION

Figs. 1 and 2 show that when tryptophan was infused in dogs over a prolonged period the tyrosine concentration remained completely unaltered in erythrocytes, plasma and its ultrafiltrate, and in cisternal cerebrospinal fluid. All the concentrations of tyrosine in the different regions of brain which were analysed in the animals killed at four hours were also within the normal range (Moir 1971*a*).

Since the work of Guroff & Udenfriend (1962), many results have suggested some inhibition of active transport systems by tryptophan associated with the entry of tyrosine to brain. The present data might be interpreted as arguing against such an inhibition, but this is not necessarily true. My data merely indicate that the equilibrium concentrations of tyrosine are not significantly altered by increased concentrations of tryptophan.

THE CEREBRAL METABOLISM OF THE BIOGENIC AMINES AFTER TRYPTOPHAN ADMINISTRATION

Tryptophan is of great interest because it is the precursor of the biogenic amine 5-hydroxytryptamine, alterations in the metabolism of which have been correlated with a variety of psychiatric and neurological illnesses and with effects of psychotropic drugs. We have previously described how the administration of tryptophan increases the rate of turnover of the 5-hydroxyindoles in brain without apparently distorting normal inter-regional variations in metabolism (Moir & Eccleston 1968; Moir 1971*a,b*). This metabolic alteration is reflected in

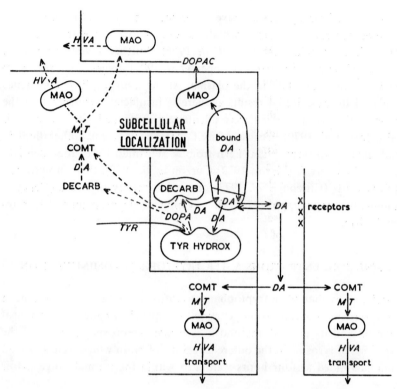

FIG. 3. Diagram of dopamine metabolism in brain showing main routes of dopamine metabolism (→) and diverted metabolism of dopa and dopamine increased by administration of tryptophan (−−−→): TYR HYDROX, tyrosine hydroxylase; DECARB, L-aromatic amino acid decarboxylase; MAO, monoamine oxidase (EC 1.4.3.4); COMT, catechol methyltransferase (EC 2.1.1.6); *TYR*, tyrosine; *DA*, dopamine; *MT*, 3-methoxytyramine; *DOPAC*, 3,4-dihydroxyphenylacetic acid; *HVA*, 4-hydroxy-3-methoxyphenylacetic acid (from Moir & Yates 1972; reproduced with permission from *Br. J. Pharmacol.*).

the concentration of 5-hydroxyindoleacetic acid in the cerebrospinal fluid. We have also shown that the administration of tryptophan alters the cerebral metabolism of dopamine (Fig. 3), although we have not yet determined whether dopamine affects steps before or after decarboxylation (Moir & Yates 1972). It seems no longer appropriate to refer generically to L-aromatic amino acid decarboxylase. Also, the regional variation in subcellular localization of the isoenzymes with their affinities for different substrates might have a considerable role in determining the locations of the different biogenic amines in the brain (Broch & Fonnum 1972; Sims *et al.* 1973). Unfortunately, we still lack comprehensive and definitive experimental evidence to show the relative importance of these decarboxylases.

CONCLUSIONS

While we can measure free tryptophan in the brain directly, we can infer much about the probable behaviour of tryptophan in brain indirectly from its concentrations in other compartments of the body which are more accessible to sampling. The concentrations of tryptophan in different metabolic compartments will undoubtedly vary in different proportions in response to many agents including drugs. The administration of tryptophan itself causes various increases in its own concentration at different sites throughout the body, increases the rate of metabolism of the 5-hydroxyindoles without disturbing the metabolism itself and interacts with the cerebral metabolism of dopamine without disturbing the equilibrium concentrations of tyrosine either in brain or in other body compartments. Analyses of the cerebrospinal fluid frequently and accurately mirror alterations of this type which occur in brain but interpretation of the data from such analyses is not always straightforward (Moir *et al.* 1970; Moir & Yates 1972) owing to the variations in the relationship between brain and cerebrospinal fluid and the lack of uniformity of concentrations in that fluid.

ACKNOWLEDGEMENT

I am grateful to all members of the MRC Brain Metabolism Unit for the continual intellectual stimuli they have provided over a number of years.

References

ASHCROFT, G. W., CRAWFORD, T. B. B., ECCLESTON, D., SHARMAN, D. F., MACDOUGALL, E. J., STANTON, J. B. & BINNS, J. K. (1966) 5-Hydroxyindole compounds in cerebrospinal fluid of patients with psychiatric or neurological diseases. *Lancet* 2, 1049–1052

BROCH, O. J. & FONNUM, F. (1972) The regional and subcellular distribution of catechol-*O*-methyl transferase in the rat brain. *J. Neurochem.* 19, 2049–2055

COPPEN, A., BROOKSBANK, B. W. L. & PEET, M. (1971) Tryptophan concentration in the cerebrospinal fluid of depressive patients. *Lancet* 1, 1393

COPPEN, A., ECCLESTON, E. G. & PEET, M. (1972) Total and free tryptophan concentration in the plasma of depressive patients. *Lancet* 1, 1415–1416

CURZON, G., JOSEPH, M. H. & KNOTT, P. J. (1972) Effects of immobilization and food deprivation on rat brain tryptophan metabolism. *J. Neurochem.* 19, 1967–1974

CURZON, G., FRIEDEL, J. & KNOTT, P. J. (1973) Effect of fatty acids on the binding of tryptophan to plasma protein. *Nature (Lond.)* 242, 199–200

ECCLESTON, D., ASHCROFT, G. W., MOIR, A. T. B., PARKER-RHODES, A., LUTZ, W. & O'MAHONEY, D. P. (1968) A comparison of 5-hydroxyindoles in various regions of dog brain and cerebrospinal fluid. *J. Neurochem.* 15, 947–957

ELWYN, D. M. (1966) Distribution of amino acids between plasma and erythrocytes in the dog. *Fed. Proc.* **25**, 854–861

FERNSTROM, J. D. & WURTMAN, R. J. (1971) Brain serotonin content: physiological dependence on plasma tryptophan levels. *Science (Wash. D.C.)* **173**, 149–152

GÁL, E. M., MORGAN, M., CHATTERJEE, S. K. & MARSHALL, F. D. (1964) Hydroxylation of tryptophan by brain tissue *in vivo* and related aspects of 5-hydroxytryptamine metabolism. *Biochem. Pharmacol.* **13**, 1639–1653

GEDDES, A. C. & MOIR, A. T. B. (1969) Removal of L-tryptophan from cerebrospinal fluid in the dog. *Br. J. Pharmacol.* **36**, 204P

GEDDES, A. C., MARTIN, M. & MOIR, A. T. B. (1973) The transport of L-tryptophan from cerebrospinal fluid in the dog. *J. Physiol. (Lond.)* **230**, 595–612

GUROFF, G. & UDENFRIEND, S. (1962) Studies on aromatic amino acid uptake by rat brain. *J. Biol. Chem.* **237**, 803–806

KNOTT, P. J. & CURZON, G. (1972) Free tryptophan in plasma and brain tryptophan metabolism. *Nature (Lond.)* **239**, 452–453

McMENAMY, R. H. & ONCLEY, J. L. (1958) The specific binding of L-tryptophan to serum albumin. *J. Biol. Chem.* **233**, 1436–1447

McMENAMY, R. H., LUND, C. C., NEVILLE, G. J. & WALLACH, D. F. H. (1960) Studies of unbound amino acid distributions in plasma erythrocytes, leucocytes and urine of normal human subjects. *J. Clin. Invest.* **39**, 1675–1689

MOIR, A. T. B. (1971*a*) Interaction in the cerebral metabolism of the biogenic amines. Effect of intravenous infusion of L-tryptophan on tryptophan and tyrosine in brain and body fluids. *Br. J. Pharmacol.* **43**, 715–723

MOIR, A. T. B. (1971*b*) Interaction in the cerebral metabolism of the biogenic amines. Effect of intravenous infusion of L-tryptophan on the metabolism of dopamine and 5-hydroxyindoles in brain and cerebrospinal fluid. *Br. J. Pharmacol.* **43**, 724–731

MOIR, A. T. B. & ECCLESTON, D. (1968) The effects of precursor loading in the cerebral metabolism of 5-hydroxyindoles. *J. Neurochem.* **15**, 1093–1108

MOIR, A. T. B. & YATES, C. M. (1970) Actions of phenelzine on the interactions of the metabolism of tryptophan and dopamine in brain. *Br. J. Pharmacol.* **40**, 563P–564P

MOIR, A. T. B. & YATES, C. M. (1970) Interaction in the cerebral metabolism of the biogenic amines. Effect of phenelzine on this interaction. *Br. J. Pharmacol.* **45**, 265–274

MOIR, A. T. B., ASHCROFT, G. W., CRAWFORD, T. B. B., ECCLESTON, D. & GULDBERG, H. C. (1970) Cerebral metabolites in cerebrospinal fluid as a biochemical approach to the brain. *Brain* **93**, 357–368

PRICE, S. A. P. & WEST, G. B. (1960) Distribution of tryptophan in the brain. *Nature (Lond.)* **185**, 470–471

SCHUBERT, J. & SEDVALL, G. (1972) Effect of amphetamines on tryptophan concentrations in mice and rats. *J. Pharm. Pharmacol.* **24**, 53–62

SIMS, K. L., DAVIS, G. A. & BLOOM, F. E. (1973) Activities of 3,4-dihydroxy-L-phenylalanine and 5-hydroxy-L-tryptophan decarboxylases in rat brain: assay characteristics and distribution. *J. Neurochem.* **20**, 449–464

TAGLIAMONTE, A., TAGLIAMONTE, P., PEREZ-CRUET, J. & GESSA, G. L. (1971) Increase of brain tryptophan caused by drugs which stimulate serotonin synthesis. *Nat. New Biol.* **229**, 125–126

YUWILER, A. (1973) Conversion of D- and L-tryptophan to the brain serotonin and 5-hydroxyindoleacetic acid and to the blood serotonin. *J. Neurochem.* **20**, 1099–1109

For *Discussion*, see pp. 230–241.

Serum free tryptophan: control of brain concentrations of tryptophan and of synthesis of 5-hydroxytryptamine

G. L. GESSA and A. TAGLIAMONTE

Institute of Pharmacology, University of Cagliari, Cagliari

Abstract   Cerebral tryptophan hydroxylase is not saturated by the concentration of tryptophan normally present in the mammalian brain and therefore the rate of synthesis of 5-hydroxytryptamine in the brain depends on the availability of this precursor.

We found that many treatments increase the concentration of tryptophan in the brain and also stimulate cerebral turnover of 5-hydroxytryptamine. Conversely, all treatments so far studied which were known to increase the synthesis of 5-hydroxytryptamine also increased the concentration of tryptophan in brain. Our results indicate that the concentration of tryptophan in brain is an index of the turnover of 5-hydroxytryptamine. In addition, the fact that tryptophan is the only amino acid present in plasma bound to serum proteins and that only free tryptophan can cross the blood–brain barrier led us to study whether free tryptophan in serum controls the cerebral concentrations of tryptophan.

Our results may be summarized as follows:—

(*a*)   The administration of exogenous L-tryptophan to rats increased the concentrations of free and total tryptophan in serum. Cerebral concentrations of tryptophan and the rate of turnover of 5-hydroxytryptamine increased proportionally and were time-related to the changes in free but not total serum tryptophan.

(*b*)   Drugs, such as salicylate, probenecid and clofibrate, which release serum tryptophan from its protein binding, increased the cerebral concentration of tryptophan and stimulated cerebral turnover of 5-hydroxytryptamine in rats.

(*c*)   Fasted rats had more free serum tryptophan (but less total serum tryptophan), more brain tryptophan and a higher rate of turnover of 5-hydroxytryptamine in the brain than fed rats.

(*d*)   Lithium and amphetamine, which increased the concentration of tryptophan and the rate of turnover of 5-hydroxytryptamine in the brain, also increased the amount of free tryptophan in serum.

(*e*)   Electroconvulsive shock and the intracerebral injection of puromycin or cycloheximide increased brain tryptophan and stimulated synthesis of 5-hydroxytryptamine without a concomitant change in free serum tryptophan.

The results indicate that the limiting step in the cerebral synthesis of 5-hydroxytryptamine is the transport of tryptophan from blood into the 5-hydroxy-tryptaminergic neuron.

The fact that in most but not all conditions studied the concentration of free serum tryptophan reflected both the concentration of brain tryptophan and the rate of turnover of 5-hydroxytryptamine indicates that free serum tryptophan is the most important—but not the only—factor controlling the transport of tryptophan into the sites of synthesis of 5-hydroxytryptamine.

The hydroxylation of tryptophan to 5-hydroxytryptophan by tryptophan hydroxylase is considered the rate-limiting step in the synthesis of 5-hydroxy-tryptamine (Green & Sawyer 1966). However, the enzyme has a Michaelis constant (K_m) for its substrate much higher than the concentration of trypto-phan normally present in the mammalian brain. Therefore, the rate of synthesis of 5-hydroxytryptamine in the brain should depend on the availability of this precursor.

These considerations led us to ask whether the drugs which have been shown to increase the rate of synthesis of this amine act by increasing tryptophan concentration in brain, and conversely if drugs which increase tryptophan concentration in brain also increase the synthesis of the amine.

After screening a large number of drugs under many conditions, we found that, with a few exceptions, any treatment which increases turnover of 5-hydroxy-tryptamine also increases the concentration of brain tryptophan. Conversely, many psychotropic drugs increased the cerebral amount of tryptophan and also increased the rate of synthesis of 5-hydroxytryptamine in the brain. The results of these experiments are summarized in Tables 1 and 2. All the treat-ments listed in Table 1 stimulate synthesis of 5-hydroxytryptamine (Reid 1970 [amphetamine]; Tozer et al. 1966 [reserpine]; Perez-Cruet et al. 1971 [lithium carbonate]; Tagliamonte et al. 1971a [dibutyryl cyclic AMP]; Reid et al. 1968 [40 °C environment]). These treatments markedly increased tryptophan concen-tration in brain—from about 60% with reserpine to 300% in animals treated with amphetamine or exposed to high environmental temperature.

Table 2 shows that other treatments and psychotropic drugs increase brain tryptophan and Table 3 indicates that they also stimulate brain synthesis of 5-hydroxytryptamine as indicated by the increases in the amounts of 5-hydroxy-indoleacetic acid. Phenelzine, a monoamine oxidase inhibitor which inhibits the formation of this acid, and fenfluramine, which has been reported to release 5-hydroxytryptamine protected from monoamine oxidase inactivation (Le Douarec et al. 1966), are two exceptions.

The increase in brain tryptophan originated neither from a non-specific stress nor from a non-specific mechanism since brain concentrations of the amino acid were influenced neither by exposure to cold nor by many psycho-

TABLE 1

Effect of different treatments known to increase the rate of synthesis of brain 5-hydroxy-tryptamine on the concentration of tryptophan in brain.

Treatment	Body temperature (°C)[a]	Brain tryptophan (µg/g)[b]	P
None	37.8 ± 0.3	4.56 ± 0.07(150)	
D-Amphetamine[c]	39.1–40.2	10.48 ± 0.67(18)	<0.001
D-Amphetamine[c]	40.5–43.1	18.40 ± 2.50(8)	<0.001
Reserpine	36.3–38.7	7.20 ± 0.26(9)	<0.001
Lithium carbonate	37.3–38.7	7.30 ± 0.28(26)	<0.001
Dibutyryl cyclic AMP[c]	35.4–39	8.25 ± 0.38(13)	<0.001
Dibutyryl cyclic AMP[c]	39.5–42.2	14.25 ± 1.08(6)	<0.001
40 °C environment	40.1–43.3	14.10 ± 0.95(12)	<0.001

[a] Values are average ± S.E. and ranges.
[b] Each value is the average ± S.E. (number of experiments). Doses and time before death were as follows: D-amphetamine, 10 mg/kg intraperitoneally (i.p.), 90 min; reserpine, 5 mg/kg i.p., 90 min; lithium carbonate, 60 mg/kg i.p. twice a day for five days, last treatment six hours before death; N^6O^2-dibutyryl cyclic AMP 100 µg/animal, 90 min (the drug, injected into the lateral ventricles, was dissolved in distilled water [5 µl] as previously described [Gessa et al. 1970]; a group of control rats injected with the same volume of distilled water showed no changes in brain tryptophan); 40 °C environment, two hours.
[c] The animals receiving these treatments were divided into two groups according to their body temperature.

TABLE 2

Effect of different treatments on tryptophan concentrations in brain.

Treatment	Body temperature (°C)[a]	Brain tryptophan (µg/g)[b]
None	37.8 ± 0.3	4.56 ± 0.07(150)
Phenmetrazine[a]	39.2–40.0	7.32 ± 0.17(10)
Phenmetrazine[b]	40.3–42.3	17.20 ± 0.95(8)
DL-Fenfluramine[a]	39.2–40.3	8.40 ± 0.65(14)
DL-Fenfluramine[b]	35.2–38.4	6.41 ± 0.28(12)
Phenelzine	37.6–38.3	6.27 ± 0.37(9)
Bulbocapnine	37.2–38.3	8.35 ± 0.31(14)
Probenecid	37.3–38.2	9.60 ± 0.74(18)
4-Hydroxybutyrate	35.1–37.1	8.86 ± 0.60(8)
ECS	37.1–38.1	8.50 ± 0.76(21)
Puromycin	36.1–37.1	9.50 ± 0.11(18)

[a] Average ± S.E. and ranges.
[b] Each value is the average ± S.E. (number of experiments). Treatments and drugs, given, intraperitoneally, were two hours before death: phenmetrazine, 20 mg/kg; DL-fenfluramine 10 mg/kg; phenelzine, 100 mg/kg; bulbocapnine, 60 mg/kg; probenecid, 300 mg/kg; 4-hydroxy-butyrate, 1.5 g/kg. ECS (electroconvulsive shocks) were produced by a constant-current stimulator which delivered 100 mA stimuli (at 50 Hz sinusoidally) lasting 0.5 s through ear clip electrodes (Tagliamonte et al. 1972). Puromycin (50 µg/animal) was injected intraventricularly (Biggio et al. 1972). All values are significantly different from control values ($P<0.001$).

TABLE 3

Effect of treatments listed in Tables 1 and 2 on brain concentrations of 5-hydroxytryptamine and 5-hydroxyindoleacetic acid.

Treatment (number of experiments)	5-Hydroxytrypt-amine $(\mu g/g)^b$	P	5-Hydroxyindole-acetic acid $(\mu g/g)^b$	P
None (150)	0.62 ± 0.03		0.72 ± 0.02	
D-Amphetamine[a] (18)	0.64 ± 0.06	N.S	0.99 ± 0.04	<0.001
D-Amphetamine[a] (8)	0.66 ± 0.07	N.S.	1.21 ± 0.08	<0.001
Reserpine (9)	0.11 ± 0.01	<0.001	1.56 ± 0.07	<0.001
Lithium carbonate (26)	0.73 ± 0.04	<0.01	1.23 ± 0.07	<0.001
Dibutyryl cyclic AMP[a] (13)	0.63 ± 0.04	N.S.	1.26 ± 0.06	<0.001
Dibutyryl cyclic AMP[a] (6)	0.64 ± 0.05	N.S.	1.54 ± 0.12	<0.001
40 °C environment (12)	0.65 ± 0.06	N.S.	1.23 ± 0.11	<0.001
Phenmetrazine[a] (10)	0.64 ± 0.02	N.S.	1.08 ± 0.08	<0.001
Phenmetrazine[a] (8)	0.67 ± 0.05	N.S.	1.48 ± 0.06	<0.001
DL-Fenfluramine [a] (14)	0.42 ± 0.03	<0.001	0.67 ± 0.05	N.S.
DL-Fenfluramine[a] (10)	0.41 ± 0.04	<0.001	0.69 ± 0.06	N.S.
Phenelzine (9)	1.12 ± 0.09	<0.001	0.32 ± 0.03	<0.001
Bulbocapnine (14)	0.60 ± 0.05	N.S.	0.89 ± 0.03	<0.001
Probenecid (18)	0.64 ± 0.02	N.S.	1.75 ± 0.09	<0.001
4-Hydroxybutyrate (8)	0.79 ± 0.03	<0.01	1.22 ± 0.10	<0.001
ECS (21)	0.65 ± 0.04	N.S.	1.40 ± 0.12	<0.001
Puromycin (18)	0.70 ± 0.06	<0.01	1.04 ± 0.05	<0.001

[a] See note c in Table 1.
[b] Each value (± s.e.) was obtained from the same animals from which the data in Tables 1 and 2 were obtained.

tropic drugs tested (such as α-methyltyrosine, morphine, apomorphine, chlorpromazine, desipramine and haloperidol). These results indicate that tryptophan concentration in the whole brain reflects the amount at the sites of synthesis of 5-hydroxytryptamine and is an index of the rate of synthesis. If we could clarify the mechanisms controlling brain tryptophan concentration, we would also throw light upon those controlling turnover of 5-hydroxytryptamine.

We observed that among treatments found to increase brain tryptophan some increased, some had no effect on and others even decreased the amount of tryptophan in plasma and, therefore, we concluded that brain tryptophan concentrations are independent of tryptophan concentration in plasma (Tagliamonte et al. 1971b,c) (Table 4). Nevertheless, since tryptophan is the only amino acid present in plasma bound to serum proteins (McMenamy & Oncley 1958) and the biological activity of many substances is related to their unbound concentration in plasma, we decided to investigate whether the quantity of tryptophan in brain was controlled by the concentration of free tryptophan in serum.

TABLE 4

Effect of treatments listed in Tables 1 and 2 on the total concentration of tryptophan in plasma.

Treatment	Plasma tryptophan (μg/ml)[b]	P
None	27.70 ± 0.52 (150)	
D-Amphetamine[a]	29.00 ± 0.64 (18)	N.S.
D-Amphetamine[a]	38.00 ± 1.50 (8)	<0.001
Reserpine	25.22 ± 1.11 (9)	N.S.
Lithium carbonate	34.78 ± 0.76 (26)	<0.001
Dibutyryl cyclic AMP[a]	26.80 ± 0.80 (13)	N.S.
Dibutyryl cyclic AMP[a]	38.60 ± 1.80 (6)	<0.001
40 °C environment	45.58 ± 1.35 (12)	<0.001
Phenmetrazine[a]	27.35 ± 1.61 (10)	N.S.
Phenmetrazine[a]	41.00 ± 2.60 (8)	<0.001
DL-Fenfluramine[a]	26.06 ± 1.75 (14)	N.S.
DL-Fenfluramine[a]	26.11 ± 0.89 (12)	N.S.
Phenelzine	29.55 ± 1.59 (9)	N.S.
Bulbocapnine	17.76 ± 0.78 (14)	<0.001
Probenecid	10.70 ± 0.70 (18)	<0.001
4-Hydroxybutyrate	22.80 ± 2.10 (8)	N.S.
ECS	29.30 ± 1.04 (21)	N.S.
Puromycin	28.60 ± 0.80 (18)	N.S.

[a] See note c in Table 1.
[b] Each value (± s.e.) was obtained from the same animals from which the data in Tables 1 and 2 were obtained.

In order to test this hypothesis, we studied (a) the effect of the administration of L-tryptophan on the concentrations of free and total serum tryptophan, of brain tryptophan and of 5-hydroxyindoleacetic acid; (b) the effect of drugs capable of releasing serum tryptophan from its protein binding on brain tryptophan content; (c) the effect of different treatments which increase synthesis of 5-hydroxytryptamine in the brain on free and total tryptophan concentrations in serum; and finally (d) whether free serum tryptophan reflected brain tryptophan concentrations and synthesis of 5-hydroxytryptamine under physiological conditions.

Table 5 shows the changes in the concentrations of free, total serum and brain tryptophan after the oral administration of L-tryptophan (50 mg/kg). The total amount of tryptophan in plasma reached a maximum value after 30 min and was still 50% above the initial value after two hours before it returned to the normal value after three hours. The changes in free tryptophan concentration in plasma were qualitatively similar but much more pronounced. Even after 180 min, when total tryptophan in plasma had returned to normal, free tryptophan was still elevated by 50%. The changes in tryptophan concentration in brain were proportional to the changes in free tryptophan but not to those

TABLE 5

Changes in concentrations of free and bound tryptophan in serum and of tryptophan in brain after the oral administration of L-tryptophan (50 mg/kg) to rats.[a]

Time after treatment (min)	n	Serum tryptophan				Brain tryptophan	
		Total (μg/ml)	Percentage increase	Free (μg/ml)	Percentage increase	Concentration (μg/g)	Percentage increase
0	48	23.70 ± 0.74	100	1.71 ± 0.04	100	3.01 ± 0.08	100
30	16	49.91 ± 1.45[b]	210	8.71 ± 0.69[b]	510	11.71 ± 0.72[b]	356
60	16	38.81 ± 1.29[b]	164	5.96 ± 0.56[b]	348	9.54 ± 0.59[b]	316
120	16	35.79 ± 1.23[b]	151	3.12 ± 0.11[b]	185	5.02 ± 0.26[b]	166
180	16	23.20 ± 1.04[c]	98	2.58 ± 0.07[b]	151	4.69 ± 0.18[b]	156

[a] Each value is the average (± s.e.); n is the number of experiments.   [b] P<0.001.   [c] Not significant with respect to control value.

TABLE 6

Effect of sodium salicylate, probenecid and clofibrate on concentrations of free and total serum tryptophan, brain tryptophan, 5-hydroxytryptamine and 5-hydroxyindoleacetic acid.[a]

Treatment[b] (number of experiments)	Dose (mg/kg)	Serum tryptophan (μg/g)		Concentrations in brain (μg/g)		
		Total	Free	Tryptophan	5-Hydroxytrypt-amine	5-Hydroxyindole-acetic acid
None (50)	—	24.61 ± 0.66	2.46 ± 0.09	3.31 ± 0.09	0.66 ± 0.06	0.73 ± 0.07
Sodium salicylate	50 (i.p)	11.75 ± 0.19[c]	3.08 ± 0.11[c]	4.15 ± 0.07[c]	0.78 ± 0.09	0.93 ± 0.05[c]
Probenecid (18)	150 (i.p.)	16.20 ± 0.25[c]	3.40 ± 0.15[c]	5.30 ± 0.30[c]	0.70 ± 0.08	1.35 ± 0.11[c]
Clofibrate (18)	200 (os)	6.85 ± 0.24[c]	4.85 ± 0.14[c]	4.37 ± 0.50[c]	0.68 ± 0.03	1.11 ± 0.08[c]

[a] Each value is the average ± s.e.   [b] Sodium salicylate, probenecid and clofibrate were injected 60, 120 and 180 min before the animal was killed, respectively.   [c] P<0.01 with respect to control value.

in total tryptophan in plasma; that is the amount of brain tryptophan rose to a maximum after 30 min and decreased over the next 150 min to a value still 50% greater than the initial value.

Salicylate, probenecid and clofibrate, which are able to release tryptophan from its binding on serum proteins (McArthur & Dawkins 1969; Tagliamonte *et al.* 1971*d*; Gessa, unpublished observations) are also capable of increasing the amount of brain tryptophan and 5-hydroxyindoleacetic acid (see Table 6). Release of serum tryptophan from its protein binding would explain both the decrease in the absolute amount of the amino acid in serum and its increase in the central nervous system. These drugs also stimulate brain synthesis of 5-hydroxytryptamine (Tagliamonte *et al.* 1971*e*, 1973).

As Table 7 shows, both D-amphetamine and lithium chloride, which stimulate synthesis of 5-hydroxytryptamine, significantly increased the amount of free serum tryptophan. Neither compound released tryptophan from its protein binding when added *in vitro* at the maximum concentration which can be reached *in vivo* after their administration. Electroconvulsive shock (ECS) and the intracerebral administration of puromycin or cycloheximide, which stimulate synthesis of the amine (Tagliamonte *et al.* 1972; Biggio *et al.* 1972), raised the quantity of brain tryptophan and 5-hydroxyindoleacetic acid but influenced neither total nor free serum tryptophan concentrations. We conclude that, under certain experimental conditions, the amount of brain tryptophan increases independently of changes in its free concentrations in serum.

To establish whether free serum tryptophan controlled brain tryptophan and synthesis of 5-hydroxytryptamine under physiological conditions, we studied the relationship between free serum tryptophan and brain tryptophan in fasted and fed rats, having previously shown that brain tryptophan concentration and turnover of the amine are higher in fasted than in fed rats.

The concentration of total tryptophan in serum is lower in rats fasted for 24 h than in rats fed for two hours (see Table 8). In contrast, the absolute concentration of free serum tryptophan is much higher in fasted than in fed rats. These changes were observed in rats eating both in the morning (1000) and in the evening (2200), in other words the variations were not circadian.

Our results provide convincing evidence that the transport of tryptophan from blood into the 5-hydroxytryptamine neuron is the most important, perhaps the only, factor limiting synthesis of the amine. The significance of this mechanism is evident if one considers that, with synthesis of the amine in the whole brain being confined to these neurons, the rate of entry of tryptophan in one hour into the neurons can be estimated in the order of mg/g tissue. However, the present data do not yet allow us to state whether this transport depends only on the concentration of free tryptophan in serum in physiological conditions.

TABLE 7

Effect of different treatments which increase brain synthesis of 5-hydroxytryptamine on concentrations of free and total serum tryptophan.[a]

Treatment[b] (number of experiments)	Dose (mg/kg)	Serum tryptophan (µg/g)		Brain concentrations (µg/g)		
		Total	Free	Tryptophan	5-Hydroxytrypt-amine	5-Hydroxyindole acetic acid
None (50)	—	24.61 ± 0.66	2.46 ± 0.09	3.31 ± 0.09	0.66 ± 0.06	0.73 ± 0.07
d-Amphetamine (10)	5 (i.p.)	30.35 ± 0.70[c]	4.45 ± 0.09[c]	5.15 ± 0.18[c]	0.70 ± 0.11	0.95 ± 0.02[c]
Lithium chloride (12)	100 (i.p.)[d]	31.35 ± 0.80[c]	4.20 ± 0.11[c]	6.12 ± 0.25[c]	0.76 ± 0.11	1.15 ± 0.08[c]
ECS (21)	—	25.50 ± 0.30	2.53 ± 0.15	6.05 ± 0.15[c]	0.66 ± 0.04	1.14 ± 0.12[c]
Puromycin (18)	0.2[e]	25.60 ± 0.80	2.18 ± 0.15	5.20 ± 0.12[c]	0.71 ± 0.05	1.06 ± 0.04[c]
Cycloheximide (42)	0.2[e]	26.4 ± 0.90	2.55 ± 0.08	5.45 ± 0.12[c]	0.75 ± 0.04[c]	1.12 ± 0.08[c]

[a] Each value is the average ± S.E. [b] D-Amphetamine, lithium chloride, puromycin and cycloheximide were injected 60, 120, 180, and 60 min before the animal was killed, respectively, and ECS was performed 60 min before the animal was killed. [c] $P<0.01$ with respect to control value. [d] Twice daily for five days. [e] Intraventricularly.

TABLE 8

Amounts of free and total tryptophan in serum and concentrations of tryptophan, 5-hydroxytryptamine and 5-hydroxyindoleacetic acid in brains of fasted and fed rats.[a]

Condition[b]	Time of death[c]	Serum tryptophan (µg/ml)		Brain tryptophan (µg/g)	Brain 5-hydroxytrypt-amine (µg/g)	Brain 5-hydroxyindole-acetic acid (µg/g)
		Free	Total			
Fasted	1200	4.56 ± 0.13	18.04 ± 0.16	5.67 ± 0.06	0.66 ± 0.03	0.61 ± 0.04
Fed	1200	2.78 ± 0.11[d]	28.21 ± 0.19[d]	3.57 ± 0.04[d]	0.64 ± 0.04	0.34 ± 0.02[d]
Fasted	2400	4.79 ± 0.09	17.52 ± 0.13	5.83 ± 0.03	0.62 ± 0.04	0.64 ± 0.03
Fed	2400	3.01 ± 0.12[d]	27.81 ± 0.21[d]	3.48 ± 0.06[d]	0.63 ± 0.02	0.41 ± 0.05[d]

[a] Each value is the average ± S.E. of at least 20 determinations. [b] Rats were fasted for 24 h or fed for two hours. [c] Rats killed at noon were those trained to eat their daily food from 1000 to 1200; rats killed at 2400 had been trained to eat their food from 2200 to 2400. [d] $P < 0.001$ with respect to fasted rats.

In at least two treatments—shock (ECS) and intracerebral injection of puromycin or cycloheximide—brain tryptophan concentration rose without a concomitant change in that of free serum tryptophan. Dibutyryl cyclic AMP, which increases the amount of brain tryptophan *in vivo*, stimulates the uptake of tryptophan by brain slices.

The involvement of other factors in the control of the concentration of tryptophan in brain is suggested by studies on the passage of amino acids from blood to brain (Guroff & Udenfriend 1962; Oldendorf 1971) or by competition studies on the uptake of amino acids by brain slices (Blasberg & Lajtha 1965) or synaptosomes (Grahame-Smith & Parfitt 1970) showing that the aromatic amino acids share the same transport mechanism from blood to brain in such a way that a high concentration of one can lower the uptake of others. Thus, the tryptophan concentration in brain might depend not only on the concentration of free serum tryptophan, but also, as suggested by Fernstrom & Wurtman (1971), on the plasma concentration of other amino acids competing for the same transport.

Since the ratio of free to bound tryptophan in serum is not constant and the free tryptophan fraction seems to be the only one biologically active, it is conceivable that the concentration of free tryptophan changes with respect to the concentration of the other circulating amino acids, including those which compete for the same transport mechanism into brain. We are still investigating the physiological significance of these other mechanisms.

References

BIGGIO, G., MEREU, G., VARGIU, L. & TAGLIAMONTE, A. (1972) . *Riv. Farmacol. Ter.* 3, 229–236
BLASBERG, R. & LAJTHA, A. (1965). *Arch. Biochem. Biophys.* 112, 361–367
FERNSTROM, J. D. & WURTMAN, R. J. (1971). *Science (Wash. D.C.)* 174, 1023–1025
GESSA, G. L., KRISHNA, G., FORN, J., TAGLIAMONTE, A. & BRODIE, B. B. (1970) Role of cyclic AMP in cell function. *Adv. Biochem. Psychopharmacol.* 3, 371–381
GRAHAME-SMITH, G. D. & PARFITT, A. G. (1970). *J. Neurochem.* 17, 1339–1353
GREEN, H. & SAWYER, J. L. (1966). *Anal. Biochem.* 15, 53–57
GUROFF, G. & UDENFRIEND, S. (1962). *J. Biol. Chem.* 237, 803–806
LE DOUAREC, J. C., SCHMITT, H. & LAUBIE, M. (1966). *Arch. Int. Pharmacodyn. Thér.* 161, 206–232
MCARTHUR, J. N. & DAWKINS, P. D. (1969). *J. Pharm. Pharmacol.* 21, 744–750
MCMENAMY, R. H. & ONCLEY, J. L. (1958). *J. Biol. Chem.* 233, 1436–1447
OLDENDORF, W. H. (1971). *Am. J. Physiol.* 221, 1629–1639
PEREZ-CRUET, J., TAGLIAMONTE, A., TAGLIAMONTE, P. & GESSA, G. L. (1971). *J. Pharmacol. Exp. Ther.* 178, 325–330
REID, W. D. (1970). *Fed. Proc.* 29, 747

REID, W. D., VOLICER, L., SMOOKLER, H., BEAVEN, M. A. & BRODIE, B. B. (1968). *Pharmacology* **1**, 329–344

TAGLIAMONTE, A., BIGGIO, G. & GESSA, G. L. (1971*d*). *Riv. Farmacol. Ter.* **2**, 251–255

TAGLIAMONTE, A., BIGGIO, G., VARGIU, L. & GESSA, G. L. (1973). *J. Neurochem.* **20**, 909–912

TAGLIAMONTE, A., TAGLIAMONTE, P., DI CHIARA, G., GESSA, R. & GESSA, G. L. (1972). *J. Neurochem.* **19**, 1509–1512

TAGLIAMONTE, A., TAGLIAMONTE, P., FORN, J., PEREZ-CRUET, J., KRISHNA, G. & GESSA, G. L. (1971*a*). *J. Neurochem.* **18**, 1191–1196

TAGLIAMONTE, A., TAGLIAMONTE, P., GESSA, R., DUCE, M., MAFFEI, C., GESSA, G. L. & CAMBA, R. (1971*e*). *Riv. Farmacol. Ter.* **2**, 207–213

TAGLIAMONTE, A., TAGLIAMONTE, P., PEREZ-CRUET, J. & GESSA, G. L. (1971*c*). *Nat. New Biol.* **229**, 125–126

TAGLIAMONTE, A., TAGLIAMONTE, P., PEREZ-CRUET, J., STERN, S. & GESSA, G. L. (1971*b*). *J. Pharmacol. Exp. Ther.* **177**, 475–480

TOZER, T. N., NEFF, N. H. & BRODIE, B. B. (1966). *J. Pharmacol. Exp. Ther.* **153**, 177–182

For *Discussion*, see pp. 230–241.

Fatty acids and the disposition of tryptophan

G. CURZON and P. J. KNOTT

Department of Neurochemistry, Institute of Neurology, London

Abstract   Synthesis of 5-hydroxytryptamine in the brain is influenced by trypto-phan concentration, since the rate-limiting enzyme, tryptophan hydroxylase, is normally unsaturated with substrate. In various circumstances, for example food deprivation and immobilization, the apparent increase in the turnover of 5-hydr-oxytryptamine in the rat brain is associated with an increase in the amount of brain tryptophan and of the small free (ultrafiltrable) fraction of plasma tryptophan.

The mechanism by which the amount of free tryptophan in the plasma increases has been investigated. Food deprivation is known to increase plasma non-esterified fatty acid (NEFA) concentration and, like tryptophan, these sub-stances are largely bound to albumin.

Recent results suggest that more NEFA in the plasma releases tryptophan from its binding to plasma albumin so that more is available to the brain, for upon addition of fatty acids to plasma *in vitro* the concentration of free trypto-phan increased. Also, drug treatments which increased the quantity of NEFA (e.g. heparin, isoprenaline, aminophylline and dopa) increased plasma free tryptophan. Conversely, insulin decreased both NEFA and free tryptophan while nicotinic acid, which also decreases NEFA, opposed the increase of free tryptophan upon food deprivation. Results are consistent with a relationship between the changes in plasma free tryptophan and fat cell cyclic AMP which mediates the lipolytic action of the catecholamines. Aminophylline is reported to increase the amount of fat cell cyclic AMP while insulin and nicotinic acid are said to decrease it. Altered tryptophan metabolism in pigs with experimental acute hepatic failure is also explicable by NEFA changes. The possible role of plasma free tryptophan in relation to stress, to the reported defective metabolism of 5-hydroxytryptamine in depression and to other pathways of tryptophan metabolism is discussed.

During the past few years, we have been investigating, in common with many others, the control of 5-hydroxytryptamine metabolism in the brain. Such investigations require methods by which the metabolism of 5-hydroxytrypt-amine or the activity of the aminergic neurons can be disturbed. However, no available experimental approach is ideal. Drugs almost invariably have side effects and may also reveal control mechanisms which, though pharmacologically significant, are not necessarily physiologically relevant. The direct stimulation of 5-hydroxytryptaminergic neurons is an attractive method but has two limitations. First, the manipulations required in sham-operations on control animals might obscure normally operative physiological mechanisms by disturbing the normal state. Secondly, physiological influences upon the metabolism of the mono-amine, which are not generated by altered 5-hydroxytryptaminergic activity, might not be revealed. Because of these problems, we concentrated upon the effects of altering environmental or nutritional parameters. The disadvantages here are the complexity of the possible consequent changes and their problema-tical relationship to 5-hydroxytryptaminergic activity. The great advantage is that any mechanisms revealed are by definition physiological mechanisms.

Therefore, we deprived rats of food (but not water) for 24 h or we immobilized them for three hours, and in both cases we observed well defined increases in the concentration of brain tryptophan. The concentration of the metabolite of the monoamine, 5-hydroxyindoleacetic acid, in the brain also increased but that of 5-hydroxytryptamine remained essentially constant (Curzon et al. 1972a). The inference that the turnover of the amine in the brain had increased was strength-ened by the observation that the concentrations of the amine increase more in deprived rats than in control rats after injection of the monoamine oxidase in-hibitor, pargyline. Perez-Cruet et al. (1972) obtained similar results. Since brain turnover of 5-hydroxytryptamine is increased by administration of tryptophan (Eccleston et al. 1965), it seems likely that the extra indole acid results from the increased amounts of brain tryptophan since tryptophan hydroxylase (the rate-limiting enzyme for synthesis of the monoamine) is apparently unsaturated with substrate under physiological conditions (Friedman et al. 1972).

Food deprivation or immobilization did not markedly alter concentrations of most other brain amino acids (Knott et al. 1973). This indicated the existence of a mechanism with some degree of specificity by which brain tryptophan (and hence turnover of 5-hydroxytryptamine) was increased in two different stressful situations.

Therefore, we investigated the mechanism by which brain tryptophan con-centration rose. Although we presume that the tryptophan is derived from plasma, we could not find a positive correlation between the amount of trypto-phan in plasma and in brain. Perez-Cruet et al. (1972) reported that, rather than rise in parallel with the concentration of tryptophan in brain, the amount in plasma progressively decreased during deprivation.

The lack of positive correlation is explicable if only a small fraction of plasma tryptophan is free and thus available to the brain. Most tryptophan in plasma is not directly available since it is bound to plasma albumin (McMenamy & Oncley 1958). Free tryptophan can be determined by ultrafiltration or by dialysis and, using the former method, we showed that the increase of brain tryptophan upon deprivation or immobilization was associated with an increase in plasma free tryptophan (Knott & Curzon 1972). Similarly, when tryptophan was administered intraperitoneally to rats (Tagliamonte *et al.* 1973), the build-up of brain tryptophan and 5-hydroxyindoleacetic acid was directly proportional to that of plasma free tryptophan.

Therefore, it is apparent that anything which can influence the equilibrium between protein-bound and free tryptophan in plasma may also influence the cerebral metabolism of 5-hydroxytryptamine. It has now been shown that this equilibrium is strongly affected by non-esterified fatty acids (NEFAs) in the plasma: the greater their concentration the greater the proportion of plasma tryptophan which is free.

NON-ESTERIFIED FATTY ACIDS AND TRYPTOPHAN IN PLASMA

It is well known that food deprivation and emotional stresses increase plasma NEFA concentrations (Dole 1956; Mallov & Witt 1961; Mayes 1962). These fatty acids are strongly bound to albumin so that only a negligible fraction is not bound under physiological conditions (Goodman 1958). Addition of oleate to bovine serum albumin in Tris buffer has been reported to interfere with tryptophan binding "to a minor extent" (McMenamy & Oncley 1958). Thus, are the increases in plasma NEFA during food deprivation or stress matched by a decrease in the binding of tryptophan to plasma albumin so that more trypto-phan is made available to the brain? The positive association between changes of plasma NEFA and of free tryptophan upon food deprivation is consistent with this hypothesis. The increase of plasma free tryptophan after injection of heparin provides further support (Knott & Curzon 1972; Curzon *et al.* 1973a). (Heparin increases plasma NEFA by releasing tissue lipase into plasma which then hydrolyses plasma glycerides.)

Direct evidence for the proposed mechanism comes from our observation that the amount of free tryptophan rises after the addition, *in vitro*, of physio-logically occurring NEFAs to rat and human plasma within the range of their physiological concentrations (Curzon *et al.* 1973a). We found positive and significant correlations between the concentrations of added linoleic, oleic and palmitic acid and those of free tryptophan for both rat and human plasma. An increase of fatty acid concentration from 0.5 to 1.0 mequiv/1 caused the concentration of the free tryptophan to double in human plasma and increase almost fourfold in rat plasma. Equilibration appears to be rapid: the increase of

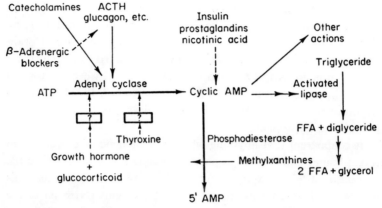

FIG. 1. Effects of drugs and hormones on lipolysis (from Robison *et al*. 1971): FFA = NEFA.

free tryptophan on adding 1.5 mequiv/1 of palmitic acid to human plasma is complete at room temperature within five minutes (the minimum time required for collection of a sufficient ultrafiltration sample).

These results suggested that many agents which affect lipolysis and plasma NEFA concentrations by acting on fat cell cyclic AMP (Robison *et al*. 1971) also influence the disposition of plasma tryptophan and thus turnover of 5-hydroxytryptamine in the brain. Therefore, we investigated the effects of injecting rats with the various substances indicated in Fig. 1 and related substances. The results are set out in Tables 1 and 2.

TABLE 1

Effect of treatments increasing plasma NEFA on disposition of plasma tryptophan and on brain tryptophan, 5-hydroxytryptamine (5HT) and 5-hydroxyindoleacetic acid (5HIAA). Values are expressed as the percentage change relative to controls.

Treatment[a]	Plasma			Brain		
	NEFA	Tryptophan		Tryptophan	5HT	5HIAA
		Free	Total			
Food deprivation, 24 h (6)	+159[d]	+150[d]	−3	+74[d]	+2	+48[d]
Heparin, 500 i.u./kg (i.v.), 15 min (6)	+88[c]	+89[c]	−18			
Isoprenaline, 0.04 mg/kg (i.v.), 5 min (6)	+92[b]	+56[c]	+10			
L-Dopa, 500 mg/kg (i.p.), 2 h (6)[e]	+132[d]	+144[d]	−36[b]	+71[d]	−15[c]	+46[c]
Aminophylline, 150 mg/kg (i.p.), 3 h (8)	+89[d]	+83[d]	+13	+150[d]	+26[d]	+157[d]

[a] Figures in brackets are numbers of animals in each group.
[b] $P < 0.05$.
[c] $P < 0.01$.
[d] $P < 0.001$.
[e] In the dopa-treated group, plasma dopa concentrations were 13.4 ± 0.3 µg/ml whereas in saline controls they were less than 1 µg/ml.

TABLE 2

Effect of treatments decreasing plasma NEFA on disposition of plasma tryptophan and on brain tryptophan, 5-hydroxytryptamine (5HT) and 5-hydroxyindoleacetic acid (5HIAA). Values are expressed as the percentage change relative to controls.

Treatment[a]	Plasma			Brain		
	NEFA	Tryptophan		Tryptophan	5HT	5HIAA
		Free	Total			
Insulin, 2 i.u./kg (i.p.), 2 h, fed rats (6)[b]	-37^d	-44^d	$+2$	-7	-14^e	-15
Insulin, 2 i.u./kg (i.p.), 2 h, fasted rats (6)[b,c]	-48^f	-45^f	$+20$	$+34^d$	$+4$	$+20$
Nicotinic acid, 50 mg/kg (i.p.), 1 h, fasted rats (6)	-63^f	-21^d	$+53^e$	-33^f	0	-15^d

[a] Figures in brackets are numbers of animals in each group.
[b] Insulin reduced plasma glucose levels 42% and 83% in fed and fasted rats respectively ($P<0.001$).
[c] Food withdrawn for 24 h for fasted rats.
[d] $P<0.05$.
[e] $P<0.01$.
[f] $P<0.001$.

EFFECTS OF AGENTS ALTERING PLASMA NEFA ON PLASMA TRYPTOPHAN AND ON TRYPTOPHAN METABOLISM IN BRAIN

Catecholamines

Plasma NEFA concentrations increase during noradrenaline infusion (Havel & Goldfien 1959); the effect is mediated through action on β receptors. The specific β agonist, isoprenaline, raises the concentration of both plasma NEFA and free tryptophan (Table 1). It is likely that episodes of acute sympathetic activity in many stress situations can change NEFA levels sufficiently to increase the availability of plasma tryptophan considerably so that the brain is flooded with tryptophan. For example, Taggart & Carruthers (1971) found grossly elevated plasma NEFA and increased plasma catecholamine concentrations in racing drivers at about the time of a race. Our *in vitro* experiments suggest that the amount of plasma free tryptophan might rise very appreciably under these conditions and that it could also be significantly affected by much milder stress stimuli. Presumably, differences of sympathetic release and of receptor sensitivity lead to large differences between such changes in different subjects.

The effect upon the NEFA–tryptophan system of the catecholamine precursor L-dopa is interesting because Weiss *et al.* (1971) showed that intraperitoneal

injection of L-dopa into rats led to increased brain tryptophan and Rivera-
Calimlim & Bianchine (1972) found that its intravenous injection into dogs led
to increased plasma NEFA. So, it is hardly surprising that we find that injection
of L-dopa also increases plasma free tryptophan. The smaller increase of brain
tryptophan might reflect competition between the two amino acids for transport
to the brain and the fall of brain 5-hydroxytryptamine might be due to displacement of the amine stores by dopamine (Ng et al. 1970). Plasma dopa
concentrations attained were considerably greater than those usually found in
Parkinsonian subjects under dopa treatment (Curzon et al. 1972b) though
comparable values are found occasionally in patients in whom tryptophan
changes might therefore result.

Purines

Aminophylline, an inhibitor of the enzymic destruction of cyclic AMP in the
fat cell, increases plasma NEFA and free tryptophan concentrations com-
parably. The cerebral amounts of tryptophan and 5-hydroxyindoleacetic acid
rose strikingly while the increase in cerebral 5-hydroxytryptamine was small
but significant. A similar mechanism might underlie the increased quantities
of the amine and its indole acid metabolite in the brain reported after injection
of caffeine (Berkowitz & Spector 1971). These findings are relevant to the
decreased depletion of the amine in the brains of rats treated with the tryptophan
hydroxylase inhibitor, α-propyldopacetamide, when aminophylline or caffeine
are also given (Corrodi et al. 1972).

Insulin and nicotinic acid

Insulin and nicotinic acid, which decrease plasma NEFA by decreasing fat
cell cyclic AMP, were also investigated (Table 2). Insulin led to comparable and
significant percentage falls of plasma NEFA and free tryptophan and to lesser
decreases of brain tryptophan, 5-hydroxytryptamine and 5-hydroxyindoleacetic
acid of which only the change in the former was significant. These findings are
at variance with those of Fernstrom & Wurtman (1971) who found that the
amounts of brain tryptophan and 5-hydroxytryptamine increased when insulin
was given to rats from whom a largely carbohydrate diet had been withdrawn
15–18 h previously. Our animals had access to a mixed diet. In an additional
experiment with rats from whom food was withdrawn 24 h before injection of
insulin plasma changes were essentially as before but now the brain changes

were more consistent with those reported by Fernstrom & Wurtman; brain tryptophan increased significantly. Interpretation of the results with insulin is not simple since insulin decreases the plasma concentrations of amino acids which interfere with the transport of tryptophan to the brain (Fernstrom & Wurtman 1972a,b). These changes tend to counteract the effect of the decrease in plasma free tryptophan. Another consideration is that, under our conditions, insulin treatment of the rats deprived of food led to a fall of plasma glucose which was much more striking than that in the fed rats. Also, convulsions were observed only in deprived animals treated with insulin.

Although injection of nicotinic acid caused a small fall of plasma NEFA, none of the other parameters measured were significantly altered. However, it completely prevented the large increase of plasma NEFA on food deprivation. The increases of plasma free tryptophan, brain tryptophan and 5-hydroxy-indoleacetic acid upon deprivation were less marked when nicotinic acid was given but were not completely prevented. This is consistent with the increase of total plasma tryptophan concentration in these rats. Plasma NEFA can also be decreased by loading with glucose (Dole 1956) and it has recently been shown that this is paralleled by a fall of plasma free tryptophan (Lipsett et al. 1973).

Thus, in general, drugs or hormones which alter plasma NEFA similarly alter plasma free tryptophan. These changes tend to lead to changes of brain tryptophan metabolism in the appropriate direction. Many studies, especially those of Tagliamonte et al. (1971), indicate that brain tryptophan determinations are now necessary when investigating drugs or procedures affecting the cerebral metabolism of 5-hydroxytryptamine. If tryptophan changes are shown, then assessment of the role of plasma NEFA also becomes necessary. An obvious example demanding study is the increase in tryptophan and 5-hydroxy-indoleacetic acid in rodent brains upon injecting amphetamines (Schubert & Sedvall 1972; Tagliamonte et al. 1971) especially as D-amphetamine raises rat plasma NEFA (Gessa et al. 1969). Also, the increase of rodent brain 5-hydroxyindoleacetic acid and/or tryptophan concentrations on treatment with inhibitors of dopamine hydroxylase (Johnson & Kim 1973), with cholino-mimetics (Haubrich & Reid 1972) and on electroconvulsive shock (Tagliamonte et al. 1972) might involve the foregoing mechanism.

EFFECT OF PLASMA NEFA ON TRYPTOPHAN METABOLISM IN STRESS AND DISEASE

Plasma NEFA changes in stress situations might differ in animals of different strains, life history or behavioural pattern. For example, rats selected for low exploratory activity show a particularly large plasma NEFA increase upon electroshock (Benes & Hrubes 1968) and might therefore also have a particularly large increase of plasma free tryptophan. Further complexities may also arise because of tryptophan destruction through the action of tryptophan oxygenase (EC 1.13.1.12) induced by a stress-provoked increase of adrenocortical secretion (Curzon & Green 1969). Together, these opposing effects upon tryptophan availability could cause qualitatively different net changes of brain tryptophan metabolism when different stresses are imposed or different groups of animals are used. The amount of tryptophan in the rat brain rose during the first 2–3 h of immobilization and then gradually declined (Curzon & M. H. Joseph, unpublished results). A gradual increase of hepatic oxygenase could be involved in the decline. Factors of this kind might explain why in earlier immobilization experiments (Curzon & Green 1969) brain 5-hydroxytryptamine gradually diminished to a minimum at six hours.

Elevated plasma concentrations of NEFAs are found in severe hepatic failure (Mortiaux & Dawson 1961). This probably explains the increase in 5-hydroxyindoleacetic acid content of cerebrospinal fluid in patients with hepatic coma (Knell et al. 1972), for using a model for human acute hepatic failure obtained by hepatic devascularization of pigs (Curzon et al. 1973b) we found that plasma NEFA and free (but not total) tryptophan were both markedly raised. Brain tryptophan concentration increased, correlating significantly with plasma free tryptophan concentration (Fig. 2), and the indole acid in brain also rose.

Could the increase in brain tryptophan have a special role in the development of hepatic coma? Direct toxicity of tryptophan itself cannot readily be invoked, since doses of DL-tryptophan of up to 15 g/day have been given by mouth in the treatment of depression (Coppen et al. 1963) and the amount of plasma free L-tryptophan in human subjects has been raised almost a hundred-fold by tryptophan infusion apparently without gross effect (Curzon, J. Friedel, B. D. Kantamaneni, M. H. Greenwood & M. H. Lader, unpublished results). However, according to Sherlock (1968), the brain may be detrimentally affected in liver disease by changes which would not affect normal subjects. Also, it could be that the greater turnover of 5-hydroxytryptamine enhances the central toxicity of other substances which accumulate in liver disease. A somewhat analogous pharmacological example is the apparent role of brain 5-hydroxytryptamine in determining the toxicity of caffeine in the mouse (Berkowitz et al. 1971).

The role of plasma NEFA in biochemical abnormalities during depression requires comment as much evidence suggests that the cerebral turnover of 5-hydroxytryptamine is defective (Coppen 1972); a recent brief report claims that plasma free tryptophan was significantly low in a group of female depressives (Coppen et al. 1972). While this would be explicable if plasma NEFA

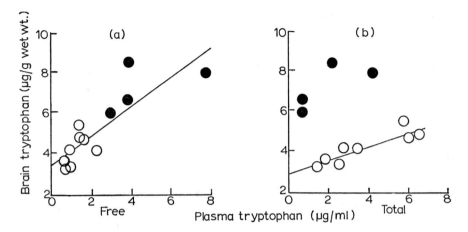

FIG. 2. Relationships between brain and plasma tryptophan concentrations. Brain tryptophan means the average concentration for hypothalamus, thalamus, caudate and cortex.

(a) A plot of brain tryptophan against plasma free tryptophan: ○, sham-operated pigs; ●, pigs with devascularized livers; r for both groups together $= 0.84$ ($n = 12$; $P<0.001$), r for sham-operated pigs $= 0.41$ ($n = 8$; n.s.), r for pigs with devascularized livers $= 0.53$ ($n = 4$; n.s.).

(b) A plot of brain tryptophan against plasma total tryptophan: ○, sham-operated pigs; ●, pigs with devascularized livers; r for both groups together $= -0.03$ ($n = 12$; n.s.), r for sham-operated pigs $= 0.89$ ($n = 8$; $P<0.001$), r for pigs with devascularized livers $= 0.74$ ($n = 4$; n.s.) (from Curzon et al. 1973b).

was low, it is, on the contrary, reported to be rather high in depression (Cardon & Mueller 1966; Van Praag & Leijnse 1966). However, the evidence of low sympathetic activity in depression (Perez-Reyes 1969) suggests that further study of these relationships will be worthwhile. Certainly, the increase of 5-hydroxyindoleacetic acid in the cerebrospinal fluid when depressed subjects simulate manic hyperactivity (Post et al. 1973) could well result from the increase of plasma NEFA on exercise (Havel et al. 1963).

COMMENT

In many circumstances, apparently, the amount of non-esterified fatty acids in plasma can alter the availability of tryptophan for brain metabolism. Further examples include the diurnal variation of plasma NEFA (Fuller & Diller 1970) which must be considered in relation to the diurnal changes of plasma tryptophan and of brain tryptophan and the 5-hydroxyindoles. As well as the diurnal changes of human plasma NEFA, Court et al. (1971) have reported fluctuations in NEFA with a period of oscillation of about two minutes and which may have considerable amplitudes and might generate similar oscillations in the plasma free tryptophan.

Thus, the availability of tryptophan to the brain for synthesis of 5-hydroxy-tryptamine and other purposes might be continuously changing. Our results suggest that often the newly synthesized amine simply spills over from replete vesicles and is destroyed by monoamine oxidase so that the amount of 5-hydroxyindoleacetic acid increases while that of the amine changes little. However, in stress situations, any depletion of the stores of the amine in brain due to firing would tend to be opposed not only by increased tryptophan hydroxylase activity (Shields & Eccleston 1972) but also by increased tryptophan availability which might therefore have homeostatic value.

Preliminary experiments by Dr C. Marsden in our laboratory suggest that the tryptophan changes are not initiated by 5-hydroxytryptaminergic activity. Thus electrical stimulation of the raphe nucleus did not increase brain trypto-phan concentration. Rats with raphe lesions and thus low turnover of the amine had slightly increased brain tryptophan concentration perhaps merely owing to decreased demands upon it for amine synthesis. These findings suggest that brain tryptophan concentration does not alter as part of a regulatory mechan-ism driven by changes in the amount of 5-hydroxytryptamine in the brain.

Other central influences might, however, be involved in the tryptophan changes. Electrical stimulation of the premammillary area of the hypothalamus increases the amount of NEFA in plasma (Barkai & Allweis 1972) thereby tending to increase the amount of brain tryptophan. Various findings are consistent with central cholinergic and noradrenergic activity having opposite effects on plasma NEFA: intraventricular noradrenaline and acetylcholine injections respectively decrease and increase sheep plasma NEFA (Clough & Thompson 1972). These results might explain the increase of brain tryptophan and 5-hydroxyindoleacetic acid when noradrenaline synthesis is inhibited by dopamine hydroxylase inhibitors (Johnson & Kim 1973) and the increase of the indole acid when brain acetylcholine levels are elevated by pilocarpine (Haubrich & Reid 1972). Stimulation of sympathetic nerves release fatty acids from adipose tissue (Rosell 1966) but sympathetic activity does not seem to be obligatory for fatty acid release as starvation increases plasma NEFA even in demedullated and noradrenaline-depleted rats (Stern & Maickel 1963).

It seems likely that plasma NEFA alters the availability of tryptophan not only to the brain but for intracellular uptake in general and, therefore, for many pathways of metabolism, for example, protein or tryptamine synthesis and degradation by liver oxygenase or aminotransferase. Increased availability of tryptophan for metabolism when free tryptophan increases upon food depriva-tion is perhaps of physiological value in opposing deficiencies which could otherwise occur more rapidly (e.g. NAD deficiency). Tryptophan availability might be a limiting factor in protein synthesis, especially in young animals (Wunner *et al*. 1966; Aoki & Siegel 1970). Also, inhibition of the synthesis of 5-hydroxytryptamine has irreversibly harmful effects in young rats (Hole 1972). Therefore, the increase of free tryptophan on food deprivation would tend to protect the developing animals against the consequences of periods of deprivation in early life.

ACKNOWLEDGEMENT

We thank the Medical Research Council for financial support.

References

AOKI, K. & SIEGEL, F. L. (1970) Hyperphenylalaninemia: disaggregation of brain polyribosomes in young rats. *Science (Wash. D.C.)* **168**, 129–130

BARKAI, A. & ALLWEIS, C. (1972) Effect of electrical stimulation of the hypothalamus on plasma levels of free fatty acids and glucose in rats. *Metabolism* **21**, 921–927

BENES, V. & HRUBES, V. (1968) Serum free fatty acids level after an experimentally induced emotional stress and its modification by nicotinic acid in rats with different characteristics of higher nervous activity. *Act. Nerv. Super.* **10**, 395–399

BERKOWITZ, B. A. & SPECTOR, S. (1971) The effect of caffeine and theophylline on the disposition of brain serotonin in the rat. *Eur. J. Pharmacol.* **16**, 322–325

BERKOWITZ, B. A., SPECTOR, S. & POOL, W. (1971) The interaction of caffeine, theophylline and theobromine with monoamine oxidase inhibitors. *Eur. J. Pharmacol.* **16**, 315–321

CARDON, P. V. & MUELLER, P. S. (1966) A possible mechanism: psychogenic fat mobilization. *Ann. N.Y. Acad. Sci.* **125**, 924–927

CLOUGH, D. P. & THOMPSON, G. E. (1972) Effect of intraventricular noradrenaline and acetylcholine on plasma unesterified fatty acid concentration in sheep. *Neuroendocrinology* **9**, 365–374

COPPEN, A. (1972) Biogenic amines and affective disorders. *J. Psychiatr. Res.* **9**, 163–175

COPPEN, A., SHAW, D. M. & FARRELL, J. P. (1963) Potentiation of the antidepressive effect of a monoamine oxidase inhibitor by tryptophan. *Lancet* **1**, 79–81

COPPEN, A., ECCLESTON, E. G. & PEET, M. (1972) Total and free tryptophan concentration in the plasma of depressive patients. *Lancet* **2**, 1415–1416

CORRODI, H., FUXE, K. & JONSSON, G. (1972) Effect of caffeine on central monoamine neurons. *J. Pharm. Pharmacol.* **24**, 155–158

COURT, J. M., DUNLOP, M. E. & LEONARD, R. F. (1971) High frequency oscillation of blood fatty acid levels in man. *J. Appl. Physiol.* **31**, 345–347

CURZON, G. & GREEN, A. R. (1969) Effects of immobilization on rat liver tryptophan pyrrolase and brain 5-hydroxytryptamine metabolism. *Br. J. Pharmacol.* **37**, 689–687

CURZON, G., JOSEPH, M. H. & KNOTT, P. J. (1972*a*) Effects of immobilization and food deprivation on rat brain tryptophan metabolism. *J. Neurochem.* **19**, 1967–1974

CURZON, G., KANTAMANENI, B. D. & TRIGWELL, J. (1972*b*) A method for the determination of dopa and 3-*O*-methyldopa in the plasma of Parkinsonian patients. *Clin. Chim. Acta* **37**, 335–341

CURZON, G., FRIEDEL, J. & KNOTT, P. J. (1973a) The effects of fatty acids on the binding of tryptophan to plasma protein. *Nature (Lond.)* **242**, 198–200

CURZON, G., KANTAMANENI, B. D., WINCH, J., ROJAS-BUENO, A., MURRAY-LYON, I. M. & WILLIAMS, R. (1973b) Plasma and brain tryptophan changes in experimental acute hepatic failure. *J. Neurochem.*, **21**, 1761–1764

DOLE, V. J. (1956) A relation between non-esterified fatty acids in plasma and the metabolism of glucose. *J. Clin. Invest.* **35**, 150–154

ECCLESTON, D., ASHCROFT, G. W. & CRAWFORD, T. B. B. (1965) 5-Hydroxyindole metabolism in rat. A study of intermediate metabolism using the technique of tryptophan loading II. *J. Neurochem.* **12**, 493–503

FERNSTROM, J. D. & WURTMAN, R. J. (1971) Brain serotonin content: increase following ingestion of carbohydrate diet. *Science (Wash. D.C.)* **174**, 1023–1025

FERNSTOM, J. D. & WURTMAN, R. J. (1972a) Elevation of plasma tryptophan by insulin in rat. *Metabolism* **21**, 337–342

FERNSTROM, J. D. & WURTMAN, R. J. (1972b) Brain serotonin content: physiological regulation by plasma neutral amino acids. *Science (Wash. D.C.)* **178**, 414–416

FRIEDMAN, P. A., KAPPELMAN, A. H. & KAUFMAN, S. (1972) Partial purification and characterization of tryptophan hydroxylase from rabbit hindbrain. *J. Biol. Chem.* **247**, 4165–4173

FULLER, R. W. & DILLER, E. R. (1970) Diurnal variation of liver glycogen and plasma free fatty acids in rats fed ad libitum or single daily meal. *Metabolism* **19**, 226–229

GESSA, G. L., CLAY, G. A. & BRODIE, B. B. (1969) Evidence that hyperthermia produced by D-amphetamine is caused by a peripheral action of the drug. *Life Sci.* **8**, 135–141

GOODMAN, D. S. (1958) The interaction of human serum albumin with long chain fatty acid anions. *J. Am. Chem. Soc.* **80**, 3892–3898

HAUBRICH, D. R. & REID, W. D. (1972) Effects of pilocarpine or arecoline administration on acetylcholine levels and serotonin turnover in rat brain. *J. Pharmacol.* **181**, 19–27

HAVEL, R. J. & GOLDFIEN, A. (1959) The role of the sympathetic nervous system in the metabolism of free fatty acids. *J. Lipid Res.* **1**, 102–108

HAVEL, R. J., NAIMARK, A. & BORCHGREVINK, C. F. (1963) Turnover rate and oxidation of free fatty acids of blood plasma in man during exercise: studies during continuous infusion of palmitate-1-C^{14}. *J. Clin. Invest.* **42**, 1054–1063

HOLE, K. (1972) Reduced 5-hydroxyindole synthesis reduces postnatal brain growth in rats. *Eur. J. Pharmacol.* **18**, 361–366

JOHNSON, G. A. & KIM, E. G. (1973) Increase of brain levels of tryptophan induced by inhibition of dopamine β-hydroxylase. *J. Neurochem.*, **20**, 1761–1764

KNELL, A. J., PRATT, O. E., CURZON, G. & WILLIAMS, R. S. (1972) Changing ideas in hepatic encephalopathy in *Eighth Symposium on Advanced Medicine* (Kneale, G., ed.), pp. 156–170, Pitman, London

KNOTT, P. J. & CURZON, G. (1972) Free tryptophan in plasma and brain tryptophan metabolism. *Nature (Lond.)* **239**, 452–453

KNOTT, P. J., JOSEPH, M. H. & CURZON, G. (1973) Effects of food deprivation and immobilization on tryptophan and other amino acids in rat brain. *J. Neurochem.* **20**, 249–251

LIPSETT, D., MADRAS, B. K., WURTMAN, R. J. & MUNRO, H. N. (1973) Serum tryptophan level after carbohydrate ingestion: selective decline in non-albumin-bound tryptophan coincident with reduction in serum free fatty acids. *Life Sci.* **12**, 57–64

MALLOV, S. & WITT, P. N. (1961) Effect of stress and tranquillisation on plasma free fatty acid level in the rat. *J. Pharmacol.* **132**, 126–130

MAYES, P. A. (1962) Blood glucose and plasma unesterified fatty acid changes induced by the stress of an emergency situation. *Experientia* **18**, 451

MCMENAMY, R. H. & ONCLEY, J. L. (1958) The specific binding of L-tryptophan to serum albumin. *J. Biol. Chem.* **233**, 1436–1447

MORTIAUX, A. & DAWSON, A. M. (1961) Plasma free fatty acid in liver disease. *Gut* **2**, 304–309

NG, K. Y., CHASE, T. N., COLBURN, R. W. & KOPIN, I. J. (1970) L-Dopa induced release of cerebral monoamines. *Science (Wash. D.C.)* **170**, 76–77

PEREZ-CRUET, J., TAGLIAMONTE, A., TAGLIAMONTE, P. & GESSA, G. L. (1972) Changes in brain serotonin metabolism associated with fasting and satiation in rats. *Life Sci.* **11**, 31–39

PEREZ-REYES, M. (1969) Differences in the capacity of the sympathetic and endocrine systems of depressed patients to react to a physiological stress. *Pharmakopsychiatr. Neuropsychopharmakol.* **2**, 245–251

POST, R. M., KOTIN, J., GOODWIN, F. K. & GORDON, E. K. (1973) Psychomotor activity and cerebrospinal fluid amine metabolites in affective illness. *Am. J. Psychiatr.* **130**, 67–72

RIVERA-CALIMLIM, L. & BIANCHINE, J. R. (1972) Effect of L-dopa on plasma free fatty acids and plasma glucose. *Metabolism* **21**, 611–617

ROBISON, G. A., BUTCHER, R. W. & SUTHERLAND, E. W. (1971) *Cyclic AMP*, Academic Press, New York & London

ROSELL, S. (1966) Release of free fatty acids from subcutaneous adipose tissue in dogs following sympathetic nerve stimulation. *Acta Physiol. Scand.* **67**, 343-351

SCHUBERT, J. & SEDVALL, G. (1972) Effect of amphetamines on tryptophan concentrations in mice and rats. *J. Pharm. Pharmacol.* **24**, 53–62

SHERLOCK, S. (1968) *Diseases of the Liver and Biliary System*, 4th edn., p. 115, Blackwell, Oxford

SHIELDS, P. J. & ECCLESTON, D. (1972) Effects of electrical stimulation of rat midbrain on 5-hydroxytryptamine synthesis as determined by a sensitive radioisotope method. *J. Neurochem.* **19**, 265–272

STERN, D. N. & MAICKEL, R. P. (1963) Studies on starvation-induced hypermobilization of free fatty acids (FFA). *Life Sci.* 872–877

TAGGART, P. & CARRUTHERS, M. (1971) Endogenous hyperlipidaemia induced by emotional stress of racing driving. *Lancet* **1**, 363–366

TAGLIAMONTE, A., TAGLIAMONTE, P., PEREZ-CRUET, J., STERN, S. & GESSA, G. L. (1971) Effect of psychotropic drugs on tryptophan concentration in the rat brain. *J. Pharmacol.* **177**, 475–480

TAGLIAMONTE, A., TAGLIAMONTE, P., DI CHIARI, G., GESSA, R. & GESSA, G. L. (1972) Increase of brain tryptophan by electroconvulsive shock in rats. *J. Neurochem.* **19**, 1509–1512

TAGLIAMONTE, A., BIGGIO, G., VARGIU, L. & GESSA, G. L. (1973) Free tryptophan in serum controls brain tryptophan level and serotonin synthesis. *Life Sci.* **12**, 277–287

VAN PRAAG, H. M. & LEIJNSE, B. (1966) Some aspects of the metabolism of glucose and of the non-esterified fatty acids in depressive patients. *Psychopharmacologia* **9**, 220–233

WEISS, B. F., MUNRO, H. N. & WURTMAN, R. J. (1971) L-Dopa: disaggregation of brain polysomes and elevation of brain tryptophan. *Science (Wash. D.C.)* **173**, 833–835

WUNNER, W. H., BELL, J. & MUNRO, H. N. (1966) The effect of feeding with a tryptophan-free amino acid mixture on rat-liver polysomes and ribosomal ribonucleic acid. *Biochem. J.* **101**, 417–428

Discussion, see overleaf

Discussion of the three preceding papers

COMPARISON OF THE EFFECTS OF CLOFIBRATE AND PROBENECID ON TRYPTOPHAN IN THE RAT

Eccleston: Current interest in drugs that influence the concentrations of tryptophan in plasma has led us to re-examine (D. Eccleston & G. W. Ashcroft, unpublished results) some results we obtained in 1964 with ethyl 2-(*p*-chlorophenoxy)-2-methylpropionate (clofibrate), a drug which is used to lower blood lipid concentrations in the prophylaxis of coronary artery disease. Rifkind (1966) showed that it lowered the amount of free fatty acid in man after Thorp (1964) found it displaced protein-bound molecules, including free fatty acids, by competition for protein-binding sites. In these experiments, the drug was given by intramuscular injection (50 mg/kg) to rats and the amounts of plasma tryptophan, brain tryptophan, 5-hydroxytryptamine and 5-hydroxyindoleacetic acid were estimated at various times afterwards. In the chronic experiments, the drug was given twice daily in the same dose and the animals killed one hour after the last dose. The experiments showed clearly a profound and rapid fall in total plasma tryptophan but that this was not reflected by a change in brain and liver concentrations of the amino acid, which were thought to be in equilibrium

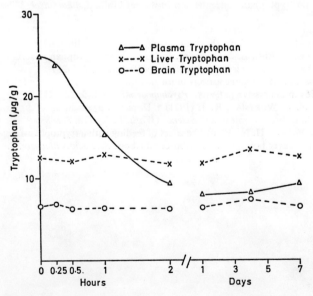

FIG. 1 (Eccleston). Concentration of plasma tryptophan in plasma liver and brain at various times after either a single dose of clofibrate (50 mg/kg) or two daily doses (50 mg/kg); the animals were killed one hour after last dose.

with the free tryptophan fraction (Fig. 1). As expected, the rate of synthesis of 5-hydroxytryptamine in brain was unchanged since the amount of 5-hydroxy-indoles was unaltered. This fall in total plasma tryptophan was encountered in the three species examined: humans, dog and rat. One trial in which we used the tryptophan content of packed cells to monitor the 'free' tryptophan in humans suggested a slight fall in the amino acid on chronic treatment with the drug.

Recently, we re-examined both the 'free' concentrations directly after admini-stration of clofibrate in the rat and the dynamic aspects using L-[³H]tryptophan. We gave the drug on two days in a single dose of 50 mg/kg and injected the animals with L-[³H]tryptophan (50 μCi; specific activity 1 Ci/mM) one hour after the second dose. We guillotined the animals 30 min later and immediately placed the heads in liquid nitrogen. We then separated the plasma at once by centrifugation at room temperature and centrifuged a portion of plasma for one minute through Amicon membranes (Centriflo 224-CF-50) to obtain the protein-free fraction. The whole separation took about 15 min from the time of killing the animal. We estimated the amount of tryptophan by the method of Denckla & Dewey (1967). Previously, we had noted that the total number of counts in plasma correlated well with the activity of [³H]tryptophan (Shields & Eccleston 1972) and so we related the specific activity to the total counts. We estimated the specific activity of tryptophan, 5-hydroxytryptamine and 5-hydroxyindoleacetic acid in brain by the method we also described in that paper.

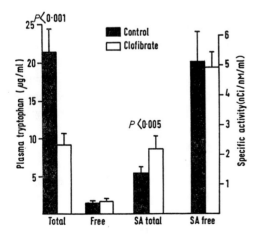

FIG. 2 (Eccleston). Concentration and specific activity of free and bound plasma tryptophan after a dose of clofibrate (50 mg/kg) followed after one hour by L-[³H]tryptophan (50 μCi); the animals were killed after 30 min.

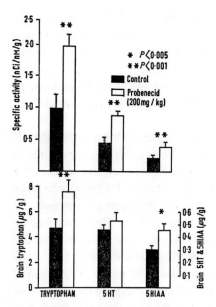

FIG. 3 (Eccleston). Concentrations and specific activity of brain tryptophan, 5-hydroxytrypt-amine (5HT) and 5-hydroxyindoleacetic acid (5HIAA) after giving probenecid (200 mg/kg) two hours before L-[³H]tryptophan (100 μCi); the animals were killed 30 min later.

The results show that there is a fall in the total amount of tryptophan in plasma but no change in that of free tryptophan and that the specific activity of the total rises significantly but that of the free is unchanged (Fig. 2). As expected, the concentration of tryptophan in the brain and the specific activity were unchanged; consequently, the specific activity of 5-hydroxytryptamine and 5-hydroxyindoleacetic acid in brain were unaltered.

These results contrast strongly with those after giving p-dipropylsulphamoyl-benzoic acid (probenecid) (200 mg/kg) to animals two hours before injecting L-[³H]tryptophan (100 μCi), and killing them 30 min later. As shown by Tagliamonte $et\ al.$ (1971), the total plasma tryptophan and brain tryptophan concentrations rise (Fig. 3); the concomitant rises of 5-hydroxytryptamine and 5-hydroxyindoleacetic acid are significant. Note the parallel increase in the specific activity of the tryptophan both in plasma and brain and a rise in the specific activity of the 5-hydroxyindoles. These results imply that probenecid does not merely block the binding sites for tryptophan in plasma as does clofibrate but that it must have some other action such as reducing the metabolism of tryptophan by the inhibition of the oxygenase (EC 1.13.1.12).

$Sharman:$ How much of the increase in 5-hydroxyindoleacetic acid after giving probenecid to a rat can be ascribed to the increase in tryptophan and how much to blockade of the efflux of this acid from the brain? After giving probenecid to different strains of rabbit, the concentration of the indole acid is frequently unchanged; whenever a change is observed, it is usually small and much delayed (Andersson & Roos 1972).

Gessa: To calculate how much of the increase is due to inhibition of transport and how much to its increased synthesis we give a monoamine oxidase inhibitor to the animals. When pargyline is given to rats 4–6 h after probenecid, the amount of the indole acid declines, and from the rate of decline we can calculate how much is due to an increase in synthesis.

Sharman: The metabolism of 5-hydroxytryptamine is said to change when mice are treated with oestrogen and progesterone (Greengrass & Tonge 1972). How do the plasma concentrations of the free fatty acids and tryptophan behave when these steroids are administered to animals?

Curzon: Values for plasma fatty acids in women on oral contraceptives are equivocal (Wynn & Doar 1966, 1969) and those for free tryptophan have yet to be determined. However, oral contraceptives increase the activity of tryptophan oxygenase presumably by elevating plasma concentrations of cortisol (Bulbrook *et al.* 1973). The metabolism of a tryptophan load is altered and a pyridoxine deficiency can develop which leads to a build-up of kynurenine (Adams *et al.* 1973). Possibly, depletion of tryptophan on account of increased oxygenase activity, and also due to an increase in the amount of kynurenine which competes with tryptophan for transport to the brain (Green & Curzon 1970), leads to a cerebral deficiency of tryptophan. This might explain why the depression caused by oral contraceptives is alleviated by pyridoxine (Adams *et al.* 1973).

Wurtman: It seems highly unlikely that changes in tryptophan oxygenase activity other than those produced by long-acting drugs such as α-methyltryptophan or by multiple treatments with large doses of glucocorticoids do modify plasma or brain concentrations of tryptophan. Certainly, there is no parallel between daily rhythmic changes in the enzyme and those in tryptophan content.

Oja: Both Professor Gessa and Dr Moir showed that some drugs increased tryptophan concentrations in plasma and in brain. These drugs have a common property—they increase the body temperature. I suggest that the increase in the concentration of tryptophan is an expression of the increased catabolism of proteins in the tissues. This view is supported by the results of electroconvulsive treatments which increase the brain tryptophan concentration and at the same time increase the catabolism of cerebral proteins. Furthermore, puromycin, applied intrathecally, increases the amount of tryptophan in the brain by decreasing the rate of synthesis of brain proteins; less tryptophan is then used for protein synthesis. Could the results with drugs be reproduced by keeping the body temperature unaltered by external cooling?

Gessa: An increase in body temperature does increase the amount of brain tryptophan, but then so do treatments which do not change body temperature (or even lower it), for instance the injection of reserpine (cf. Tables 1–3, pp. 209, 210).

Munro: I must emphasize that the methods used to determine concentrations of free and bound tryptophan differ from group to group. We have published data on free and bound tryptophan in the serum (Lipsett *et al.* 1973). We carefully went through McMenamy's paper (McMenamy *et al.* 1961) and concluded that the meticulous conditions he used provided the best conditions representative of *in vivo* serum.

They dialysed the plasma at 37 °C against a buffer which contained many of the small molecular constituents of serum. This seems important, because under these conditions, the figures for the amount of free tryptophan in plasma are about 25%, instead of the 10–20% (cf. Moir p. 197) that others have been finding. We did think that the ultrafiltration method was probably subject to more variability because of packing of the albumin at the point at which the fluid passed through the Diaflo membrane. One problem is to determine the equilibrium *in vivo*. P. Farrell (unpublished data, 1973) has injected [^{14}C]-tryptophan into patients with a dialysis loop attached arteriovenously to their forearms and finds 25% free tryptophan in plasma.

Moir: The method I used was principally that of filtration through the simple dialysis membrane, akin to the method that McMenamy eventually developed in his initial dialysis experiments (McMenamy *et al.* 1961). However, in control experiments on dogs, we used a fluid for dialysis that had the ionic composition of plasma and also contained tyrosine (Geddes *et al.* 1973). We tried various concentrations of tyrosine in the presence of tryptophan. We found the amount of free tryptophan in canine plasma is much the same by either method. Farrell's results raise similar questions to those that Dr Eccleston's data do, for instance: does the labelled material mix with the total tryptophan or is it being concentrated to some extent within the free tryptophan? I suspect the latter is true.

Munro: Did you have any difficulty in measuring the concentration of tryptophan in erythrocytes? We find that it is virtually impossible to recover tryptophan added to erythrocytes that have been lysed because of a substance which suppresses the fluorescence of the fluor which is used in the method of Denckla & Dewey (1967).

Moir: After precipitating the proteins with trichloroacetic acid, we use Hess & Udenfriend's method (1959) rather than that of Denckla & Dewey to estimate the tryptophan concentration. We have had no difficulties with this method.

Curzon: [*Note added in proof*]: A relevant point in the determination of free tryptophan is that since plasma lipase can hydrolyse glycerides *in vitro* (Grossman *et al.* 1954) the amount of free tryptophan can also increase *in vitro*. These changes, of course, cannot influence brain tryptophan. Their magnitude rises with the time required for determinations, with temperature and with substrate availability in plasma.

Munro: We all habitually talk of brain tryptophan and brain 5-hydroxytryptamine, but we should remember that the brain cells in which the tryptophan is distributed mostly do not synthesize its metabolite. We are assuming that changes in total tryptophan are necessarily those at the site of synthesis of the neurohormone.

Wurtman: A corollary concerns the locus of the accumulation of cerebral 5-hydroxytryptamine in animals which have received 5-hydroxytryptophan. After this treatment, virtually the whole brain emits the yellow fluorescence characteristic of 5-hydroxytryptamine. The capacity of cells to take up 5-hydroxytryptophan and decarboxylate it is widespread and by no means limited to those few neurons that normally contain the monoamine. Can we then deduce anything about the rate-limiting step in the synthesis of 5-hydroxytryptamine by comparing the increases in the brain amine after treatment with tryptophan or 5-hydroxytryptophan? After all, many fewer cells in the brain will hydroxylate and decarboxylate tryptophan than will just decarboxylate 5-hydroxytryptophan. So, of course, the initial rates of increase in brain 5-hydroxytryptamine concentration will differ in animals receiving one amino acid or the other.

Moir: That is possible but only relevant to the rat data. Different species provide different evidence. It is qualitative evidence from both rat and dog that consistently suggests that the hydroxylation is the rate-limiting step.

Wurtman: Most people agree that hydroxylation is the rate-limiting step in 5-hydroxytryptaminergic neurons. The question is whether this assumption is really proved by the different rates at which brain 5-hydroxytryptamine concentrations increase after treatment with tryptophan or 5-hydroxytryptophan.

Gessa: Without exception we found that whenever the brain concentration of tryptophan increases, the turnover of 5-hydroxytryptamine also increases. Therefore, I suggest that brain tryptophan concentration reflects the synthesis of the amine. Possibly, there are situations where the synthesis of the amine increases while the brain concentration of tryptophan does not.

Fernstrom: The best evidence that the increase in tryptophan in the whole brain also occurs specifically in 5-hydroxytryptamine neurons comes from the work of Aghajanian & Asher (1971). They reported that single cells in the raphe nucleus of the brain stem fluoresce more brightly when animals are injected with a large dose of tryptophan.

Dr Moir, you used a brain homogenate for your determination of the concentrations of free and bound tryptophan in the rat brain. How do you exclude the possibility of a non-specific association of tryptophan with proteins or particles resulting from homogenization?

Moir: I am not sure what 'free' tryptophan represents in the brain. Before I undertook this series of experiments I felt that at that time we only knew as much about the total amounts of tryptophan in the brain after homogenizing the tissue as we knew a few years ago about the total amount of tryptophan in whole blood and more recently in whole plasma. Admittedly, this treatment disrupts many membranes but it should not liberate the membrane-bound or particulate-bound tryptophan. I exclude these by my use of 'free'. Perhaps most of this free tryptophan is concentrated inside synaptosomes or in parts of cells; I do not know. Certainly, it does not seem to be bound in the same way that some tryptophan is bound in plasma to albumin.

Munro: Badawy & Smith (1972) also identified some binding of tryptophan to protein in the liver. I wonder what correction is necessary for plasma trapped in the organ. Albumin is being produced in the liver, so possibly the cell sap or the total homogenate contains significant amounts of albumin.

Moir: Amphetamine causes a substantial change in the concentration of free tryptophan but leaves the total amount of plasma tryptophan unaltered. Initially, I postulated that this also happened in brain but that the changes were obscured by estimating total tryptophan in brain homogenate after an acid precipitation.

Gál: I have been considerably refreshed by the remarks of Drs Curzon, Eccleston, Grahame-Smith and Kaufman. We should not be too preoccupied with K_m of L-tryptophan obtained with 6,7-dimethyltetrahydropterin as cofactor and the presumptive unsaturation of the enzyme. Dr Kaufman's classical studies on phenylalanine 4-hydroxylase with 6,7-dimethyltetrahydropterin served the purpose of some kinetic measurements *in vitro*, yet we all know that biopterin and possibly other factors in the brain bring the value of K_m of L-tryptophan to about 50 μM; intracellularly it may be even 20 μM.

In the rat brain, at least, the concentrations of 5-hydroxytryptamine and tryptophan are 2–3 and 30–40 μM, respectively. If we assume that the value of K_m is about 20 μM *in vivo*, we might as well forget about the concept of unsaturation of tryptophan 5-hydroxylase.

I am glad to hear that Professor Gessa has retracted the statement that concentration of tryptophan rather than tryptophan hydroxylation was rate-limiting (Tagliamonte *et al.* 1973).

Gessa: I still believe that tryptophan hydroxylase is not the rate-limiting step in the synthesis of 5-hydroxytryptamine but the transport of tryptophan into the monoamine neuron. The fact that any treatment capable of increasing the concentration of brain tryptophan also increases the rate of formation of brain 5-hydroxytryptamine clearly indicates that the tryptophan hydroxylase is not saturated *in vivo*.

Gál: Of course, if you supply more substrate you will find, within limits, more neurotransmitter is synthesized.

Wurtman: Does that generalization always hold? Is it true that adding more tyrosine will increase the synthesis of brain catecholamines?

Gál: There is sufficient free tyrosine in the brain; consequently, a 10–20% increase of it might not be significant.

Grahame-Smith: The fact that you can raise the concentration of 5-hydroxytryptamine in the brain by giving tryptophan might be therapeutically significant in regard to the manipulation of brain 5-hydroxytryptamine concentrations in the treatment of affective disorders.

Fernstrom: Dr Gál, how do you know that a 20% change of brain 5-hydroxytryptamine is not important?

Gál: What kind of behavioural concomitants can you relate to a 10–20% increase? Only when the concentration of the amine reaches about 10 µM in the presence of a monoamine oxidase inhibitor will one find behavioural changes.

Fernstrom: No experiment has yet been done to change monoamine concentrations within the physiological range to see if there is any concomitant change in behaviour, so you cannot assert that it will not change behaviour.

Wurtman: What's more, J. Harvey and his associates (personal communication) demonstrated a diurnal rhythm in pain sensitivity in animals that is inversely related to the rhythm in telencephalic 5-hydroxytryptamine content.

Gál: I should be the last to deny that it is related to pain, sleep and sex, but we know from our behavioural studies (unpublished findings, 1970) that a 10, 20 or even 25% increment in the amine concentration did not make much difference. When monoamine levels reach 30 µM *in vivo* in presence of a monoamine oxidase inhibitor, the polysomes disaggregate.

Carlsson: On the whole, I agree with Dr Gál. Despite the tremendous biochemical work now being done, isn't it time to start to do some simple experiments on the behavioural pharmacology of tryptophan? To my knowledge, nobody has recorded any change in behaviour within the dose range that increases the synthesis of the neurotransmitter—up to 300 mg/kg (intraperitoneally) in the rat. If the dose is raised from 300 to 1000 mg/kg, behavioural depression is observed, but that is at a time when tryptophan hydroxylase is already saturated. That means that the gross behavioural change that you can easily see after administration of tryptophan is probably not related to synthesis of 5-hydroxytryptamine at all.

Curzon: We have been infusing tryptophan into normal human volunteers, so that plasma free tryptophan rises up to 100 times normal without any detectable behavioural change (Curzon, J. Friedel, Greenwood & Lader, unpublished results). But what happens when the synthesis of 5-hydroxytryptamine is already deficient for some pathological reason and the normal behavioural requirement of the amine is not being synthesized? Maybe then small changes in the availability of tryptophan significantly affect behaviour.

Gál: I concur with Dr Curzon. Certain behavioural functions or symptoms in tryptophan-deficient animals can be eliminated by administration of 5-hydroxytryptophan (unpublished findings, 1970).

Curzon: In a long-term tryptophan-deficiency experiment, not only is 5-hydroxytryptamine synthesis diminished but other aspects of metabolism may be radically altered, for example protein synthesis. Not surprisingly, treatment of such animals with 5-hydroxytryptophan did not normalize their behaviour (Boullin 1963).

Gessa: May I explain a possible artifact in measuring turnover without considering the existence of free and bound tryptophan in plasma? If one measures the rate of synthesis of 5-hydroxytryptamine by intravenously injecting labelled tryptophan and if the injected amino acid does not readily mix with total plasma tryptophan, it will enter the brain with higher specific activity than that calculated from total plasma tryptophan.

Wurtman: In the process of trying to determine whether substrate-induced changes in this amine are important, we are fortunate in that, instead of having to administer an amino acid or other 'dirty' (i.e. non-specific) drug (see p. 128) to an animal, we can simply give groups of animals single meals consisting of one of two diets that simulate normal human diets: one, which lacks protein, will cause a major increase in the amount of amine in the brain and the other, which is rich in protein, will not. Then we may ask whether these naturally induced changes in brain 5-hydroxytryptamine within neurons that normally contain this amine are associated with behaviour that we can examine. We cannot do this experiment with brain catecholamines; at least, I am not aware of any simple way of modifying the brain concentrations of catecholamines without disrupting the feedback systems that normally are responsible for maintaining these concentrations.

Concerning the possibility of stress artifacts, we have considerable evidence that the acute consumption of our experimental diets is not stressful; for example, daily plasma corticosterone rhythms persist.

In standard nutrition experiments with diets containing 0–40% protein, a correlation of brain tryptophan or 5-hydroxyindoles with plasma total tryptophan concentration is successful some of the time, but attempts to correlate brain tryptophan with plasma free tryptophan almost always fail. However, if one plots for each animal the brain concentration of tryptophan against the ratio in plasma of the concentration of tryptophan to that of its major competitors for brain uptake (i.e. tyrosine, phenylalanine, isoleucine, leucine and valine), one obtains very strong evidence of correlation ($r = 0.9$) (Fernstrom & Wurtman 1972).

A second experimental situation is starvation; animals are fasted for 22 h. In this condition, the increase in brain tryptophan correlates well with the rise in plasma free tryptophan; plasma total tryptophan does not rise (cf. Curzon and Gessa, this volume).

A third situation to be explained is the increase in brain tryptophan that follows the administration of various drugs like salicylate that clearly have dramatic effects on the capacity of tryptophan to be bound to albumin (cf. Gessa, this volume).

Before we can hope to apply Ockham's razor and devise the simplest model for the control of brain tryptophan concentration, we simply need more data. As Dr Munro mentioned (p. 20), the metabolism of the branched chain amino acids that compete with tryptophan for brain uptake differs considerably from that of the aromatic amino acids. The latter are metabolized principally within the liver; the former in skeletal muscle. The effect of the 22 h fast on the metabolism (and plasma levels) of branched chain amino acids might differ markedly from its effect on plasma tryptophan (or tyrosine or phenylalanine), and might be responsible for the rise in brain tryptophan. If we are to resolve the different models discussed here, it is imperative that we have more data. I propose four possible correlations be tested: brain tryptophan and 5-hydroxyindole concentrations against (i) plasma total tryptophan; (ii) plasma free tryptophan; (iii) the ratio of plasma total tryptophan to the competing amino acids; and (iv) the ratio of free tryptophan to competing amino acids. Meanwhile, all of us who work on 5-hydroxytryptamine in brain must adopt the convention of providing specific details about the diets that our animal or human subjects consume chronically and, more important, in the hours preceding our experiments.

Sourkes: In regard to the tryptophan content of the diet, we should realize that commercial animal foods may have the same protein content from lot to lot but nevertheless change in composition depending on current market prices for the constituent grains, animal-protein supplements and other ingredients. I can foresee a time when journal editors will reject papers which do not specify, say, the tryptophan content of the particular diet used.

Moir: In relation to the use of the specific activity of total tryptophan in plasma, is the method of Lin *et al.* (1969) to calculate turnover rates valid? I know it gives the correct answer, but is this fortuitous? There are two separate relationships: that of free tryptophan in plasma to tryptophan in the brain and that of the tryptophan in the brain to that which is used to synthesize biogenic amines. For instance, one apparently does not use the specific activity of tryptophan in the brain to indicate the specific activity of the precursors for 5-hydroxytryptamine but the specific activity of total tryptophan in plasma. Logically, this does not seem correct.

Gál: No. We cannot estimate the turnover of 5-hydroxytryptamine or the formation of 5-hydroxyindoleacetic acid precisely unless we simultaneously determine the specific activity of tryptophan in the brain.

Wurtman: Most people inject labelled tryptophan and then kill the animals shortly thereafter when the concentration of radioactive tryptophan in the brain is rising (i.e. Hyyppa *et al.* 1973). From that they attempt to estimate the initial rate of synthesis of the amine.

Clearly, what we really need to know is not how much 5-hydroxytryptamine is present in the brain, but how much of the amine is being released into the synapses and whether changes in this rate affect synaptic function.

References

ADAMS, P. W., WYNN, V., ROSE, D.P., SEED, M., FOLKARD, J. & STRONG, R. (1973) Effect of pyridoxine hydrochloride (vitamin B) upon depression associated with oral contraception. *Lancet* 1, 897–904

AGHAJANIAN, G. K. & ASHER, I. M. (1971) Histochemical fluorescence of raphe neurons: selective enhancement by tryptophan. *Science (Wash. D.C.)* 172, 1159–1161

ANDERSSON, H. & ROOS, B. E. (1972) 5-Hydroxyindoleacetic acid and homovanillic acid in cerebrospinal fluid and brain of different rabbit breeds after treatment with probenecid. *J. Pharm. Pharmacol.* 24, 165–166

BADAWY, A. A. & SMITH, M. J. (1972) Changes in liver tryptophan and tryptophan pyrrolase activity after administration of salicylate and tryptophan to the rat. *Biochem. Pharmacol.* 21, 97–101

BOULLIN, D. J. (1963) Behaviour of rats depleted of 5-hydroxytryptamine by feeding a diet free of tryptophan. *Psychopharmacologia* 5, 28–38

BULBROOK, R. D., HAYWARD, J. L., HERIAN, M., SWAIN, M. C., TONG, D. & WANG, D. Y. (1973) Effect of steroidal contraceptives on levels of plasma androgen sulphate and cortisol. *Lancet* 1, 628–631

DENCKLA, W. D. & DEWEY, H. K. (1967) The determination of tryptophan in plasma, liver and urine. *J. Lab. Clin. Med.* 69, 160–169

FERNSTROM, J. D. & WURTMAN, R. J. (1972) Brain serotonin content: physiological regulation by plasma neutral amino acids. *Science (Wash. D.C.)* 178, 414–416

GEDDES, A. C., MARTIN, M. J. & MOIR, A.T.B. (1973) The transport of L-tryptophan from cerebrospinal fluid in the dog. *J. Physiol. (Lond.)* 230, 595–612

GREEN, A. R. & CURZON, G. (1970) The effect of tryptophan metabolites on brain 5-hydroxytryptamine metabolism. *Biochem. Pharmacol.* 19, 2061–2068

GREENGRASS, P. M. & TONGE, S. R. (1972) Effects of oestrogen and progesterone on brain monoamines: interactions with psychotropic drugs. *J. Pharm. Pharmacol.* 24 (Suppl.), 149P

GROSSMAN, M. I., PALM, M., BECKER, G. H. & MOELLER, H. C. (1954) Effect of lipemia and heparin on free fatty acid content of rat plasma. *Proc. Soc. Exp. Biol. Med.* **87**, 312–315

HESS, S. M. & UDENFRIEND, S. (1959) A fluorimetric procedure for the measurement of tryptamine in tissues. *J. Pharmacol. Exp. Ther.* **127**, 175–177

HYYPPA, M. T., CARDINALI, D. P., BAUMGARTEN, H. G. & WURTMAN, R. J. (1973) Rapid accumulation of H3-serotonin in brains of rats receiving intraperitoneal H3-tryptophan: effects of 5,6-dihydroxytryptamine or female sex hormones. *J. Neural Transm.*, **34**, 111

LIN, R. C., COSTA, E., NEFF, N. H., WANG, C. T. & NGAI, S. H. (1969) *In vivo* measurement of 5-hydroxytryptamine turnover rate in the rat brain from the conversion of C^{14}-tryptophan to C^{14}-5-hydroxytryptamine. *J. Pharmacol. Exp. Ther.* **170**, 232–238

LIPSETT, D., MADRAS, B. K., WURTMAN, R. J. & MUNRO, H. N. (1972) Serum tryptophan level after carbohydrate ingestion: selective decline in non-albumin-bound tryptophan coincident with reduction in serum free fatty acids. *Life Sci.* **12**, 57–64

MCMENAMY, R. H., LUND, C. C., VAN MARCKE, J. & ONCLEY, J. L. (1961) The binding of L-tryptophan in human plasma at 37 °C. *Arch. Biochem. Biophys.* **93**, 135–139

RIFKIND, B. M. (1966) Effect of CPIB ester on plasma free fatty acid levels in man. *Metabolism* **15**, 673–675

SHIELDS, P. J. & ECCLESTON, D. (1972) Effects of electrical stimulation of rat midbrain on 5-hydroxytryptamine synthesis as determined by a sensitive radioisotope method. *J. Neurochem.* **19**, 265–272

SHIELDS, P. J. & ECCLESTON, D. (1973) Evidence for the synthesis and storage of 5-hydroxytryptamine in two separate pools in the brain. *J. Neurochem.* **20**, 881–888

TAGLIAMONTE, A., TAGLIAMONTE, P., PEREZ-CRUET, J., STERN, S. & GESSA, G. L. (1971) Effect of psychotropic drugs on tryptophan concentration in rat brain. *J. Pharmacol. Exp. Ther.* **177**, 475–480

TAGLIAMONTE, A., BIGGIO, G., VARGIU, L. & GESSA, G. L. (1973) Increase of brain tryptophan and stimulation of serotonin synthesis by salicylate. *J. Neurochem.* **20**, 909–912

THORP, J. M. (1964) in *Absorption and Distribution of Drugs* (Binns, T. B., ed.), p. 64, Livingstone, Edinburgh & London

WYNN, V. & DOAR, J. W. H. (1966) Some effects of oral contraceptives on carbohydrate metabolism. *Lancet* **2**, 715–719

WYNN, V. & DOAR, J. W. H. (1969) Some effects of oral contraceptives on carbohydrate metabolism. *Lancet* **2**, 761–765

Monoamine metabolites in the cerebrospinal fluid: indicators of the biochemical status of monoaminergic neurons in the central nervous system

MARIN BULAT*

'*Ruder Bošković*' *Institute, Zagreb, Yugoslavia*

Abstract The origin of substances in the cerebrospinal fluid of animals and man is poorly understood today. In animals, 5-hydroxyindoleacetic acid and 4-hydroxy-3-methoxyphenylacetic acid (homovanillic acid) in cerebral fluid derive from the metabolism of 5-hydroxytryptamine and dopamine, respectively, in the brain. Measurements of the amounts of both acids in patients' spinal fluid by lumbar puncture are often made on the assumption that they reflect the metabolism of the two monoamines in the brain. This need not be so.

  We have investigated three potential sources of 5-hydroxyindoleacetic acid in feline spinal fluid: brain, blood and spinal cord. When the amount of the acid in the fluid of cisterna magna was increased, that in the spinal fluid remained unchanged. Increasing the concentration of the acid in the blood does not result in more acid in the spinal fluid. Thus, it appears that the 5-hydroxyindoleacetic acid in the spinal fluid originates neither in the brain nor in the blood. Changes in the concentration of the acid in the spinal cord are, however, followed by corresponding changes in the spinal fluid and in the perfusate of the spinal subarachnoid space. Concentrations of the acid in the spinal fluid and perfusate are linearly related to those in the spinal cord. This indicates that exchange of 5-hydroxyindoleacetic acid between spinal cord and spinal fluid is a process of passive diffusion. Efflux of the acid from the spinal cord and spinal fluid can be restricted by inhibitors of active transport; this implies an active transport process. We suggest, therefore, that 5-hydroxyindoleacetic acid in the spinal fluid originates in the spinal cord and reflects the metabolism of 5-hydroxytryptamine in the spinal tissue. We are also trying to determine whether homovanillic acid in spinal fluid is derived from brain, blood or spinal cord. Whether a given substance in the spinal fluid reflects biochemical changes in the brain or those in the spinal cord depends primarily on the concentration of this substance in the cisternal fluid, its source in the spinal cord and its active transport out of the spinal fluid.

The investigation of biochemical changes in the brain of man *in vivo* is a hard task on account of the inaccessibility of human nervous tissue. Tissue taken

---

* *Present address:* Department of Pharmacology, Chicago Medical School, Chicago, Illinois.

from the central nervous system *post mortem* is drastically changed by autolytic processes so that results obtained from it should be interpreted with great caution. Analysis of some body fluids (e.g. blood and urine) is of little value since these fluids contain the substances derived predominantly from peripheral organs and not from the central nervous system. The cerebrospinal fluid is in intimate contact with the central nervous system and might reflect the biochemical processes in that system. Recent results support such a role of cerebrospinal fluid although some uncertainties and confusion still remain.

Experiments in animals have shown that 5-hydroxyindoleacetic acid (5HIAA) and homovanillic acid in ventricular and cisternal fluid are metabolites of 5-hydroxytryptamine and dopamine, respectively, in the brain (Eccleston *et al.* 1968; Guldberg & Yates 1968; Bowers 1970). These findings have opened a new and promising avenue in clinical research, namely that, in a search for the biochemical basis of some mental and neurological diseases, these monoamine metabolites have frequently been measured in patients' spinal fluid by lumbar puncture as indicators of 5-hydroxytryptamine and dopamine metabolism in the brain (cf. Gottfries *et al.* 1969; Moir *et al.* 1970; Van Praag *et al.* 1970; Bowers 1972). Do the concentrations of these two acids in the spinal fluid reflect the metabolism of the parent amines in the brain? We have examined the origin of 5HIAA in the spinal fluid of the lumbar region (lumbar fluid) (Bulat & Živković 1971). Furthermore, we were interested to discover whether and how the changes of 5-hydroxytryptamine metabolism in the spinal cord are reflected in the spinal fluid.

5-Hydroxyindoleacetic acid in the spinal fluid has three potential sources: the brain, the spinal cord and blood. We investigated these three possibilities in cats.

If the acid in the spinal fluid originates from the brain tissue, then a rise in concentration of 5HIAA in the fluid of cisterna magna (cisternal fluid) should result in a corresponding increase in the spinal fluid (Bulat & Živković 1971). Accordingly, we injected 0.5 µg of 5HIAA into cisterna magna and followed the concentration of 5HIAA in the cisternal and spinal fluid at different time intervals (see Fig. 1). After the injection, the concentration of the acid in the cisternal fluid increases strikingly; the effect lasts for about two hours. In contrast, the amount of the acid in the spinal fluid is unchanged even after seven hours. Clearly, 5HIAA does not reach the spinal fluid from the cisterna magna and we conclude that changes in 5-hydroxytryptamine metabolism in the brain are not reflected in the spinal fluid.

5-Hydroxytryptamine in the spinal cord could be the source of 5HIAA in the spinal fluid. If so, the acid should appear in perfusate of the spinal subarachnoid space (Živković & Bulat 1971b). We perfused the spinal subarachnoid space from thoracic to sacral region with artificial cerebrospinal fluid and

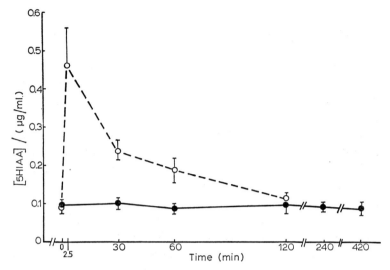

FIG. 1. The concentration of 5HIAA in the cisternal fluid (○) and spinal fluid (●) in μg/ml plotted against time. 5HIAA (0.5 μg) was injected intracisternally during the first 30 s of the experiment and samples of cisternal and spinal fluid were taken at various time intervals after the injection (from Bulat & Živković 1971).

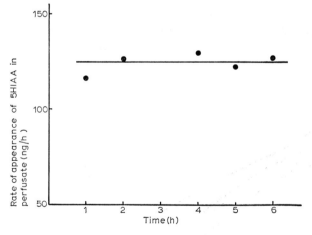

FIG. 2. The rate of appearance of 5HIAA in samples of perfusate (in ng/h) of the spinal subarachnoid space.

measured the amount of 5HIAA in samples of perfusate collected at one hour intervals (Fig. 2). The rate of appearance of the acid in the perfusate remains relatively constant over six hours of perfusion. This indicates that 5HIAA in the perfusate and probably in the spinal fluid originates in the spinal cord.

2-(5-Hydroxyindole)ethanol, a metabolite of 5-hydroxytryptamine formed in the spinal cord, also enters the perfusate of spinal subarachnoid space (Bulat *et al.* 1970).

How are the changes in 5HIAA concentration in the spinal cord reflected by the concentration of 5HIAA in the perfusate? To answer this question, we followed the concentration of 5HIAA in the spinal cord and in the perfusate after application of reserpine (Fig. 3). We found that the concentration of the acid in the spinal cord rises after reserpine treatment for three hours before decreasing towards control levels after nine hours. A corresponding increase in the amount of the acid in the perfusate follows the increase in the spinal cord but one hour later. If the graph of the concentration of 5HIAA in the perfusate is shifted to the left by a time interval of one hour, the changes in the 5HIAA concentration in the spinal cord parallel those in the perfusate (Fig. 3). This time lag is probably due to the fact that the acid is formed from the 5-hydroxytryptamine in the grey matter of spinal cord and needs a certain time to diffuse into perfusate across the white matter located in the periphery and across the pia-glial membrane (Bulat & Živković 1971).

If concentrations of 5HIAA in the spinal cord (see Fig. 3) are plotted against those in perfusate a linear relationship is obtained (Fig. 4; correlation coefficient 0.91; $P < 0.01$). Thus, any variation in the concentration of 5HIAA in the spinal cord after administration of reserpine is linearly related to corresponding changes in the perfusate. We may conclude, therefore, that metabolism of 5-hydroxytryptamine in the spinal cord is accurately reflected in the perfusate and probably in the spinal fluid (see later).

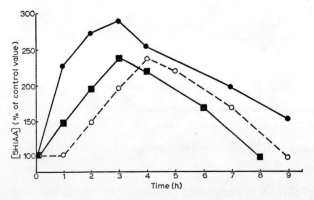

FIG. 3. The variation of the concentration of 5HIAA (expressed as a percentage of control values) in the spinal cord (●) and in the perfusate of the spinal subarachnoid space (○) after injection of reserpine (2 mg/kg i.v.). If the concentration of 5HIAA in the perfusate is plotted against $(t-1)$ h, the values marked ■ are obtained (from Bulat & Živković 1971).

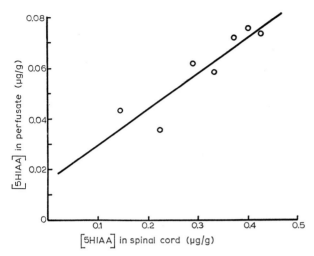

FIG. 4. The relationship between the concentration of 5HIAA in the spinal cord (in μg/g) and that in the perfusate in (μg/g) of superfused spinal cord after injection of reserpine.

The active transport mechanism which removes 5HIAA from the ventricular fluid into blood can be inhibited by probenecid, whereupon the concentration of the acid rises in ventricular fluid and consequently in cisternal fluid (Guldberg *et al.* 1966). This active transport is supposedly located in the choroid plexuses (Ashcroft *et al.* 1968) or possibly in the ependyma of brain ventricles. Since neither the choroid plexus nor ependyma is present in the spinal subarachnoid space, how is 5HIAA removed from spinal fluid? We placed an extradural thread ligature in the lower region of the thoracic spinal cord and measured the concentration of 5HIAA in the spinal cord, spinal and cisternal fluid after application of probenecid: Fig. 5 shows the dramatic increase in the concentration of the acid in all three fluids. This increase of 5HIAA in the cisternal fluid supports previous findings that the acid is removed from the ventricular fluid to blood by active transport. The supplementation of the 5HIAA in the spinal cord and spinal fluid indicates that the acid is also actively transported from the spinal cord and spinal fluid to blood (Živković & Bulat 1971*a*). Since the spinal fluid was separated from the rest of the cerebrospinal fluid system by a ligature, it is clear that the build up of 5HIAA in the spinal fluid is the consequence of an increase in the amount of the acid in the spinal cord and hence that 5HIAA in the spinal fluid originates in the spinal cord (as I suggested before).

In a similar experiment in which we administered dinitrophenol to cats, we found an increase in the amount of 5HIAA in the spinal cord and in the spinal and cisternal fluids of a similar magnitude to that observed after probenecid treatment (see Fig. 5). While probenecid is a competitive inhibitor of the active

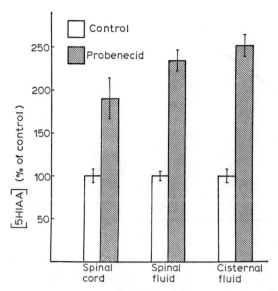

FIG. 5. The concentration of 5HIAA (expressed as a percentage of control) in the spinal cord, spinal fluid and cisternal fluid in saline (control) and probenecid-treated cats. Probenecid was given intraperitoneally three hours (200 mg/kg) and one hour (100 mg/kg) before the samples of spinal and cisternal fluid and spinal cord were taken for analysis. The value in animals treated with saline (control animals) was taken as 100%. The values of [5HIAA] represent means ± s.e.m. of four experiments.

transport of 5HIAA, dinitrophenol is known to block the energy supply for active transport processes dependent on ATP derived from oxidative phosphorylation. Thus, energy for the active transport of 5HIAA from the spinal cord and cerebrospinal fluid to blood seems to be derived from ATP.

After successive applications of probenecid, we measured 5HIAA in the spinal cord and spinal fluid over five hours. During that time, we observed a linear increase in the amount of 5HIAA in the spinal cord followed by a similar linear increase in the spinal fluid. In Fig. 6, the concentration of 5HIAA in the spinal cord is plotted against that in the spinal fluid for the successive time intervals after administration of probenecid: the linear relationship is clear (correlation coefficient 0.88; $P < 0.05$). Thus, the build up of 5HIAA in the spinal cord is followed by a corresponding rise of 5HIAA in the spinal fluid. We conclude that 5HIAA in the spinal fluid accurately reflects the metabolism of 5-hydroxy-tryptamine in the spinal cord, as does the amount of 5HIAA in the perfusate of the spinal subarachnoid space (see Fig. 4). These results show that the quantity of 5HIAA in the cerebrospinal fluid reflects the biochemical status of 5-hydroxytraminergic neurons in the central nervous system. Any change in the rate of turnover of 5-hydroxytryptamine in the central nervous system

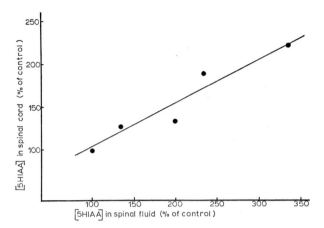

FIG. 6. The relationship between the concentration of 5HIAA (expressed as a percentage of control) in the spinal cord and spinal fluid in saline- (100%) and probenecid-treated cats. Probenecid was given intraperitoneally at the beginning of the experiment (200 mg/kg). Two hours later and subsequently every hour 100 mg/kg of probenecid was injected in the same way. Spinal cord and spinal fluid were taken in different cats 1, 2, 3 and 5 h after the first injection of probenecid. Each point represents mean of about four experiments.

should be mirrored by corresponding changes in the quantity of 5HIAA in the cerebrospinal fluid. This justifies the calculation of the turnover of 5-hydroxy-tryptamine in central nervous system of man and animals *in vivo* on the basis of the amount of 5HIAA in cerebrospinal fluid.

The direct dependence of the concentrations of 5HIAA in perfusate (Fig. 4) and in spinal fluid (Fig. 6) on the concentrations of 5HIAA in the spinal cord means that egress of 5HIAA from the spinal cord into spinal fluid or perfusate is under control of the gradient of 5HIAA concentration in the spinal cord. This suggests that the acid enters the spinal fluid or perfusate by a passive diffusional process. I should mention that the penetration of 5-hydroxytrypt-amine into the central nervous system from both the cerebrospinal fluid (Bulat & Supek 1966; Bulat & Supek 1968a) and blood (Bulat & Supek 1967; Bulat & Supek 1968b) is also a passive diffusional process.

Is 5HIAA in cerebrospinal fluid 'contaminated' by blood 5HIAA? One reason why the acid hardly crosses the blood–cerebrospinal fluid and blood–brain barriers (Ashcroft *et al.* 1968) might be the active transport which pumps the acid out of the cerebrospinal fluid and central nervous system into blood. If so, inhibition of this active transport by probenecid should facilitate the penetration of 5HIAA across these barriers. If the acid is given intravenously during perfusion of the spinal subarachnoid space, we observe only a relatively small increase of 5HIAA in the perfusate during the first 30 min (Fig. 7).

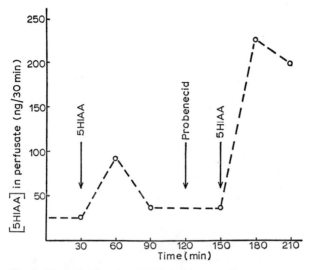

FIG. 7. The concentration of 5HIAA (in ng/30 min) in samples of perfusate of spinal sub-arachnoid space collected during 30 min intervals. Injection of 5HIAA (1 mg/kg i.v.) or probenecid (200 mg/kg i.p.) is indicated by an arrow.

However, if the acid is given after application of probenecid a striking increase in its concentration in perfusate is found, which lasts at least 60 min.

In another experiment, we applied 5HIAA intravenously and measured the concentration of the acid in the spinal fluid, in the cisternal fluid, and in the spinal cord of cats pretreated with saline or probenecid. These pretreated cats, which received an intravenous injection of saline, served as our controls. Fig. 8 shows that, after intravenous application of 5HIAA, there is no increase in the amount of the acid in the spinal fluid, in the cisternal fluid and in the spinal cord of the saline-pretreated cats. However, when animals are pretreated with probenecid, a striking increase is observed in the spinal and cisternal fluids after intravenous injection of 5HIAA. This suggests that 5HIAA penetrates the blood–cerebrospinal fluid barrier after pretreatment with probenecid. At the same time, we found no significant increase in the amount of 5HIAA in the spinal cord (Fig. 8). This does not mean, however, that 5HIAA cannot penetrate the blood–cord and blood–brain barriers, for we observed an increase of 5HIAA in the central nervous system after a high intravenous dose of 5HIAA (20 mg/kg).

Our results (Figs. 7 and 8) indicate that the active transport of 5HIAA from cerebrospinal fluid to blood counteracts the penetration of 5HIAA across the blood–cerebrospinal fluid barrier. Since probenecid increases the concentration of free penicillin in the blood by competitive binding to plasma proteins, it

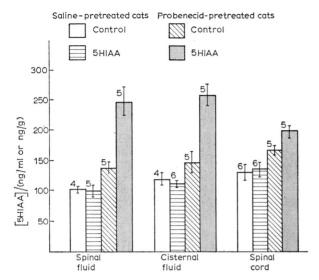

FIG. 8. The concentration of 5HIAA in the spinal and cisternal fluid (in ng/ml), and in the spinal cord (in ng/g) of cats. The animals were pretreated intraperitoneally with saline or probenecid (200 mg/kg). After 30 min, the saline-pretreated cats were intravenously treated with saline or 5HIAA (1 mg/kg). The probenecid-pretreated animals were simultaneously given intravenous injection of saline or of 5HIAA (1 mg/kg). The spinal fluid, cisternal fluid and spinal cord were taken for analysis 30 min after the second injection. Means of [5HIAA] ± S.E.M. with number of experiments are shown (from Bulat & Živković 1973).

might act similarly on 5HIAA given intravenously. In such a case, the increased penetration of 5HIAA across the barrier may be ascribed, at least in part, to the raised concentration of free 5HIAA in the blood (Bulat & Živković 1973). Since the concentration of the acid in the blood is low or unmeasurable and, under control conditions, it does not penetrate the blood–cerebrospinal fluid barrier (Fig. 8), we conclude that 5HIAA in cerebrospinal fluid is not 'contaminated' by blood 5HIAA (Bulat & Živković 1971).

Can the concentration of 5HIAA in the spinal fluid be used as an indicator of 5-hydroxytryptamine metabolism in the brain or in the spinal cord? Our results show clearly that 5HIAA does not reach spinal fluid when applied to cisternal fluid (Fig. 1). Since no anatomical barrier exists between cisternal fluid and spinal fluid, some physiological factors might be responsible for the inability of 5HIAA to reach the lumbar region from the cisterna magna. We know that the circulation or flow of cerebrospinal fluid from the cisterna magna to the lumbar sac along the spinal subarachnoid space is practically non-existent (Grundy 1962; Davson 1967), so we can assume that 5HIAA moves from the cisternal fluid to the lumbar fluid only by a slow diffusional process. However, on its way into the lumbar sac, 5HIAA is actively removed into blood (Fig. 5) so that its concentration should decrease with the distance from the cisterna magna to the lumbar sac. This hypothesis is supported by our experiments (Živković & Bulat, unpublished results) in which the spinal subarachnoid space was perfused from the cisterna magna to the lumbar sac with an artificial cerebrospinal fluid which contained 5HIAA and inulin or dextran blue. Calculation of the respective clearance values showed that 5HIAA is cleared much faster than either inulin or dextran blue from the spinal subarachnoid space. Thus, the active transport located in the spinal subarachnoid space might be a physiological barrier which prevents access of 5HIAA from the cisternal into the spinal fluid. We conclude, therefore, that 5HIAA in the spinal fluid cannot be derived from the metabolism of 5-hydroxytryptamine in the brain. On the other hand, our results (Figs. 5 and 6) clearly show that the concentration of 5HIAA in the spinal fluid reflects metabolic changes of 5-hydroxytryptamine in the spinal cord. Since the spinal cord is in intimate contact with the spinal fluid, it can constantly supply the spinal fluid with the metabolite.

Recently, Weir *et al.* (1973) perfused the subarachnoid space with labelled 5HIAA from an inflow needle placed in the basal cistern to an outflow needle introduced percutaneously in feline lumbar sacs. The specific activity of 5HIAA in the cisterna magna fluid rose to a steady value during the first hour of perfusion, while that in lumbar outflow fluid reached a plateau between the first and second hours. In the course of experiment the specific activity of 5HIAA was 5–6 times higher in the perfusion fluid taken from the cisterna magna than in the lumbar outflow fluid. This indicates only that 5HIAA is effectively removed from the perfusion fluid during its flow along the spinal subarachnoid space. However, incorrectly assuming that normal flow of cerebrospinal fluid along spinal subarachnoid space equals the rate of bulk production of cerebrospinal fluid and using numerous approximations Weir *et al.* (1973) concluded that about 40% of total 5HIAA in the lumbar fluid is derived from the brain tissue.

In almost all studies on patients, it has been taken for granted that the 5HIAA in the spinal fluid is the result of the metabolism of 5-hydroxytryptamine in the brain. However, Eccleston *et al.* (1970) tried to establish the cerebral origin of the 5HIAA in the spinal fluid. They injected tryptophan into the bones of patients and noted an increase in the amount of tryptophan in the spinal fluid which for several hours preceded the increase of 5HIAA in the same fluid. They explained this delay as the time of diffusion of 5HIAA from the brain to the spinal fluid. We believe that this time delay can be simply explained as the time needed for tryptophan to reach a sufficiently high concentration in the spinal cord for the enzymic formation of 2-(5-hydroxyindole)ethanol, 5-

hydroxytryptamine and 5-hydroxyindoleacetic acid, and for the diffusion of the acid from the grey matter of spinal cord to spinal fluid (see before). Thus, there is no need to postulate an extraspinal source of 5HIAA to explain the results of Eccleston et al. (1970). On the other hand, Curzon et al. (1971) described a significant fall of 5HIAA in the spinal fluid in patients with restricted communication between the spinal fluid and other parts of the cerebrospinal fluid system. They suggested that the 5HIAA in the spinal fluid derives from both brain and spinal cord. In contrast, Post et al. (1973) detected no significant changes in the amount of 5HIAA in the spinal fluid of patients with blockage of spinal fluid communication, interpreting this to mean that the 5HIAA in the spinal fluid derives exclusively from the spinal cord. In view of some of the difficulties in determining the amount of 5HIAA in patients' spinal fluid (Wilk & Green 1972), an improvement in the methods for measuring 5HIAA concentrations in the spinal fluid might resolve this obvious controversy.

Cerebral dopamine and homovanillic acid are mainly concentrated in the striatum. After making nigrostriatal lesions in animals, Papeschi et al. (1971) observed a decrease of dopamine and homovanillic acid in the striatum and also a drop in homovanillic acid in the cisternal fluid, and concluded that the concentration of homovanillic acid in the cisternal fluid reflects the metabolism of dopamine in the brain. It is frequently claimed that the presence of dopamine in the spinal cord is not proved and that, therefore, homovanillic acid cannot be formed in the spinal tissue. The concentration of homovanillic acid is extremely diminished in the spinal fluid of patients with a restriction (Curzon et al. 1971) and a blockage (Post et al. 1973) of spinal fluid communication with other parts of the cerebrospinal fluid system. These arguments are usually used as support to the idea that the quantity of homovanillic acid in the spinal fluid reflects the biochemical status of dopaminergic neurons in the brain.

The origin of homovanillic acid in the spinal fluid is still not clarified in our opinion. Recent evidence (Atack et al. 1972) shows that dopamine is present in the spinal cord and that after application of L-dopa its concentration rises in the spinal tissue (Baker & Anderson 1970). This indicates that homovanillic acid might be formed in the spinal cord. We have started experiments in the cats to verify whether this is so and whether it enters the spinal fluid. However, although some existing methods for determining homovanillic acid (e.g. that of Gerbode & Bowers 1968) are satisfactory for measuring its concentration in cisternal fluid and striatum of cats and in the ventricular fluid of patients, they are not sufficiently sensitive for the precise determination of homovanillic acid in the spinal fluid of patients and cats. Bowers (1972) has stressed that baseline values of homovanillic acid concentration in the spinal fluid of patients lie at the lower limit of the method and are not probably reliable. Moreover,

Kirschberg *et al.* (1972) have shown that several acid metabolites of catechol-amines suppress the intensity of fluorescence of homovanillic acid. These data raise doubts about some studies of patients whose concentrations of homovanillic acid in the spinal fluid have been reported to be lower than those of control subjects.

If homovanillic acid is formed in the spinal cord, it is in a good strategic position to enter the adjacent spinal fluid easily by diffusion, as we have shown for 5HIAA. In contrast, homovanillic acid might hardly reach the spinal fluid from the cisterna magna especially if active transport removes it from the spinal subarachnoid space, as for 5HIAA. (Both homovanillic acid and 5HIAA are actively transported from ventricular fluid to bloodstream [Ashcroft *et al.* 1968].) The considerable effect of active transport on the concentration of homovanillic acid in various compartments of the cerebrospinal fluid can be illustrated by the observation (Ashcroft *et al.* 1968) that the concentration of homovanillic acid in the ventricular fluid (1.56 µg/ml) is about 20 times higher than in the cisternal fluid (0.076 µg/ml). Clearly, the cerebrospinal fluid is effectively cleansed of homovanillic acid during its passage from the lateral ventricle to the cisterna magna. Moreover, the distance from the cisterna magna to the lumbar sac is greater than to the lateral ventricle, and the flow of cerebrospinal fluid along the spinal subarachnoid space is insignificant in comparison with flow of cerebrospinal fluid in ventricles (Davson 1967). All these considerations suggest that homovanillic acid hardly reaches the spinal fluid from the cisterna magna. We believe that the diagnostic significance of homovanillic acid in the spinal fluid will remain uncertain until the possibility of the formation of homovanillic acid in the spinal cord and the potential access of homovanillic acid from the cisternal fluid into the spinal fluid have been clarified.

On the basis of our results and a consideration of the dynamics of cerebrospinal fluid flow, we suggest that certain substances in the spinal fluid have biochemical autonomy in the spine, that is to say they originate in the spinal cord, can pass into the spinal fluid and be transported into the spinal bloodstream. Several factors might control whether these substances in the spinal fluid reflect the biochemical processes in the brain or those in the spinal cord. If the concentration of a given substance is high in the cisternal fluid and negligible or non-existent in the spinal cord, we believe that this substance can successfully diffuse from the cisternal fluid into the spinal fluid and so reflect biochemical changes in the brain. In contrast, if a substance has a plentiful source in the spinal cord, its concentration in the spinal fluid should reflect biochemical processes in the spinal cord. Active transport systems for the substances operating in the spinal subarachnoid space, could effectively counteract their diffusion from the cisterna magna into the lumbar sac.

At the present time, the origin and significance of the substances in the spinal fluid are poorly understood; this kind of research is only just beginning. In view of the increasing use of spinal fluid analysis for diagnosis and prognosis in medicine, further experiments are urgently required to gain insight into basic laws which control the origin and fate of substances in the spinal fluid.

ACKNOWLEDGEMENTS

I thank Dr B. Živković for allowing me to report some of our unpublished results. This investigation was supported by the Fund for Scientific Activities of SR Croatia and by NIH PL 480 Research Agreement (No. 01–015–1). Figs. 1 and 3 are reproduced by permission of the American Association for the Advancement of Science and Fig. 8 is reproduced from *J. Pharm. Pharmacol.* with permission.

References

ASHCROFT, G. W., DOW, R. C. & MOIR, A. T. B. (1968) The active transport of 5-hydroxy-indol-3-ylacetic acid and 3-methoxy-4-hydroxyphenylacetic acid from a recirculatory perfusion system of the cerebral ventricles of the unanaesthetized dog. *J. Physiol. (Lond.)* **199**, 397–425

ATACK, C. V., LINDQVIST, M. & MAGNUSSON, T. (1972) Single-sample determination of several biogenic amines and related compounds in *V Int. Congr. Pharmacol.* (San Francisco), Abstracts, p. 11

BAKER, R. G. & ANDERSON, E. G. (1970) The effects of L-3,4-dihydroxyphenylalanine on spinal reflex activity. *J. Pharmacol. Exp. Ther.* **173**, 212–223

BOWERS, M. B. JR. (1970) 5-Hydroxyindoleacetic acid in the brain and cerebrospinal fluid of the rabbit following administration of drugs affecting 5-hydroxytryptamine. *J. Neurochem.* **17**, 827–828

BOWERS, M. B. JR. (1972) Clinical measurements of central dopamine and 5-hydroxytryptamine metabolism: reliability and interpretation of cerebrospinal fluid acid monoamine metabolite measures. *Neuropharmacology* **11**, 101–111

BULAT, M. & SUPEK, Z. (1966) Fate of intracisternally injected 5-hydroxytryptamine in the rat brain. *Nature (Lond.)* **211**, 637–638

BULAT, M. & SUPEK, Z. (1967) The penetration of 5-hydroxytryptamine through the blood–brain barrier. *J. Neurochem.* **14**, 265–271

BULAT, M. & SUPEK, Z. (1968a) Mechanism of 5-hydroxytryptamine penetration through the cerebrospinal fluid–brain barrier. *Nature (Lond.)* **218**, 72–73

BULAT, M. & SUPEK, Z. (1968b) Passage of 5-hydroxytryptamine through the blood–brain barrier, its metabolism in the brain and elimination of 5-hydroxyindoleacetic acid from the brain tissue. *J. Neurochem.* **15**, 383–389

BULAT, M. & ŽIVKOVIĆ, B. (1971) Origin of 5-hydroxyindoleacetic acid in the spinal fluid. *Science (Wash. D.C.)* **173**, 738–740

BULAT, M. & ŽIVKOVIĆ, B. (1973) Penetration of 5-hydroxyindoleacetic acid across the blood–cerebrospinal fluid barrier. *J. Pharm. Pharmacol.* **25**, 178–179

BULAT, M., ISKRIĆ, S., STANČIĆ, L., KVEDER, S. & ŽIVKOVIĆ, B. (1970) The formation of
5-hydroxytryptophol from exogenous 5-hydroxytryptamine in cat spinal cord *in vivo.*
J. Pharm. Pharmacol. **22**, 67–68

CURZON, G., GUMPERT, E. J. W. & SHARPE, D. M. (1971) Amine metabolites in the lumbar
cerebrospinal fluid of humans with restricted flow of cerebrospinal fluid. *Nat. New Biol.*
231, 189–191

DAVSON, H. (1967) *Physiology of the Cerebrospinal Fluid,* Churchill, London

ECCLESTON, D., ASHCROFT, G. W., MOIR, A. T. B., PARKER-RHODES, A., LUTZ, W. & O'MAHO-
NEY, D. P. (1968) A comparison of 5-hydroxyindoles in various regions of dog brain and
cerebrospinal fluid. *J. Neurochem.* **15**, 947–957

ECCLESTON, D., ASHCROFT, G. W., CRAWFORD, T. B. B., STANTON, J. B., WOOD, D. & McTURK,
P. H. (1970) Effect of tryptophan administration on 5-HIAA in cerebrospinal fluid in
man. *J. Neurol. Neurosurg. Psychiatr.* **33**, 269–272

GERBODE, F. A. & BOWERS, M. B. JR. (1968) Measurements of acid monoamine metabolites in
human and animal cerebrospinal fluid. *J. Neurochem.* **15**, 1053–1055

GOTTFRIES, C. G., GOTTFRIES, J. & ROOS, B. E. (1969) Homovanillic acid and 5-hydroxy-
indoleacetic acid in the cerebrospinal fluid of patients with senile dementia, presenile
dementia and parkinsonism. *J. Neurochem.* **16**, 1341–1345

GRUNDY, H. F. (1962) Circulation of cerebrospinal fluid in the spinal region of the cat.
J. Physiol. (Lond.) **163**, 457–465

GULDBERG, H. C., ASHCROFT, G. W. & CRAWFORD, T. B. B. (1966) Concentrations of 5-
hydroxyindolylacetic acid and homovanillic acid in the cerebrospinal fluid of the dog
before and during treatment with probenecid. *Life Sci.* **5**, 1571–1575

GULDBERG, H. C. & YATES, C. M. (1968) Some studies of effects of chlorpromazine, reserpine
and dihydroxyphenylalanine on the concentrations of homovanillic acid, 3,4-dihydroxy-
phenylacetic acid and 5-hydroxyindol-3-ylacetic acid in ventricular cerebrospinal fluid of
the dog using the technique of serial sampling of the cerebrospinal fluid. *Br. J. Pharmacol.
Chemother.* **33**, 457–471

KIRSCHBERG, G. J., COTE, L. J., LOWE, Y. H. & GINSBURG, S. (1972) Interference with the
fluorimetric assay for homovanillic acid caused by acid metabolites of catecholamines.
J. Neurochem. **19**, 2873–2876

MOIR, A. T. B., ASHCROFT, G. W., CRAWFORD, T. B. B., ECCLESTON, D. & GULDBERG, H. C.
(1970) Cerebral metabolites in cerebrospinal fluid as a biochemical approach to the brain.
Brain **93**, 357–368

PAPESCHI, R., SOURKES, T. L., POIRIER, L. J. & BOUCHER, R. (1971) On the intracerebral origin
of homovanillic acid of the cerebrospinal fluid of experimental animals. *Brain Res.* **28**,
527–533

POST, M. R., GOODWIN, F. K. & GORDON, E. (1973) Amine metabolites in human cerebrospinal fluid: effects of cord transection and spinal fluid block. *Science (Wash. D. C.)* **179**,
897–899

VAN PRAAG, H. M., KORF, J. & PUITE, J. (1970) 5-Hydroxyindoleacetic acid levels in the cerebrospinal fluid of depressive patients treated with probenecid. *Nature (Lond.)* **225**, 1259–
1260

WEIR, R. L., CHASE, T. N., NG, L. K. Y. & KOPIN, I. J. (1973) 5-Hydroxyindoleacetic acid in
spinal fluid: relative contribution from brain and spinal cord. *Brain Res.* **52**, 409–412

WILK, S. & GREEN, J. P. (1972) On the measurement of 5-hydroxyindoleacetic acid in cerebrospinal fluid. *J. Neurochem.* **19**, 2893–2895

ŽIVKOVIĆ, B. & BULAT, M. (1971a) Inhibition of 5-hydroxyindoleacetic acid transport from the
spinal fluid by probenecid. *J. Pharm. Pharmacol.* **23**, 539–540

ŽIVKOVIĆ, B. & BULAT, M. (1971b) 5-Hydroxyindoleacetic acid in spinal cord and spinal
fluid. *Pharmacology* **6**, 209–215

Discussion

Sandler: We must be very careful about extrapolating perfusion data from immobilized cats to humans. Post *et al.* (1973*a*) recently measured the amounts of homovanillic acid and 5HIAA in the cerebrospinal fluid of mildy depressed patients and then persuaded them to simulate manic activity—running through corridors, banging drums, shouting and whistling. After this activity, significantly higher values of both acids were present; exercise alone caused a similar rise. The authors did not fully interpret this. I believe that what they were observing was just a mixing process. It is well known that the concentration of homovanillic acid, for instance, is normally considerably higher in ventricular and cisternal fluids than in lumbar fluid.

Bulat: Post *et al.* (1973*a*) also found no change in the concentration of proteins in the spinal fluid of patients before and after the patient simulated manic hyperactivity. Since spinal fluid contains twice as much protein as does cisternal fluid, this finding indicates that the two fluids do not mix; if they did, the resulting protein concentration in the spinal fluid should be the mean of these two concentrations. During simulated mania, turnover of 5-hydroxy-tryptamine in the central nervous system probably increased. The observed increase in the amount of 5HIAA in the spinal fluid might be the consequence of more 5HIAA in the spinal cord.

Moir: Many years ago, Sweet *et al.* (1949) demonstrated that sodium enters the cerebrospinal fluid through the brain and the choroid plexuses. In humans, the time taken for radioactive sodium to enter the lumbar fluid from the ventricular fluid can be reduced by increasing activity (Fotherby *et al.* 1963). Possibly, activity enhances mixing of the ventricular and spinal fluids. We have noticed that the concentration of 5HIAA in the lumbar cerebrospinal fluid of patients under strict bed rest is somewhat lower than in those who are allowed light activity (Moir *et al.* 1970). It has also been reported (Ashcroft *et al.* 1966) that air encephalography causes an immediate elevation of 5HIAA concentration in lumbar fluid.

Curzon: I would like to believe the mixing hypothesis, because we find low concentrations of homovanillic acid and 5HIAA in the lumbar sac of severely akinetic patients and not only in basal ganglia disease where the production of homovanillic acid might be low. For example, chair-bound patients with multiple sclerosis had very low concentrations of homovanillic acid and 5HIAA in the cerebrospinal fluid (Curzon 1973). However, these results might also be explained by akinetic subjects having decreased neuronal turnover in brain and spine and therefore lower spinal concentrations of the metabolites.

Moir: I agree with Dr Sandler that it is dangerous to relate results from an experimental animal to the human because there are vast differences particularly with regard to acid transport systems from cerebrospinal fluid and also the physical dimensions and relationships of the cerebrospinal fluid systems. For example, the cat has less openings from its cerebral ventricular system. The work of Davson *et al.* (1962) on the acid uptake system suggests that, in the cat, the choroid plexus in the lateral ventricles are very active but that perhaps the third ventricle is relatively inactive. In the rabbit, the activity in taking up acids seems to be more evenly spread and the cisternal region is perhaps more active (Pollay & Davson 1963; Pullar 1971). We studied the dog most intensively and found that between the third ventricle and the cisterna magna the concentration of 5HIAA dropped markedly owing to an active transport system at that site (Ashcroft *et al.* 1968). The human data indicate that the removal of 5HIAA is localized within the region of the cisterna magna. What are the concentrations of 5HIAA in the cat at these different sites in the cerebrospinal fluid system?

Bulat: The concentrations of 5HIAA in the cisternal and spinal fluids of the cat are about 100 ng/ml. We have just started experiments to find out the mean concentration of 5HIAA in ventricular fluid. As far as I know, there is no evidence that 5HIAA in the spinal fluid of humans originates in the brain. I can see no valid argument against the idea that 5HIAA in the spinal fluid of man originates exclusively or predominantly in the spinal cord. The human spinal cord is long; the cisterna magna and the lumbar sac are far apart. So, 5HIAA has to diffuse a long way from the brain through the cisterna magna to reach the lumbar sac. During its diffusion along spinal subarachnoid space the acid is probably removed into blood by active transport. In contrast, 5HIAA in the spinal cord has to diffuse a short distance to enter the spinal fluid.

Carlsson: Has the spinal fluid of a patient with a complete spinal transection ever been analysed? One or two months after a transection of the human spinal cord, there should be no more production of either 5-hydroxyindoleacetic acid or homovanillic acid below the lesion.

Bulat: According to Post *et al.* (1973*b*), patients whose spinal cords have been transected in some way, such as in a car accident, have lower concentrations of homovanillic acid in the spinal fluid than do control subjects. However, at the same time, the communication between the spinal fluid and cisternal fluid was usually blocked in those patients. It is hard to evaluate the significance of these results since the method used for determination of homovanillic acid in the spinal fluid is not sensitive enough.

In some preliminary experiments, we injured feline spinal cords in the thoracic region and after about three weeks we found that not only 5-hydroxytryptamine and 5HIAA in the cord below the lesion but also 5HIAA in the spinal fluid had almost completely disappeared. At the same time, the concentration of 5HIAA in the cisternal fluid was unchanged. From these chronic experiments we gather that the concentration of 5HIAA in the spinal fluid reflects the metabolism of 5-hydroxytryptamine in the spinal cord but not in the brain.

Grahame-Smith: If the 5HIAA in spinal fluid reflects all the 5-hydroxytryptamine released as a neurotransmitter, measurements of the acid as an index of the neurotransmitter function of 5-hydroxytryptamine are fine. But if only a portion of the monoamine synthesized is used as a neurotransmitter, the rest being metabolized intraneuronally by monoamine oxidase without being functionally active, then the amount of 5HIAA will not necessarily indicate the functional activity of 5-hydroxytryptaminergic neurons.

Bulat: I don't believe that the concentration of 5HIAA in the central nervous system and cerebrospinal fluid can be taken as a specific indicator of the 5-hydroxytryptamine acting as neurotransmitter at the postsynaptic membrane but rather as an index of the total biochemical status of 5-hydroxytryptaminergic neurons.

Wurtman: Most of us tend to forget 2-(5-hydroxyindole)ethanol (5-hydroxytryptophol), the reduction product from 5-hydroxytryptamine. It has been suggested that, in the catabolism of catecholamines, urinary 1-(4-hydroxy-4-methoxyphenyl)-1,2-ethanediol (4-methoxy-3-hydroxyphenylglycol; MHPG), a similar reduction product, or its sulphate specifically represents catecholamines that are released from the brain. We have data suggesting that this is not so and that most MHPG is derived from peripheral sources (i.e., sympathetic neurons and adrenal medulla). Thus, the intravenous injection of 6-hydroxydopamine, which destroys sympathetic neurons without affecting brain catecholamines, depresses urinary concentrations of MHPG in the rat (Hoeldtke *et al.* 1973). The intracisternal administration of 6-hydroxydopamine does not have this effect.

Bulat: According to our calculations, about 20% of the 5HIAA formed in the spinal cord diffuses into the spinal fluid and about 80% is transported into the bloodstream. Finally, all the 5HIAA and (5-hydroxyindole)ethanol formed in the central nervous system are excreted in urine. However, I do not believe that either of these two metabolites in urine reflect the metabolism of 5-hydroxytryptamine in the central nervous system because this constitutes only a small percentage of the total 5-hydroxytryptamine metabolism in the body.

Eccleston: As long ago as 1960 Dr Sharman studied cerebrospinal fluid in patients. We are still trying to evaluate the relative contributions from cord and brain to the concentrations of metabolites in the lumbar cerebrospinal fluid. We still do not know why the concentrations differ so widely between patients.

Wurtman: There are two families of 5-hydroxytryptaminergic neurons: those whose axons descend into the spinal cord and those whose axons ascend into the brain. If we consider the response of these neurons to circulating inputs such as tryptophan or various drugs, we might assume as a first approximation that the biology of both the descending and ascending 5-hydroxytryptamine neurons is the same and thus that their responses will be similar. This formulation is based on the assumption that the synaptic connections of a particular neuron do not influence its response to circulating factors. It allow us to believe that, for example, the release of 5HIAA into the cerebrospinal fluid from axons of descending, spinal 5-hydroxytryptamine neurons parallels the turnover of 5-hydroxytryptamine within the brain itself. I suspect that these assumptions are not valid, and that not all monoaminergic neurons respond in the same way to a given drug or other circulating input (e.g. tryptophan). If the carotid sinus in a rabbit is denervated, the animal develops profound hypertension, whereupon there is a marked increase in activity of the descending bulbospinal noradrenergic neurons that terminate at the thoracic cord. This is manifest by the increased activity of tyrosine hydroxylase and by the greater turnover of noradrenaline (Chalmers & Wurtman 1971). No similar changes are found in the cervical cord or in most of the brain. Can you explain this dichotomy? Is it reasonable to hope that changes in *lumbar* fluid will reveal *brain* responses when we are dealing with the effects of tryptophan on 5-hydroxytryptaminergic neurons and not with responses to neuronal influences?

Bulat: I agree with your interpretation. In addition, some pathological processes could affect either descending or ascending 5-hydroxytryptaminergic neurons located at distant places so that 5HIAA in the spinal fluid cannot reflect the state within the brain.

Sandler: We know that administration of L-dopa alone increases the concentrations of homovanillic acid in the cerebrospinal fluid whereas L-dopa together with a peripheral decarboxylase inhibitor barely affects the concentration. How do we interpret this? Is this finding compatible with the idea that homovanillic acid concentration reflects intracerebral dopamine metabolism?

Carlsson: Exogenous dopa is decarboxylated in the cerebral or spinal capillaries and a peripheral decarboxylase inhibitor will cause the concentration of homovanillic acid in the spinal fluid to go down (cf. Tissot 1970). This has nothing to do with physiology because in all probability dopa is not formed in the capillaries and does not occur in the blood plasma normally.

Sandler: You seem to think that the dopa treatment of Parkinson's disease is a special case and that the spinal capillaries may not participate in the formation of homovanillic acid in the subject not treated with dopa. Yet it is likely that we take in far more L-dopa with our food than had previously been thought possible. I am not only thinking of broad beans but of the recent observations of Dr Wurtman and his colleagues that cereals may contain unexpectedly high amounts of L-dopa.

Sharman: Atack (1973) discovered traces of dopamine in the spinal cord. Using gas chromatography, we have found only minute quantities of homovanillic acid in the spinal cord of rabbit. This may come from the cerebrospinal fluid by diffusion.

Sourkes: Dr Carlsson, since you found some dopa in the spinal cord using decarboxylase inhibitors, we must assume that this dopa is accumulating in noradrenergic fibres, for dopaminergic fibres do not seem to have been identified in the spinal cord. You suggest that some of the dopa is eventually metabolized to homovanillic acid which could reach the cerebrospinal fluid, in other words, under natural conditions homovanillic acid in the cerebrospinal fluid is derived mainly from regions of the brain, but in experiments in which decarboxylase inhibitors are used some of that acid in the cerebrospinal fluid of spinal compartments might originate in the spinal cord itself.

Dr Bulat, Table 1 shows the concentrations of tryptophan and 5HIAA in human cerebrospinal fluid. The only compartment that needs explanation is the so-called mixed one. In patients about to undergo pneumoencephalography, a lumbar puncture is performed and some cerebrospinal fluid is withdrawn at once. Air is then injected (for radiographic visualization), and the fluid that escapes is what we call mixed cerebrospinal fluid. It includes fluid from the subarachnoid space in the lumbar region but also in the thoracic, cervical and probably sometimes cisternal regions. The values in the table show that tryptophan does not display as sharp a gradient as 5HIAA. That for homovanillic acid is even steeper (Sourkes 1973*a*).

We asked ourselves whether there is any correlation at all between the tryptophan and 5HIAA concentrations in the cerebrospinal fluid. A positive correlation would favour the idea that the amount of tryptophan in the central nervous system determined the amount of 5-hydroxytryptamine there. However, the study of Young *et al.* (1973) reveals no such relationship. Of course, Young

TABLE 1 (Sourkes)

Concentration gradients in the human cerebrospinal fluid.

Compartment	Tryptophan[a]		5HIAA[b]	
	No.	Concentration (ng/ml)	No.	Concentration (ng/ml)
Ventricular	8	860	20	105
Cisternal	—	—	2	86
Mixed	7	603	18	37
Lumbar	34	531	17	29

[a] From Young *et al.* (1973)
[b] For original references see collected data in Sourkes (1973*a*) and Garelis & Sourkes (1973).

TABLE 2 (Sourkes)

Concentrations of 5HIAA and homovanillic acid (in ng/ml) in the cerebrospinal fluid of patients with obstructed circulation of that fluid.

Site of blockage	5HIAA			Homovanillic acid		
	Cisternal	Mixed	Lumbar	Cisternal	Mixed	Lumbar
Complete blockage of both foramina of Monro[a]		35	26		<10	19
Complete block at T_2–T_3[a]	86		32	188		
Partial blocks at T_1, T_3 and T_8[a]		54	22		57	<10
Complete block at T_8–T_{10}[b]	80		32	46		0

[a] From Garelis & Sourkes (1973).
[b] From Young et al. (1973).

et al. (1973) were measuring the concentrations of 'spent' metabolites, and the absence of a correlation does not rule out the possibility of a positive relationship between the concentrations in the cerebrospinal fluid with regard to their intracellular concentrations.

In a few of our patients the circulation of the cerebrospinal fluid was obstructed. Values for 5HIAA and homovanillic acid for these patients are set out in Table 2. The first case was dramatic: the foramina of Monro were blocked so that fluid was essentially unable to escape from the lateral ventricles. The concentration of homovanillic acid in the lumbar fluid was more than two standard deviations below the mean value for that compartment, while the concentration of 5HIAA was normal. The second case had a complete block of circulation at T_2–T_3; the concentrations of homovanillic acid in the cisternal fluid and of 5HIAA in cisternal and lumbar fluids were normal. In the third case, where there were partial blocks at T_1, T_3 and T_8 (and which became complete a few days after our samples were drawn), concentration of homovanillic acid in the lumbar fluid, that is below the blockages, was very low but in the mixed fluid, which was drawn from the region above the blockage, the concentration was within the normal range. By contrast, the amount of 5HIAA in lumbar fluid in this case was almost normal. In the last case (complete blockage of the subarachnoid space from the upper level of T_{10} to the lower level of T_8), we were able to obtain both lumbar and cisternal samples of the fluid. Homovanillic acid was undetectable in the lumbar sample and lower in the cisternal sample than in other such samples that we have analysed. The values for 5HIAA both above and below the blockage in this patient are certainly normal and thus indicate clearly that the lumbar sac is not entirely dependent upon circulation of the cerebrospinal fluid from above for its content of 5HIAA.

We have good evidence that all (or nearly all) the homovanillic acid comes from the caudate nucleus (Sourkes 1973*b*), part of which lines portions of the lateral ventricles. On the other hand, 5HIAA may be contributed to the cerebrospinal fluid at many levels of the central nervous system.

Bulat: Even if 20–40% of 5HIAA in the spinal fluid derives from the brain, that still does not mean that 5HIAA in the spinal fluid reflects the metabolism of 5-hydroxytryptamine in the brain (cf. my comment on p. 258/259). We have no doubt about the origin of 5HIAA in the spinal fluid of the cat. Of course, we have to be careful in extrapolating this to humans. Recently, we found high concentrations of both 5-hydroxytryptamine (about 600 ng/g) and 5HIAA (about 500 ng/g) in the human lumbosacral cord, *post mortem*, which is close to the spinal fluid in the lumbar sac. Thus, we believe that also in humans 5HIAA in the spinal fluid does reflect the metabolism of 5-hydroxytryptamine in the spinal cord.

Sourkes: I am not disputing any of your data, but we still do not have sufficient information to state categorically that all 5HIAA in the lumbar cerebrospinal fluid is unequivocally derived from the spinal cord. In order to state that, we need more information about the transport mechanisms for removal of the monoamine metabolites from the cerebrospinal fluid, and particularly whether there are separate systems for homovanillic acid and 5HIAA.

Sharman: Dr Bulat, I once performed an experiment to see how much homovanillic acid came out of the caudate nucleus of the sheep into a perfusate of the lateral ventricle. If the release of homovanillic acid from the spinal cord bore the same relation to the concentration of the acid in the tissue as that observed with the caudate nucleus and the concentration in the cord was as high as 10 ng/g, then one would have to collect a sample for about 3–4 h before it could be estimated by fluorimetry.

References

ASHCROFT, G. W., CRAWFORD, T. B. B., ECCLESTON, D., SHARMAN, D. F., MACDOUGALL, E. J., STANTON, J. B. & BINNS, J. K. (1966) 5-Hydroxyindole compounds in the cerebrospinal fluid of patients with psychiatric or neurological diseases. *Lancet* **2**, 1049–1052

ASHCROFT, G. W., DOW, R. C. & MOIR, A. T. B. (1968) The active transport of 5-hydroxyindol-3-ylacetic acid and 3-methoxy-4-hydroxyphenylacetic acid from a recirculatory perfusion system of the cerebral ventricles of the unanaesthetized dog. *J. Physiol. (Lond.)* **199**, 397–425

ATACK, C. V. (1973) The determination of dopamine by a modification of the dihydroxyindole fluorimetric assay. *Br. J. Pharmacol.* **48**, 699–714

CHALMERS, J. P. & WURTMAN, R. J. (1971) Participation of central noradrenergic neurons in arterial baroreceptor reflexes in the rabbit. *Circ. Res.* **28**, 480–491

CURZON, G. (1973) Involuntary movements other than Parkinsonism: biochemical aspects. *Proc. R. Soc. Med.* **66**, 873–876

DAVSON, H., KLEEMAN, C. R. & LEVIN, E. (1962) Quantitative studies of the passage of different substances out of the cerebrospinal fluid. *J. Physiol. (Lond.)* **161**, 126–142

FOTHERBY, K., ASHCROFT, G. W., AFFLECK, J. W. & FORREST, A. D. (1963) Studies on sodium transfer and 5-hydroxyindoles in depressive illness. *J. Neurol. Neurosurg. Psychiatr.* **26**, 71–73

GARELIS, E. & SOURKES, T. L. (1973) The sites of origin in the central nervous system of monoamine metabolites measured in human cerebrospinal fluid. *J. Neurol. Neurosurg. Psychiatr.* **36**, 625–629

HOELDTKE, R., ROGAWSKI, M. & WURTMAN, R. J. (1973) Effect of selective destruction of central and peripheral catecholamine-containing neurons with 6-hydroxydopamine on catecholamine excretion in the rat. *Br. J. Pharmacol.*, in press

MOIR, A. T. B., ASHCROFT, G. W., CRAWFORD, T. B. B., ECCLESTON, D. & GULDBERG, H. C. (1970) Cerebral metabolites in cerebrospinal fluid as a biochemical approach to the brain. *Brain* **93**, 357–368

POLLAY, M. & DAVSON, H. (1963) The passage of certain substances out of cerebrospinal fluid. *Brain* **86**, 137–150

POST, R. M., KOTIN, J., GOODWIN, F. K. & GORDON, E. K. (1973a) Psychomotor activity and cerebrospinal fluid amine metabolites in affective illness. *Am. J. Psychiatr.* **130**, 67–72

POST, R. M., GOODWIN, F. K. & GORDON, E. (1973b) Amine metabolites in human cerebrospinal fluid: effects of the cord transection and spinal fluid block. *Science (Wash. D.C.)* **179**, 897–899

PULLAR, I. A. (1971) The accumulation of $[^{14}C]$5-hydroxyindol-3-ylacetic acid by the rabbit choroid plexus *in vitro*. *J. Physiol. (Lond.)* **216**, 201–211

SOURKES, T. L. (1973a) Enzymology and sites of action of monoamines in the central nervous system. *Adv. Neurol.* **2**, 13–36

SOURKES, T. L. (1973b) On the origin of homovanillic acid in the cerebrospinal fluid. *J. Neural Transm.* **34**, 153

SWEET, W. H., SELVERSTONE, B., SOLOMON, A. & BAKOY, L. (1949) Studies of formation and absorption of constituents of cerebrospinal fluid in man. *J. Clin. Invest.* **28**, 814

TISSOT, R. (1970) L-Dopa and decarboxylase inhibitor (IDC): biochemical and clinical implications in L-*Dopa and Parkinsonism* (Barbeau, A. & McDowell, F. H., eds.), pp. 80–86, Davis, Philadelphia

YOUNG, S. N., LAL, S., MARTIN, J. B., FORD, R. M. & SOURKES, T. L. (1973) 5-Hydroxyindoleacetic acid, homovanillic acid and tryptophan levels in CSF above and below a complete block of CSF flow. *Psychiatr. Neurol. Neurochir.*, in press

Do tryptophan concentrations limit protein synthesis at specific sites in the brain?

SAMUEL H. BARONDES

Department of Psychiatry, University of California, San Diego, La Jolla, California

Abstract   Since physiological variations in cerebral concentrations of tryptophan have been found to limit the biosynthesis of 5-hydroxytryptamine, it is
possible that synthesis of protein in the brain might be limited as well. No
definitive studies of this problem have been made. In the absence of direct support for this hypothesis and in the light of a review of the relevant literature, it
seems unlikely that this hypothesis holds true for the whole brain; but the situation in 5-hydroxytryptaminergic neurons is obscure.

Although the aromatic amino acids of brain have received extraordinary attention because they are precursors of noradrenaline and 5-hydroxytryptamine,
proteins are quantitatively more important products. Since about 10% of brain
weight is protein and since each aromatic amino acid comprises about 5% of
the total amino acid content of the protein, about 5 mg of each aromatic amino
acid is incorporated in protein per g of brain. In contrast, the amines noradrenaline and 5-hydroxytryptamine are only $1/10^4$ as abundant since they are
found in concentrations of about 0.5 µg/g brain. Even if one takes into account
the more rapid turnover of amines, the proteins remain the major product
derived from aromatic amino acids in brain. Half-lives of brain proteins are very
varied but we may consider them to be of the order of several days whereas half-
lives of total brain amines have been estimated to be about several hours (Costa
& Neff 1970). If amines turn over about 25–50 times as fast as proteins, then
400–200 times as much aromatic amino acid is used for brain protein synthesis
as for amine synthesis.

Such considerations of the other possible fates of aromatic amino acids
might be of little concern were it not for the indications that tryptophan is
rate limiting for the biosynthesis of 5-hydroxytryptamine under normal physiological circumstances (Fernstrom & Wurtman 1971). Since amines and proteins

might be rival products from tryptophan, this observation raises the possibility that tryptophan is limiting for protein synthesis as well in the brain.

That amino acids may be the limiting factor in protein synthesis has been clearly established in certain extreme situations. For example, the division of cultured mammalian cells can be terminated by complete removal of the amino acid from the medium, since amino acids are absolutely required for a net increase in the total protein necessary for cell division. Yet complete absence of extracellular amino acids is an artificial and extreme situation which is never approached in an intact mammal. My purpose in this paper is to consider the possibility that amino acids, particularly tryptophan, regulate protein synthesis in brain in less extreme circumstances.

Unfortunately, definitive answers to the questions raised by that possibility still elude us. On the basis of both indirect and direct evidence, I shall argue that tryptophan does not regulate overall brain protein synthesis under conditions when it might limit biosynthesis of 5-hydroxytryptamine. Opinions about its potential role in regulating protein synthesis on ribosomes in 5-hydroxytryptaminergic cell bodies and in mitochondria at 5-hydroxytryptaminergic nerve endings must be withheld until this is studied directly.

SEGREGATION OF AMINE BIOSYNTHESIS AND PROTEIN SYNTHESIS

The synthesis of amines is anatomically segregated from most of the protein synthesis in the brain in two ways. First, amine synthesis is confined to a few neurons of brain, for all the other neurons as well as the glia lack the aromatic amino acid hydroxylases necessary for competition with the protein-synthesizing apparatus for aromatic amino acids. Secondly, the bulk of amine biosynthesis takes place in nerve terminals whereas protein is synthesized in neurons almost exclusively in nerve cell bodies. Yet the amine-synthesizing enzymes are also found in nerve cell bodies where they themselves are made. Although the relative concentrations of these enzymes in cell bodies and nerve endings is not known accurately, an estimate of their activity may be derived from histochemical fluorescence studies of the relative concentrations of amines at these two sites. These studies indicate that nerve cell bodies contain about 1% of the amine concentration of the nerve endings (Hillarp *et al.* 1966). Since the amine-synthesizing system is relatively dilute in the nerve cell body, the finding that a reduction in brain concentrations of tryptophan leads to a measurable reduction in brain 5-hydroxytryptamine tells us only that in nerve endings tryptophan may be limiting for amine synthesis; in no way does it indicate that tryptophan in the nerve cell body is limiting for protein synthesis, even in 5-hydroxytryptaminergic neurons. However, if protein synthesis at nerve endings is significant for regulation of nerve ending function, the possibility of vigorous competition in aminergic nerve endings demands serious attention. Accordingly, I shall now consider studies of protein synthesis at nerve endings.

PROTEIN SYNTHESIS AT NERVE ENDINGS

It seems virtually certain that mitochondria in nerve endings share with other mitochondria the capacity to synthesize a small amount of the total mitochondrial protein. The possible existence of other types of protein synthesis in the nerve endings has also commanded considerable attention because of speculations about their possible role in regulating synaptic function. Attempts have been made to study this directly by incubating subcellular fractions enriched in nerve endings with radioactive amino acids.

These fractions enriched in nerve endings incorporate amino acids into protein (Austin et al. 1970). Because the fractions are contaminated by other protein-synthesizing particles, this observation has been difficult to interpret. To evaluate the relative role of contaminants, Gambetti et al. (1972) examined the localization of the radioactive amino acid incorporated by these fractions by autoradiography with the electron microscope. They showed that almost half the radioactive amino acid incorporated is found in membrane-bound vesicles which contain ribosomes and which are contaminants of nerve-ending fractions. Another 20% of the radioactivity was associated with mitochondria either contaminating the fraction or contained within the nerve-ending particles. The remainder was found in association with the membranes of nerve-ending particles. Unfortunately, these difficult studies did not establish that the fraction of the radioactivity associated with the nerve-ending membranes was incorporated into protein rather than some other product. While the results suggest the existence of a non-mitochondrial protein-synthesizing system at nerve endings, they do not conclusively prove this.

Another approach which has been devised to search for non-mitochondrial protein synthesis at nerve endings is based on polyacrylamide gel electrophoresis of the radioactive products of in vitro incubation of nerve-ending fractions with radioactive amino acids. The finding of unique protein products associated with the nerve-ending fraction and not with microsomal or mitochondrial fractions which would contain the major contaminants would be strong evidence for specific nerve ending protein synthesis. Two such studies have been reported. In one, Ramirez et al. (1972) observed chloramphenicol-sensitive incorporation of amino acids into proteins in nerve-ending fractions. Since chloramphenicol inhibits only mitochondrial protein synthesis rather than microsomal protein synthesis in mammals, Ramirez *et al.* compared the radioactive products of the protein-synthesizing system of the nerve-ending fraction with those of a mitochondrial fraction derived from brain. Striking differences were found between the radioactive products separated by polyacrylamide gel electrophoresis of the radioactively labelled proteins synthesized *in vitro* by the two fractions. These results suggest that a unique protein-synthesizing system exists in nerve endings which has the chloramphenicol sensitivity of mitochondrial protein-synthesizing systems but which incorporates amino acids into a special and restricted group of proteins. Using a similar strategy, Gilbert (1972) has identified a protein-synthesizing system sensitive to cycloheximide and insensitive to chloramphenicol in nerve-ending fractions. Since microsomal protein synthesis is sensitive to cycloheximide and insensitive to chloramphenicol, the nature of the radioactive products made by the nerve-ending fraction were compared with those made by the microsomal fraction. Again differences were found, suggesting a special protein-synthesizing system of a microsomal type in nerve endings.

Because of the impurity of nerve-ending fractions and the absence of detectable ribosomes in these structures, it is difficult to accept the existence of non-mitochondrial protein-synthesizing systems in nerve endings despite this evidence. Nevertheless, the two strategies which have been employed—electron microscope autoradiography and identification of unique products of the protein-synthesizing system of the nerve-ending fraction—are consistent with the existence of a special protein-synthesizing system in the nerve ending fraction. This could, of course, be associated with the postsynaptic component of the synaptosomal particles and, therefore, be segregated from competition with amine-synthesizing enzymes in aminergic nerve terminals. Nevertheless, it is still possible that an important type of protein synthesis occurs at this site which could be in direct competition with amine synthesis. Certainly, the mitochondrial protein-synthesizing apparatus in the abundant mitochondria in nerve endings could be competing uniquely with the amine-synthesizing apparatus for available aromatic amino acid. However, the significance of this competition is unclear.

SEGREGATION OF AMINO ACID POOLS USED FOR PROTEIN SYNTHESIS

In addition to the gross anatomical segregation of amine synthesis and protein synthesis, the amino acid pools for protein synthesis are segregated in another manner whose anatomical basis is obscure. This may or may not play a role in reducing competition for aromatic amino acids by the two major metabolic pathways under discussion.

The existence of functionally segregated pools of amino acids in brain was first clearly indicated in studies of incorporation of radioactive glutamic acid into radioactive glutamine and into the overall radioactive glutamic acid pool of brain (Berl & Clark 1969). Measurement of the specific activity of total brain glutamine and glutamic acid after administration of radioactive glutamic acid showed that the overall specific activity of glutamine was four times as great as that of glutamic acid. This indicated that some of the exogenously administered radioactive glutamic acid was preferentially used for glutamine synthesis rather than absorbed into the overall glutamic acid pool.

Segregation of amino acids for protein synthesis has been observed in several systems. In one study, Hider et al. (1969) incubated intact muscle for 30 min with [^{14}C]leucine. They replaced the extracellular amino acid with [^{3}H]leucine and observed the incorporation over the next 30 min. Between 30 and 40 min after the start of the experiment the intracellular amino acid pool was relatively rich in [^{14}C]leucine and relatively poor in [^{3}H]leucine. Yet in this interval, there was considerable incorporation of [^{3}H]leucine into protein compared with the incorporation of [^{14}C]leucine. This indicates that amino acids taken directly from the medium are used for protein synthesis in preference to the amino acids from the overall intracellular pool.

In another elegant study, Righetti et al. (1971) showed that some amino acids derived from protein degradation are used for protein synthesis in preference to the amino acids in the overall intracellular pool. The intracellular pools of free lysine, leucine and phenylalanine in HeLa cells were equilibrated for 40 min with radioactive amino acids added to the medium. Righetti et al. kept the radioactive amino acids in the medium thereafter and followed the radioactivity in the intracellular amino acid pools throughout. After the 40 min preincubation, they induced the synthesis of ferritin, which was initially absent from the HeLa cells, by adding iron to the medium. Three hours later, they isolated ferritin by immunoprecipitation, using a specific antibody to this pure protein, and they determined the specific activity of each of the amino acids incorporated into the induced ferritin. The specific activity of leucine in ferritin was 30% lower than that of leucine in the intracellular pool while the specific activity of phenylalanine was 62% lower. In contrast, the specific activity of lysine in ferritin was the same as that in the intracellular amino acid pool. The unlabelled leucine and phenylalanine used for ferritin synthesis must, therefore, have come to a large extent from degradation of cellular proteins. These had a low specific activity since the cells had been labelled for only a short period of time. The experiments indicate that leucine and phenylalanine derived from protein degradation are relatively sequestered into the pool for protein synthesis rather than equilibrating freely in the amino and pool of the cell. In contrast, lysine appears to equilibrate freely.

These and other studies indicate that the overall amino acid content of the cell is not a true reflection of the material available for protein synthesis. Rather, recently accumulated amino acids from the medium as well as amino acids directly derived from degradation of proteins are used preferentially. Estimation of the specific activity of the precursor amino acid in a cell, therefore, does not reflect the specific activity of the amino acids which are being converted into protein. The problem is further complicated by the evidence of segregation of the amino acid pools used for amine biosynthesis. Whether proteins and amines draw upon the same or different segregated pools is a matter for further investigation.

DIRECT MEASUREMENT OF TRYPTOPHYL-tRNA

Since the amino acid pool for protein synthesis is segregated from the overall amino acid pool of the cell, the size of the cellular amino acid pool may not accurately reflect the amount of amino acid available for protein synthesis. Despite this potential impediment, the amount of a given amino acid which is available for protein synthesis can be estimated by determining the amount of this amino acid bound to its transfer RNA (tRNA) in a given circumstance. Since aminoacyl-tRNA is the direct precursor of protein, we can determine the amount of tryptophan available for protein synthesis in brain by measuring the amount of tryptophyl-tRNA (not the acceptor but the amount of tryptophan actually bound into it) in the brain.

A simplified method for this determination has been devised by Johnson & Chou (1973). They have studied the concentration of tryptophyl-tRNA in immature and adult mouse brain. Since the rate of protein synthesis in immature brain has been estimated to be as much as ten times greater than in the adult (Johnson & Luttges 1966), the concentrations of tryptophyl-tRNA were determined in brains of immature and adult mice. Johnson & Chou (1973) found that the concentration of tryptophyl-tRNA in immature brain is not strikingly different from that in the adult. From these findings, it might be argued that tryptophyl-tRNA is not limiting for adult brain protein synthesis since similar concentrations are sufficient for the greater rates of protein synthesis in the immature brain. Although such an argument must be qualified in many ways, direct measurements of tryptophyl-tRNA should prove more useful than studies of amino acid concentration for an analysis of the possible limiting role of tryptophan in protein synthesis. To encourage studies with this method I shall briefly describe its principle.

Transfer RNA can be isolated fairly simply. The amount of an amino acid bound to its specific tRNA *in vivo* can be determined because the tRNA which has an amino acid bound to it does not lose its acceptor activity when treated with periodate (Johnson & Chou 1973). In contrast, free tRNA is inactivated by periodate and can no longer accept an amino acid. Therefore, isolation and treatment of tRNA with periodate destroys the acceptor activity of the tRNA not bound to an amino acid but not that of the tRNA which is attached to an amino acid. After periodate treatment, the amino acid can be readily released from its tRNA and can be replaced by a radioactive amino acid with an appropriate enzyme. In this way, one can conveniently determine how much tRNA was bound to amino acid *in vivo*, since the free tRNA was inactivated by the periodate treatment.

This method could be useful for determining the relationship between changes in brain tryptophan concentrations and the amount of tryptophan available for brain protein synthesis. Of course, a reduction in the amount of tryptophyl-tRNA might not indicate a reduction in brain protein synthesis since it might not be limiting in this range. Nevertheless, a constant tryptophyl-tRNA concentration would argue against regulation of brain protein synthesis by the change in the amount of tryptophan which was produced.

TRYPTOPHAN CONCENTRATION AND POLYSOMES

So far, I have concentrated on the potential regulation of rates of brain protein synthesis by precursor availability. Yet, in the usual circumstances, it seems clear that the synthesis of a specific protein is regulated by the presence (and state of activity) of the messenger RNA (mRNA) which directs its synthesis. By inference, the overall rate of protein synthesis is generally thought to be limited by the availability of the total mRNA of the cell rather than of precursor. Evidence suggests that, in the regulation of protein synthesis, tryptophan acts not as a limiting precursor but rather by regulating either the synthesis or degradation of mRNA.

A general measure of availability of mRNA can be obtained from the percentage of polysomes in an extract of tissue; mRNA binds to several ribosomes thereby generating a polysome which is held together by the strand of mRNA. This polysome sediments more rapidly than free ribosomes unassociated with mRNA. By determining the ratio of polysomes to monosomes in a particular extract it is possible to estimate the amount of messenger RNA in the preparation. It is therefore of great interest that reduction of hepatic tryptophan concentration by food deprivation or administration of an amino acid mixture devoid of tryptophan leads to a reduction of liver polysomes in adult mice (Sidransky *et al.* 1968). The effect seems to be relatively specific for tryptophan since deletion of other essential amino acids from the diet does not have this effect. Since tryptophan appears to regulate the proportion of liver polysomes, it might act directly in the regulation of either the synthesis or degradation of many mRNAs.

Similar results were found in immature (seven-day-old) rat brains after the brain concentration of tryptophan had been reduced by injection of large amounts of phenylalanine which competes with tryptophan for entry into the brain. In the seven-day-old rat brains, the concentration of free tryptophan fell from 0.55 to 0.31 mg/100 g tissue after treatment with phenylalanine and there was a concomitant marked decrease in polysomes (Aoki & Siegel 1970). Unexpectedly, the effect of reduced tryptophan on brain polysomes was not observed in four-week-old rats. To further complicate matters, it should be noted that whole brain (rather than 5-hydroxytryptaminergic neurons) was examined. It is certainly possible that alteration of the amount of tryptophan in 5-hydroxytryptaminergic neurons could have a marked effect on polysomes and protein synthesis but that this effect is obscured by examining whole brain. Were there a special uptake mechanism for this amino acid in 5-hydroxytryptaminergic cell bodies, the issue would be even further complicated. In view of Roberts' findings (see pp. 299-318), interpretation must be particularly cautious.

HOW TOLERABLE IS A REDUCTION IN BRAIN PROTEIN SYNTHESIS?

Regulation of protein synthesis is a major method of controlling cell function and is generally conducted by altering the rate of synthesis or state of activity of specific mRNAs which direct the synthesis of specific proteins. If tryptophan concentrations limited overall protein synthesis, might this not pose a great danger to the control of this critical process?

Although rigorous proof is required before the heretical notion that tryptophan concentrations regulate brain protein synthesis under physiological circumstances is accepted, some evidence indicates that a substantial and general reduction in brain protein synthesis for many hours might be quite tolerable. For example, inhibition of about 80% of brain protein synthesis for several hours by injections of acetoxycycloheximide, which interferes with the protein-synthesizing apparatus, has no effect on the ability of mice to perform, learn or remember a discrimination task (Barondes 1970). Only more extreme inhibition has a detectable effect on memory formation, which may be one of the most important regulatory processes in brain. Thus, brain function can continue despite far greater reduction in the rate of brain protein synthesis than one would expect by simply reducing tryptophan concentrations. Presumably, this is possible because of mechanisms for structural alterations of pre-existing proteins in brain which provide a large repertoire of regulatory processes which are independent of brain protein synthesis (Barondes & Dutton 1972). The concentrations of specific proteins might be regulated through control of the rates of degradation (Schimke & Doyle 1970) as well as rates of synthesis. In the case of brain cell suspensions, inhibition of protein synthesis with puromycin was accompanied by inhibition of degradation as well (Gilbert & Johnson 1972), a finding which indicates that a reduction in rate of brain protein synthesis (by reducing the amount of a critical precursor) might not be devastating.

CONCLUSIONS

We must draw separate conclusions for whole brain and for amine-synthesizing neurons. Given the complexities and heterogeneity of the system, all conclusions must be made with caution in the absence of direct and detailed studies.

For whole brain we can conclude that (1) competition for tryptophan between amine synthesis and protein synthesis is obviated by the segregation of amine synthesis in the small population of 5-hydroxytryptaminergic neurons, primarily in their nerve terminals; (2) evidence suggests that the amount of tryptophyl-tRNA is not strictly limiting for protein synthesis in the adult brain over the expected physiological range, implying that tryptophan is not rate limiting for brain protein synthesis; (3) in the immature brain, tryptophan depletion (by administration of phenylalanine) has an effect on brain polysomes whose mechanism is unknown, but in the adult brain this effect is not detectable.

For 5-hydroxytryptaminergic neurons we conclude that (1) competition between protein synthesis and 5-hydroxytryptamine synthesis is to some extent segregated because amine synthesis is concentrated at nerve endings whereas protein synthesis is concentrated in nerve cell bodies; (2) there is amine synthesis in nerve cell bodies which could compete with protein synthesis, although there is no direct evidence to support this; (3) possibly the concentration of tryptophan significantly affects synthesis or degradation of mRNA (as measured by the proportion of polysomes) in the nerve cell bodies of these neurons; (4) the active amine synthesis at nerve terminals suggests potentially significant competition with mitochondrial protein synthesis at 5-hydroxytryptaminergic terminals. If other protein-synthesizing systems exist at nerve terminals, they too might have to compete.

ACKNOWLEDGEMENTS

I thank the National Institutes of Mental Health and the Alfred P. Sloan Foundation for financial support.

References

AOKI, D. & SIEGEL, F. L. (1970) Hyperphenylalaninemia: disaggregation of brain poly-ribosomes in young rats. *Science (Wash. D.C.)* **168**, 129–130

AUSTIN, L., MORGAN, I. G. & BRAY, J. J. (1970) in *Protein Metabolism of the Nervous System* (Lajtha, A. ed.), pp. 271–290, Plenum, New York

BARONDES, S. H. (1970) Cerebral protein synthesis inhibitors block long-term memory. *Int. Rev. Neurobiol.* **12**, 177

BARONDES, S. H. & DUTTON, G. R. (1972) in *Basic Neurochemistry* (Albers, R. W., Siegel, G. J., Katzman, R. & Agranoff, B. W., eds.), pp. 229–244, Little Brown, Boston

BERL, S. & CLARK, D. D. (1969) in *Handbook of Neurochemistry*, Vol. 2 (Lajtha, A., ed.), pp. 447–472, Plenum, New York

COSTA, E. & NEFF, N. H. (1970) in *Handbook of Neurochemistry*, Vol. 4 (Lajtha, A., ed.), pp. 45–90, Plenum, New York

FERNSTROM, J. D. & WURTMAN, R. J. (1971) Brain serotonin content: physiological dependence on plasma tryptophan levels. *Science (Wash. D.C.)* **173**, 149–151

GAMBETTI, P., AUTILIO-GAMBETTI, L. A., GONATAS, N. K. & SHAFER, B. (1972) Protein synthesis in synaptosomal fractions. *J. Cell Biol.* **52**, 526–535

GILBERT, J. M. (1972) Evidence for protein synthesis in synaptosomal membranes. *J. Biol. Chem.* **247**, 6541–6550

GILBERT, B. E. & JOHNSON, T. C. (1972) Protein turnover during maturation of mouse brain tissue. *J. Cell. Biol.* **53**, 143–147

HIDER, R. C., FERN, E. B. & LONDON, D. R. (1969) Relationship between intracellular amino acids and protein synthesis in the extensor digitorum longus muscle of rats. *Biochem. J.* **114**, 171–178

HILLARP, N. A., FUXE, K. & DAHLSTROM, A. (1966) in *Mechanisms of Release of Biogenic Amines* (Von Euler, U. S., Rosell, S. & Uvnas, B., eds.), pp. 31–57, Pergamon Press, New York

JOHNSON, T. C. & CHOU, L. (1973) Level and amino acid acceptor activity of mouse brain t-RNA during neural development. *J. Neurochem.* **20**, 405–414

JOHNSON, T. C. & LUTTGES, M. W. (1966) The effects of maturation on *in vitro* protein synthesis by mouse brain cells. *J. Neurochem.* **13**, 545–552

RAMIREZ, G., LEVITAN, I. B. & MUSHYNSKI, W. (1972) Highly purified synaptosome membranes from rat brain; incorporation of amino acids into membrane proteins *in vitro*. *J. Biol. Chem.* **247**, 5382

RIGHETTI, P., LITTLE, E. P. & WOLF, G. (1971) Reutilization of amino acids in protein synthesis in HeLa cells. *J. Biol. Chem.* **246**, 5724–5732

ROBERTS, S. (1974) *This Volume*, pp. 299–318

SCHIMKE, R. J. & DOYLE, D. (1970) Control of enzyme levels in animal tissues. *Annu. Rev. Biochem.* **39**, 777

SIDRANSKY, H., SARMA, D. S. R., BONGIORNO, M. & VERNEY, E. (1968) Effect of dietary tryptophan on hepatic polyribosomes and protein synthesis in fasted mice. *J. Biol. Chem.* **243**, 1123–1132

Discussion

Wurtman: In contrast to the brain, the liver receives a portal circulation in which the concentrations of amino acids vary markedly according to whether the animal has eaten recently. As Dr Munro showed, the concentrations of amino acids coming into the liver after protein is consumed can increase 5–10 fold. This increase apparently generates the daily rhythms in hepatic polysome aggregation (Fishman *et al.* 1969) and in the activity of tyrosine aminotransferase (Wurtman *et al.* 1968*a*), and other rhythms in the hepatic synthesis of protein.

We have found that the concentrations of most amino acids in human (Wurtman *et al.* 1968*b*) and in rat serum (Fernstrom *et al.* 1971) undergo characteristic daily rhythms. These rhythms are of considerably smaller amplitude—about 20–100% of the daily nadir—than the rhythms in amino acid concentrations in the portal vein, and bear some relationship to the total quantity of the particular amino acid in the body: rhythms in the relatively scarce amino acids (e.g., the aromatic and branched-chain compounds) are of greater amplitude than, for example, alanine and glutamic acid. The blood amino acid rhythms are also generated largely by food consumption and disappear completely after people have fasted for a few days. The most important factor in their production seems to be the post-prandial secretion of insulin (Wurtman 1970); this hormone causes major shifts in the distribution of amino acids between the plasma and most tissues.

As Dr Fernstrom pointed out (cf. p. 162), a diet-induced change of 40–60% in the amount of tryptophan in the brain is sufficient to induce a change in the rate of 5-hydroxytryptamine synthesis. However, it appears that the post-prandial increase in brain tryptophan is not sufficient to change the state of aggregation of brain polysomes: Drs Weiss and Munro and I were unable to detect any daily rhythm in the aggregation of brain polysomes, an observation which suggests that, under normal conditions, brain protein synthesis is not limited by amino acids. The point is that it is essential to establish the normal dynamic range of amino acid concentrations to which a tissue is exposed, and then to ask whether fluctuations within that range are sufficient to produce changes in the use of the amino acids for protein synthesis or the production of low molecular weight derivatives like monoamines.

Munro: The tryptophan content of the brain rises as the polysomes disaggregate after administration of dopa; the increase is independent of or greater than the changes in the plasma which tend to be in the same direction (Weiss *et al.* 1971). This implies that lack of tryptophan in the brain is not the effector of this cerebral disaggregation.

Many problems of amino acid supply are still unresolved. First, the phenomenon of tryptophan supply and polysome disaggregation (Fleck *et al.* 1965) is only demonstrable under certain circumstances, such as fasting in the rat. By preparing the animals a different way, we can produce changes in the relative amounts of free amino acids in organs. Once the animals have different proportions of amino acids in the liver pool, the polysome patterns can be changed by administration of other amino acids, such as threonine (Pronczuk *et al.* 1970). Secondly, in studies on perfused livers when the input of amino acids can be much better controlled, a change in the amino acid composition of the perfusate can vary polysome aggregation. Some workers are surprised by the amounts of amino acids needed—ten times those in plasma—to cause such aggregation. However, changes of this magnitude occur in the portal vein after feeding.

Roberts: I shall refer to polysome aggregation and disaggregation later (pp. 299–318) but I will just comment that I agree that the polysomal disaggregation which is seen after administration of amino acids in large quantities appears to be largely due to factors other than a decline in tryptophan concentrations.

Barondes: The studies I quoted on tryptophan deprivation on hepatic polysomes certainly suggest that tryptophan concentrations are critical.

Roberts: You emphasized (p. 269) that different amino acids appear to be differently used for protein synthesis as a result of the probable compartmentation of amino acids arising from protein degradation (Righetti *et al.* 1971). What was neglected in that study was the fact that the different amino acids which are presented to the protein-synthesizing systems are being degraded at different rates. The finding that phenylalanine is incorporated at 30% the rate of lysine, compared to 60% for certain other amino acids might simply reflect different rates of degradation of these amino acids.

Munro: I agree; we should be measuring charged tRNA. Most of the compartmental data to which you referred, Dr Barondes, are drawn from *in vitro* experiments. Are the amino acids being incorporated on the surface of the slice or in the piece of tissue which is being examined? The pool which dilutes the radioactivity is the pool of the entire tissue. Consequently, protein synthesis appears to have a higher specific activity than the pool has, because the pool is largely non-exchanging owing to premature *in vitro* death of the tissue.

Barondes: Hider *et al.* (1969) drew their data (see p. 269) from an *in vitro* situation in which contact between muscle and medium is not ideal, but the segregation of pools that they observed cannot be dismissed. The cells in contact with the medium do show this phenomenon. In the experiment with HeLa cells (Righetti *et al.* 1971), these tissue culture cells are in good contact with the medium, so your criticism is not applicable.

Dr Munro, do you know if dopa attaches to tRNA?

Munro: We have not ruled out this possibility but it seems unlikely. The ligases (the activating enzymes) are reasonably specific.

Wurtman: Dopa is not incorporated into protein.

Barondes: That is to be expected. Was this examined in a population which is highly enriched in dopaminergic or adrenergic cell bodies?

Wurtman: The disruption of polysomes by dopa causes a marked cessation in protein synthesis that lasts about two hours and is entirely reversible. This is a convenient way of investigating the importance of the reduction of protein synthesis.

Munro: Since dopa disaggregates polysomes through dopamine formation, we come full circle to the question you originally asked: do these amines have any effect on protein synthesis?

Barondes: Do you mean that if the decarboxylase is inhibited, the condition of polysomes is not affected?

Lajtha: As yet, there is no evidence for or against the existence of precursor compartments for protein synthesis in the living brain. There have been some reports that extracellular amino acids are preferentially used for protein synthesis in rat muscle (Hider *et al.* 1969), in rat pancreas (Van Venrooij *et al.* 1972) and in embryonic chick cartilage (Adamson *et al.* 1972). In our recent studies with brain slices (D. Dunlop and I, unpublished results, 1973), we found no evidence for the preferential use of extracellular amino acids, either in the kinetics of incorporation or in double-labelling experiments.

Munro: Not all the evidence favours compartmentation. For example, London's group (Fern *et al.* 1971) claim to have demonstrated compartmentation in the gut mucosal cells, namely that there is a preferential external source of amino acids for mucosal cell protein synthesis, but Alpers & Thier (1972) contradict this.

Lajtha: You mentioned that protein synthesis might be ten times more active at birth than in maturity. This was deduced from most *in vitro* data (which we have summarized in Lajtha & Marks [1971], p. 570). The changes in the living brain are somewhat less (Lajtha & Dunlop 1973); synthetic activity decreases by a factor of three and turnover is halved. Synthesis decreases more than breakdown because of the net deposition of proteins during growth. The reason for a greater decrease *in vitro* than *in vivo* is that in the young brain *in vivo* the rate of incorporation and of incorporation in slices are almost identical whereas in the adult animal the rate of incorporation by slices is at best only 20% of that of the living brain; the changes in slices are much greater during development because of the greater *post-mortem* changes in adult slices.

Barondes: I agree that the estimates of the relative rates of protein synthesis at different ages are not completely accurate. The point I wanted to make is that to facilitate interpretation it would help to know the specific activity of aminoacyl-tRNA.

Lajtha: I am not convinced that this would be a significant help in determining true rates. Certainly, tRNA is more of a precursor than is free amino acid. However, if we assume that general free amino acid specific activity is meaningless because the free amino acids are compartmented, we can also assume that the general specific activity of the tRNA is meaningless because tRNA itself may be compartmented. Having said that, let me say that it seems unlikely that tRNA would be compartmented because of the very high rates of turnover or the tRNA pool. With tRNA present at very low concentrations, the whole tRNA content of the tissue must be used up for protein synthesis in less than a minute. With such rapid utilization of tRNA pools, it is difficult to see how they could be sharply compartmented.

Barondes: But aminoacyl-tRNA remains the immediate precursor. I agree that in a tissue like brain with multiple cell types this estimate could also be misleading. Specific activity of aminoacyl-tRNA might be much higher in some cells than others, but they might be making less protein. Still, an estimation based on the specific activity of aminoacyl-tRNA is more likely to reflect the true rate of protein synthesis *in vivo*.

Roberts: I tend to agree with Dr Lajtha. On scrutiny, the data which suggest a profound decline in protein synthesis in the brain during development become less and less convincing. I am not even sure that it is halved. The *in vitro* results can largely be explained by the fact that brain polysomes or protein-synthesizing systems in the developing animals are much less stable than those in immature animals.

To use tRNA as an index of the precursor for protein synthesis instead of the amino free acid pool is ill-advised because there is not only the possible compartmentation of tRNA but also the turnover of the aminoacyl-tRNA.

Barondes: If one measures the specific activity of tRNA for a given amino acid at various times during the course of incorporation and integrates this over the duration, one has a direct measure of the specific activity of the precursor pool that is being used to make protein. I have more confidence in estimates based on this than in those in which the specific activity of the precursor pool is estimated from the specific activity of the free amino acid in the preparation.

Wurtman: In normal conditions, what fraction of the brain tRNA for any particular amino acid is saturated with its amino acid?

Barondes: According to Johnson & Chou (1973) more than half the brain tRNA capable of binding to tryptophan is covalently bound to this amino acid. Other tRNAs are in the same range.

Wurtman: Is there experimental evidence that a selective reduction in the charging of one tRNA can by itself suppress protein synthesis?

Munro: In the red cell, Hunt *et al.* (1969) slowed the time of production of one globin chain by withdrawing threonine from the medium. They did not measure charging, but presumably there must have been a deficiency in the number of charged tRNA molecules.

Barondes: Withdrawal of an amino acid is a radical manipulation which can only be achieved *in vitro*. In starvation, the concentrations of circulating amino acids appear to be maintained by catabolism of liver protein. Certainly, the free amounts never come close to zero.

Mandel: Dr Barondes, even in the best preparation of nerve endings, contamination is at least 5%. Such contamination might explain data obtained for protein synthesis.

Barondes: Gambetti *et al.* (1972) attempted to settle this by electron microscope autoradiography and found that mitochondria in nerve endings incorporate amino acid into protein. Whether there is extra-mitochondrial synthesis of protein is unclear.

Mandel: Different pools of tRNA are found in the mitochondria, the nuclei, the cytoplasm and in the polysomes. It is not obvious that the specific activities of the free tRNA, the polysomal tRNA, and the different 'iso' tRNAs are the same.

We have developed a simple method for the isolation of aminoacyl-tRNAs: we run the glycerol–0.25M-sucrose–0.14M-NaCl (13 : 1 : 16 v/v) soluble fraction through a Sephadex column in a slightly acidic medium. The aminoacyl-tRNA is eluted readily.

One difficulty with electron microscope autoradiography is that one must use extremely thin slices and highly radioactive material. Thus, the method cannot be very sensitive: it is difficult to follow the trace, and to discover whether and where the trace is incorporated.

Barondes: I agree, but Gambetti *et al.* found enough disintegrations to count easily and they used an elegant method to localize individual grains. I believe that they provided strong evidence for protein synthesis in the mitochondria contained within nerve endings. They also found grains associated with plasma membranes of synaptosomes but did not prove that these grains are incorporated into protein. Had they shown that the grains were no longer seen after inhibition of protein synthesis with both cycloheximide and chloramphenicol, they would have proved their point.

Mandel: It is possible that *peptides* could be synthesized at the nerve endings? A few months ago we found a relatively high incorporation of amino acids in the phosphopeptide fraction.

Barondes: It is evident in microbial systems that synthesis of polypeptide antibiotics is not directed by mRNA and that ribosomes are not required. I believe the synthesis of the tripeptide, thyrotropin-releasing factor, is similar. In most studies of protein synthesis, these compounds would not be studied because they are not precipitable with trichloroacetic acid. In the autoradiography studies, they would probably have been washed out during fixation, but this is not certain.

Wurtman: Thyrotropin-releasing factor (pyroglutamylhistidylprolinamide) is apparently synthesized by enzymes, not polysomes. Reichlin and his co-workers have obtained evidence for the existence of a 'TRF–synthetase' complex.

Fernstrom: Vasopressin, an octapeptide hormone secreted from the posterior pituitary, is released by neurons whose nerve endings terminate in this organ but whose cell bodies lie in the supraoptic nucleus. Vasopressin is not synthesized in the posterior pituitary (i.e. in nerve endings) (Sachs *et al.* 1969) but rather in the supraoptic nucleus (cell bodies). Sachs *et al.* indicate that it is synthesized on polysomes, but their evidence is not compelling.

Lajtha: In addition to the classical pathways for polypeptide synthesis, possibly biologically active peptides can be formed by breakdown of pre-existing hormones or other precursor materials. For example, we discovered that the factor inhibiting the release of the melanocyte-stimulating hormone, namely, prolylleucylglycinamide, can be formed by splitting of the *N*-terminal tripeptide from oxytocin. (Breakdown of oxytocin also yields a peptide *stimulating* the release of the melanocyte-stimulating hormone.) The tripeptide can then be inactivated by brain peptidase with release of proline and leucine (Marks *et al.* 1973). Specific peptidases, therefore, can form and break down active peptides.

References

ADAMSON, L. F., HERINGTON, A. C. & BARNSTEIN, J. (1972) Evidence for the selection by the membrane transport system of intracellular or extracellular amino acids for protein synthesis. *Biochem. Biophys. Acta* **282**, 352–365

ALPERS, D. H. & THIER, S. O. (1972) Role of the free amino acid pool of the intestine in protein synthesis. *Biochem. Biophys. Acta* **262**, 535–545

FERN, E. B., HIDER, R. C. & LONDON, D. R. (1971) Studies *in vitro* on free amino acid pools and protein synthesis in rat jejunum. *Eur. J. Clin. Invest.* **1**, 211–215

FERNSTROM, J. D., LARIN, F. & WURTMAN, R. J. (1971) Daily variations in the concentrations of individual amino acids in rat plasma. *Life Sci.* **10**, 813–820

FERNSTROM, J. D., MADRAS, B. K., MUNRO, H. N. & WURTMAN, R. J. (1974) Nutritional control of the synthesis of 5-hydroxytryptamine in the brain, *This Volume*, pp. 153–166

FISHMAN, B., WURTMAN, R. J. & MUNRO, H. N. (1969) Daily rhythms in hepatic polysome profiles and tyrosine transaminase activity: role of dietary protein. *Proc. Natl. Acad. Sci. U.S.A.* **64**, 677–682

FLECK, A., SHEPHERD, J. & MUNRO, H. N. (1965) Protein synthesis in rat liver: influence of amino acids in diet on microsomes and polysomes. *Science (Wash. D.C.)* **150**, 628–629

GAMBETTI, P., AUTILIO-GAMBETTI, L. A., GONATAS, N. K. & SHAFER, B. (1972) Protein synthesis in synaptosomal fractions. *J. Cell Biol.* **52**, 526–535

HIDER, R. C., FERN, E. B. & LONDON, D. R. (1969) Relationship between intracellular amino acids and protein synthesis in the extensor digitorum longus muscle of rats. *Biochem. J.* **114**, 171–178

HUNT, R. T., HUNTER, A. R. & MUNRO, A. J. (1969) The control of haemoglobin synthesis: factors controlling the output of α and β chains. *Proc. Nutr. Soc.* **28**, 248–254

JOHNSON, T. C. & CHOU, L. (1973) Level and amino acid acceptor activity of mouse brain t-RNA during neural development. *J. Neurochem.* **20**, 405–414

LAJTHA, A. & DUNLOP, D. (1973) Alterations of protein metabolism during development of the brain in *Drugs and the Developing Brain* (Vernadakis, A. & Weiner, N., eds.), Plenum Press, New York

LAJTHA, A. & MARKS, N. (1971) Protein turnover in *Handbook of Neurochemistry*, vol. 5 (Lajtha, A. ed.), pp. 551–629, Plenum Press, New York

MARKS, N., ABRASH, L. & WALTER, R. (1973) Degradation of neurohypophyseal hormones by brain extracts and purified brain enzymes. *Proc. Soc. Exp. Biol. Med.* **42**, 455–460

PRONCZUK, A. W., ROGERS, Q. R. & MUNRO, H. N. (1970) Liver polysome patterns of rats fed amino acid imbalanced diets. *J. Nutr.* **100**, 1249–1258

RIGHETTI, P., LITTLE, E. P. & WOLF, G. (1971) Reutilization of amino acids in protein synthesis in HeLa cells. *J. Biol. Chem.* **246**, 5724–5732

ROBERTS, S. (1974) Effects of amino acid imbalance on amino acid utilization, protein synthesis and polyribosome function in cerebral cortex, *This Volume*, pp. 299–318

SACHS, H., FAWCETT, P., TAKABATAKE, Y. & PORTANOVA, R. (1969) Biosynthesis and release of vasopressin and neurophysin. *Recent. Prog. Horm. Res.* **25**, 447–392

VAN VENROOIJ, W. J., POORT, C., KRAMER, M. F. & JANSEN, M. T. (1972) Relationship between extracellular amino acids and protein synthesis *in vitro* in the rat pancreas. *Eur. J. Biochem.* **30**, 427–433

WEISS, B. F., MUNRO, H. N. & WURTMAN, R. J. (1971) L-Dopa: disaggregation of brain polysomes and elevation of brain tryptophan. *Science (Wash. D.C.)* **173**, 833–835

WURTMAN, R. J. (1970) Diurnal rhythms in mammalian protein metabolism in *Mammalian Protein Metabolism* (Munro, H. N. ed.), vol. 4, ch. 36, Academic Press, New York

WURTMAN, R. J., SHOEMAKER, W. J. & LARIN, F. (1968a) Mechanism of the daily rhythm in hepatic tyrosine transaminase activity: role of dietary tryptophan. *Proc. Natl. Acad. Sci. U.S.A.* **59**, 800–807

WURTMAN, R. J., ROSE, C. M., CHOU, C. & LARIN, F. (1968b) Daily rhythms in the concentrations of various amino acids in human plasma. *N. Engl. J. Med.* **279**, 171–175

Aromatic amino acid supply and brain protein synthesis

S. S. OJA, P. LÄHDESMÄKI* and M.-L. VAHVELAINEN+

*Department of Biomedicine, University of Tampere, Finland, *Department of Biochemistry, University of Oulu, Finland and + Department of Physiology, University of Helsinki, Finland*

Abstract The possibility that the intracellular concentration of amino acids regulates the rate of brain protein synthesis has been studied for phenylalanine, tyrosine and tryptophan. Hyperphenylalaninaemia decreased the exchange of tyrosine between the plasma and brain *in vivo*. Reciprocal competitive inhibition was observed between the aromatic amino acids in the influx into brain cortex slices. The kinetics of the incorporation of tritium-labelled aromatic amino acids into protein and of the incorporation of [³H]phenylalanine into polyphenylalanine and phenylalanyl-tRNA were studied in cell-free brain preparations. Phenylalanine competitively inhibited the incorporation of tyrosine into protein, and similarly tyrosine and tryptophan competitively inhibited the incorporation of phenylalanine into polyphenylalanine and phenylalanyl-tRNA. Extrapolation of these results suggests that neither the transport systems at the brain cell membranes nor the protein-synthesis systems inside the cells are saturated with aromatic amino acids. Alterations in extracellular concentrations could thus influence the influx of aromatic amino acids into brain cells and alterations in intracellular concentrations influence brain protein synthesis.

The possible role of amino acids in the regulation of protein synthesis in the brain is at present far from being resolved. Variations in the intracellular concentrations of amino acids can obviously influence the rate of protein synthesis in mammalian cells (Eagle *et al.* 1959; Morgan *et al.* 1971). Protein synthesis in the brain is also sensitive to changes in metabolite concentrations which lead to an amino acid imbalance (Peterson & McKean 1969; Agrawal *et al.* 1970). The nature of these relationships and the site of action at the molecular level have not yet been defined. A clear understanding of these phenomena is, however, necessary for a knowledge of the regulation of protein metabolism in brain. In this paper, we shall summarize the results of our investigation into the influence of varying intracellular concentrations of the aromatic amino acids, phenylalanine, tyrosine and tryptophan, on each other's polymerization into cerebral proteins, and of varying extracellular concentrations of the aromatic amino acids on their penetration across brain cell membranes in adult rats both *in vivo* and *in vitro*.

METHODS

Studies in vivo

The determination of the transport rates of tyrosine between plasma and brain and of the metabolic rates of brain proteins *in vivo* have been described in detail before (Oja 1967; Lindroos & Oja 1972). A catenary three-compartment model, open at the plasma end, was used to describe the transfer of L-[^{3}H]-tyrosine from plasma to brain and to brain proteins, and *vice versa*. The model was then resolved with digital computers. The alterations in the specific radioactivity of tyrosine in the compartments were simulated by best-fit, third order, polynomials (Lindroos & Oja 1972).

Influx of aromatic amino acids into brain cortex slices

The experimental conditions are described in detail elsewhere (Vahvelainen & Oja 1972). Cortical slices from rat brain were incubated in Krebs–Ringer phosphate medium (5 ml; pH 7.4) for five minutes at 37 °C under oxygen with glucose (10 µmol/ml), ^{3}H-labelled L-amino acid (1 µCi/ml) and various amounts of the corresponding unlabelled amino acid. When the influxes of the two aromatic amino acids were to be determined simultaneously, one amino acid was ^{14}C-labelled (0.1 µCi/ml).

We assumed that the rate of influx of amino acids into brain cortex slices could be described by the equation:

$$v = \frac{V_{\max} S}{S + K_m} + K_D S$$

where v is the velocity of influx, $V_{\max}$ the maximal velocity of the saturable transport, K_m the transport constant equivalent to the Michaelis constant, K_D the diffusion constant and S the amino acid concentration in the medium. In order to calculate these constants, we first separated the saturable transport from the non-saturable transport due to diffusion. After a successful subtraction of the non-saturable transport, the remaining saturable transport alone should yield a straight line in a plot of velocity v against v/S. We programmed a digital computer to search by trial and error for an estimated value for K_D which best satisfied the above stipulation. The other calculation procedures have been previously described (Vahvelainen & Oja 1972).

Incorporation of aromatic amino acids into protein in cell-free brain preparations

The incorporation of [^{3}H]phenylalanine, [^{3}H]tyrosine and [^{3}H]tryptophan into the protein of rat brain homogenates has been studied (Oja 1972). Samples (1 ml) were incubated at 37 °C (pH 7.4) for five minutes with 1–10 µCi of ^{3}H-labelled aromatic amino acids and various amounts of unlabelled aromatic amino acids.

The synthesis of protein, polyphenylalanine and phenylalanyl-tRNA was also studied with subcellular fractions from the brain. We prepared free ribosomes and cell sap according to the method of Andrews & Tata (1971) and determined the RNA content of the ribosomal preparations by Ragnotti's method (1971). We measured the protein content of the cell sap spectro-photometrically at 280 nm or as described by Lowry *et al.* (1951).

The standard incubation mixture contained Tris–HCl buffer (50 µmol; pH 7.4), KCl (100 µmol), MgCl$_2$ (10 µmol), 2-mercaptoethanol (10 µmol), ATP (3 µmol), GTP (1 µmol), creatine phosphate (5 µmol) and creatine phosphokin-ase (10 µg) in a final volume of 1 ml. Cell sap (2 mg protein), ribosomes (50–100 µg RNA) and each of the other 20 protein amino acids (100 nmol) were in-cluded in the study of the incorporation of [^{3}H]phenylalanine, [^{3}H]tyrosine or [^{3}H]tryptophan (2 µCi) into protein. In some incubations, we used an excess of one of the other protein amino acids. To study the synthesis of polyphenylalanine, we included polyuridylic acid (200 µg) in the reaction mixture. Only cell sap (6 mg protein) was added to the standard mixture when we examined the formation of [^{3}H]phenylalanyl-tRNA from [^{3}H]phenylalanine (2 µCi). We studied the syn-thesis of protein and polyphenylalanine from phenylalanyl-tRNA with [^{3}H]-phenylalanyl-tRNA, prepared as above. [^{3}H]Phenylalanyl-tRNA, ribosomes and 20 protein amino acids were added, as described, to the standard incubation mixture. We used puromycin (2 µmol/ml), tyrosyl-tRNA or tryptophyl-tRNA as an inhibitor of polyphenylalanine synthesis. The reaction rates were almost constant during incubation for 30 min.

We prepared the protein and polyphenylalanine samples according to the method of Andrews & Tata (1971) and the phenylalanyl-tRNA samples as described by Fangman & Neidhardt (1964). All the samples were dissolved in 1M-NaOH and mixed with a Beckman solubilizer (Bio-Solv BBS-2). We deter-mined the radioactivity of the samples as before (Lähdesmäki & Oja 1972).

RESULTS

Intraperitoneally administered phenylalanine (3 μmol/g) increased the plasma concentrations of phenylalanine and tyrosine more than six times and three times respectively (Table 1). The cerebral concentration of phenylalanine rose by a factor of about 2.5 and that of tyrosine by about 30%. Loading with tyrosine had no effect on phenylalanine concentrations while the plasma concentration of tyrosine doubled and its cerebral concentration rose by 40%. Despite the increase in the plasma and brain concentrations of tyrosine, the tyrosine exchange rate between plasma and brain diminished in hyperphenyl-alaninaemic rats. The exchange rates between the free and protein-bound compartments of the brain in both experimental groups were significantly lower ($P < 0.01$) than in the control group.

TABLE 1

Concentrations of free phenylalanine and tyrosine in plasma and brain, and exchange rates of tyrosine between plasma and brain and between free and protein-bound tyrosine in brain in adult rats *in vivo*.

| Treatment | Concentration | | | | Exchange rates of tyrosine | |
| | Phenylalanine | | Tyrosine | | | |
	Plasma (nmol/ml)	Brain (nmol/g)	Plasma (nmol/ml)	Brain (nmol/g)	Plasma/brain (nmol min^{-1} g^{-1})	Free/protein-bound (nmol min^{-1} g^{-1})
Controls	195 ± 12	134 ± 6	107 ± 7	92 ± 4	6.51 ± 0.89	3.31 ± 0.41
Phe (3 μmol/g)	1278 ± 93	317 ± 20	341 ± 35	118 ± 4	4.69 ± 0.42	2.20 ± 0.15
Tyr (3 μmol/g)	194 ± 10	128 ± 8	203 ± 18	127 ± 6	7.28 ± 0.99	2.19 ± 0.12

The control animals received [³H]tyrosine (0.5 μCi/g; intraperitoneally) without any other treatment; the second and the third group of animals had received phenylalanine or tyrosine (3 μmol/g; intraperitoneally) 30 min earlier. The rats were killed at various intervals from 5 to 45 min after the administration of the label. The results (± S.E.M.) are given per g of brain fresh weight or per ml of plasma. Data from Lindroos & Oja (1972).

Fig. 1 depicts the influx of aromatic amino acids into slices of brain cortex as a function of the concentration of the medium. It also shows that the influx of one aromatic amino acid may be effectively inhibited by another aromatic amino acid. Table 2 lists the results of our calculations of Michaelis constants (K_m) and of the maximal velocities (V_{max}) of active transport. The K_D constants were on average 0.10 min^{-1}. These comprise both the diffusion into extra-cellular spaces of slices and the non-saturable penetration of the labelled amino acid into cells. The maximal velocity of saturable transport of tryptophan was significantly lower ($P < 0.01$) than that of phenylalanine or tyrosine, but the

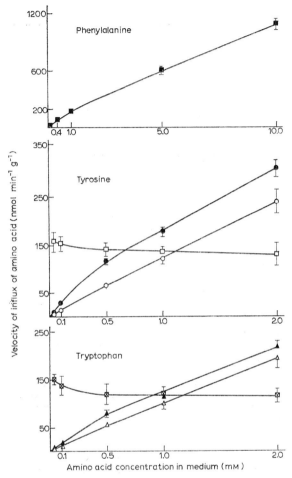

FIG. 1. Velocity of influx of phenylalanine, tyrosine and tryptophan into slices of adult rat brain cortex. Influx of [³H]phenylalanine (■) at various concentrations of phenylalanine in the incubation medium. Influx of [³H]tyrosine at various concentrations of tyrosine in the incubation medium with (○) and without (●) phenylalanine (1 μmol/ml), and the influx of [¹⁴C]phenylalanine (□) in the presence of various concentrations of tyrosine. Influx of [³H]tryptophan at various concentrations of tryptophan in the incubation medium with (△) and without (▲) phenylalanine (1 μmol/ml) and the influx of [¹⁴C]phenylalanine (⊠) in the presence of varying concentrations of tryptophan. Results (± S.D.) are given per unit of fresh weight of unincubated slices without any subtractions for extracellular space.

affinity of tryptophan for the transport systems was greater. Phenylalanine seems to be a competitive inhibitor of the influx of other aromatic amino acids across brain cell membranes, shown by the significant rise ($P < 0.01$) in the K_m values for tyrosine and tryptophan in the presence of phenylalanine.

288 S. S. OJA et al.

TABLE 2

Apparent Michaelis constants (K_m) and maximal velocities of transport (V_{max}) of aromatic amino acids in brain slices of adult rats.

Amino acid transported	Inhibitor amino acid	K_m/mM	$V_{max}/nmol\ min^{-1}\ g^{-1}$
Phe	None	0.68 ± 0.10	132 (121–145)
Tyr	None	0.57 ± 0.10	162 (144–182)
Tyr	Phe	1.38 ± 0.24	100 (81–123)
Trp	None	0.37 ± 0.07	55 (50–60)
Trp	Phe	1.92 ± 0.70	55 (32–95)

The results with their 95 % confidence limits are calculated from the experiments depicted in Fig. 1. They are given per fresh weight of incubated slices.

TABLE 3

Effect of one aromatic amino acid on the incorporation into protein of another.

Inhibitor amino acid	Rate of incorporation of (% of control)		
	Phenylalanine	Tyrosine	Tryptophan
None	100.0 ± 5.8	100.0 ± 4.7	100.0 ± 6.1
Phe	—	74.0 ± 8.3^a	103.2 ± 2.2
Tyr	61.5 ± 7.3^b	—	99.0 ± 2.3
Trp	80.1 ± 6.9^a	94.6 ± 7.5	—

^{a} $P<0.05$.
^{b} $P<0.01$.

Subcellular preparations from rat brain were incubated with ^{3}H-labelled aromatic amino acid (0.1 µmol/ml) and with an excess of 1 µmol/ml of some other aromatic amino acid. Results ($\pm$ s.e.m.) are given as percentages of the corresponding control incubation without the inhibitor amino acid for seven experiments.

Table 3 shows that a tenfold excess of tyrosine or tryptophan inhibited the incorporation of [^{3}H]phenylalanine into the protein of subcellular preparations from rat brain. The incorporation of [^{3}H]tyrosine was also inhibited by phenylalanine at least. Phenylalanine and tyrosine did not affect the incorporation of [^{3}H]tryptophan. The incorporation of phenylalanine was inhibited by 23 % by puromycin (2 µmol/ml), but not by tryptophyl-tRNA or tyrosyl-tRNA. An excess of 1 µmol/ml of any of the other protein amino acids—glutamic acid, glutamine, aspartic acid, asparagine, histidine, valine, leucine, isoleucine, alanine, glycine, threonine, serine, cysteine, methionine, proline, lysine or arginine—did not inhibit the incorporation of phenylalanine, tyrosine or tryptophan.

The incorporation of aromatic amino acids into protein of brain homogenates seems to obey the Michaelis kinetics reasonably well with respect to the incuba-

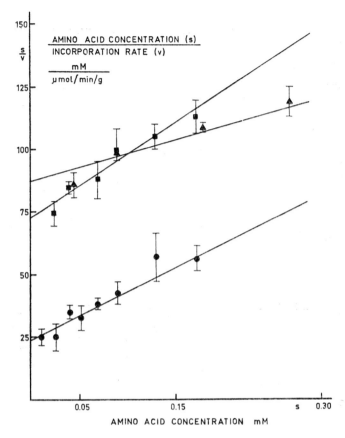

FIG. 2. Ratio of the concentration of phenylalanine (■), tyrosine (●) and tryptophan (▲) in the incubation medium (mM) to the rate of incorporation of the respective amino acid into protein (μmol min⁻¹ [g protein]⁻¹) as a function of its concentration (S/v against S). Each point (± s.e.m.) is the mean of four or five experiments. Data are from Oja (1972).

tion medium concentration (Fig. 2). The apparent K_m value for tryptophan was about seven times as great and that for phenylalanine more than twice as great as that for tyrosine (Table 4). The K_m for tyrosine increased significantly ($P <$ 0.01) in the presence of an excess of phenylalanine, as did the K_m for phenylalanine in the presence of tryptophan. These obviously represent cases of competitive inhibition, because the apparent affinity of tyrosine and tryptophan diminishes in the presence of the inhibitor amino acid. On the other hand, V_{max} for tyrosine diminished significantly ($P < 0.01$) in the presence of an excess of tryptophan, while the K_m remained unaltered. The type of inhibition provoked by tyrosine in the incorporation of phenylalanine could not be determined, since

TABLE 4

Apparent Michaelis constants (K_m) and maximal velocities (V_{max}) of incorporation into protein for phenylalanine, tyrosine and tryptophan.

Inhibitor amino acid	Incorporation into protein					
	Phenylalanine (5)		*Tyrosine (5)*		*Tryptophan (4)*	
	K_m/m$_M$	V_{max}/nmol min^{-1} g^{-1}	K_m/m$_M$	V_{max}/nmol min^{-1} g^{-1}	K_m/m$_M$	V_{max}/nmol min^{-1} g^{-1}
None	0.29 ± 0.03	4.0 ± 0.4	0.13 ± 0.02	5.5 ± 0.3	0.87 ± 0.10	10.0 ± 1.1
Phe	—	—	0.27 ± 0.02	6.1 ± 0.3	—	—
Tyr	0.33 ± 0.09	3.4 ± 0.1	—	—	—	—
Trp	0.49 ± 0.02	4.8 ± 0.8	0.13 ± 0.001	2.4 ± 0.1	—	—

Constants of control incubations are calculated from experiments described in Fig. 2. Similar incubations were performed in the presence of 1.7m$_M$-tyrosine, -tryptophan or -phenylalanine as an inhibitor amino acid. Means ($\pm$ s.e.m.) are given. The number of experiments is given in parentheses. Data are from Oja (1972).

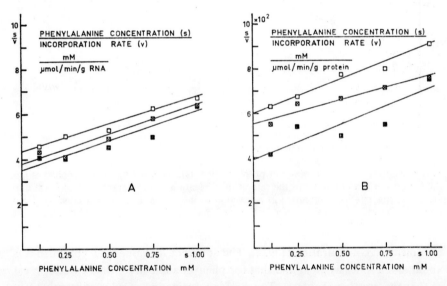

FIG. 3. Ratio of the concentration of phenylalanine in the incubation medium (m$_M$) to the rate of incorporation of phenylalanine into (A) polyphenylalanine (nmol min^{-1} [g RNA]$^{-1}$) or (B) phenylalanyl-tRNA (nmol min^{-1} [g protein]$^{-1}$) as a function of the concentration of phenylalanine (S/v against S). Incorporation of [^{3}H]phenylalanine with an excess (5 µmol/ml) of tyrosine (□) or tryptophan (⊠) and without inhibitors (■). Each point is the mean of six or seven experiments.

TABLE 5

Apparent Michaelis constants (K_m) and maximal velocities (V_{max}) of incorporation into poly-phenylalanine and phenylalanyl-tRNA for phenylalanine.

	Incorporation of [^3H]phenylalanine into			
Inhibitor amino acid	Phenylalanyl-tRNA		Polyphenylalanine	
	K_m/mM	V_{max}/nmol min^{-1} (g protein)$^{-1}$	K_m/mM	V_{max}/nmol min^{-1} (g RNA)$^{-1}$
None	1.3	3.16	1.3	382
Tyr	1.8	3.27	1.8	415
Trp	2.6	4.69	1.5	380

The constants were graphically determined from the experiments described in Fig. 3.

tyrosine did not produce statistically significant differences in either K_m or V_{max}. The K_m and V_{max} values for tryptophan were not determined in the presence of phenylalanine or tyrosine since these amino acids did not inhibit the incorporation of tryptophan to any significant degree.

The polyuridylic acid-directed synthesis of [^3H]polyphenylalanine and the formation of [^3H]phenylalanyl-tRNA from [^3H]phenylalanine were both inhibited by tyrosine and tryptophan (Fig. 3). The apparent K_m values for phenylalanine increased in the presence of the inhibitor amino acids, while the V_{max} values did not decrease (Table 5). This indicates that this inhibition is probably competitive.

DISCUSSION

The intracellular concentration of amino acids in the brain seems to be at least partially determined by transport processes. For instance, the concentration of various amino acids in vivo to a certain degree parallels the extent to which they accumulate in brain slices in vitro (Lajtha 1967; Battistin et al. 1969). Extrapolation of our results suggests that the mechanisms for active transport of aromatic amino acids are not saturated by extracellular amino acids. Let us assume, on the basis of the plasma data (Joosten et al. 1966), that the physiological extracellular concentrations of these amino acids in vivo vary from 0.05 to 0.1 μmol/ml. The calculated transport constants for phenylalanine, tyrosine and tryptophan are several times greater than these extracellular concentrations. This implies that the alterations in the extracellular concentrations affect the supply of aromatic amino acids to brain tissue. The loading of the organism with aromatic amino acids increased both their cerebral concentrations and exchange rates between plasma and brain *in vivo*. It must be admitted, however, that the increases were less than were predicted from the studies on brain slices.

The aromatic acids strongly inhibited each other's transport into brain slices. This competitive inhibition suggests that they share a common transport mechanism. The present study shows that *in vivo*, too, phenylalanine inhibited the exchange of tyrosine between plasma and brain, an inhibition which was not fully compensated by the elevated plasma and brain concentrations of tyrosine. The inhibition provoked by phenylalanine in the transport of other aromatic amino acids may be one reason for the disturbances of brain development encountered in phenylketonuria. In both hyperphenylalaninaemia and hypertyrosinaemia, the composition of the intracerebral pool of free amino acids is altered (McKean *et al.* 1968; Diasamidze 1970). This imbalance could also explain the inhibition of cerebral protein synthesis (Lindroos & Oja 1972).

From our present results, it is clear that the protein-synthesizing systems are not saturated with aromatic amino acids. The K_m values for phenylalanine, tyrosine and tryptophan in the process of incorporation into protein were all higher than the brain concentrations of these amino acids (Carver 1965). This relationship signifies that alterations in the concentrations of the intracellular amino acids affect the rates of synthesis of brain proteins. Not too much trust should be placed, however, in the comparisons between the calculated K_m values and the brain concentrations. Obviously, the simple concept of an intracellular amino acid pool freely available for translational events in brain protein synthesis is not tenable. We must emphasize the complexities involved in the compartmentation of amino acids in brain tissue and the unknown preferences or discriminations in protein synthesis with respect to precursors originating from different sources (for example, amino acid supply from plasma, from protein breakdown or from synthesis *in situ*). The phenomenological nature of the present K_m constants should also be kept in mind. The apparent K_m and V_{max} constants calculated cannot be given full significance in a multicompartment system such as the incorporation of amino acids into protein. Furthermore, the affinities of aromatic amino acids for protein synthesis systems may be totally different *in vivo* than in our preparations *in vitro*. It should be noted, for instance, that the present K_m values derived from experiments with homogenates and those derived with subcellular fractions differed by a magnitude.

The incorporation of phenylalanine and tyrosine into protein was inhibited by other aromatic amino acids. The concentrations of the inhibitor amino acids needed were so high that *in vivo* they are only met in animals with inborn errors of amino acid metabolism. Previous analyses have suggested that only a single rate-limiting step in the complex protein synthesis is affected (Oja 1972). The present studies with phenylalanine suggest that this step is most probably the formation of the aminoacyl-tRNA. The reciprocal inhibition of incorporation seems to be confined to aromatic amino acids only since all aliphatic amino acids were ineffective. The greater structural dissimilarity between the tryptophan molecule and the phenylalanine and tyrosine molecules presumably also renders the incorporation of tryptophan insensitive to the effects of excessive amounts of phenylalanine and tyrosine.

ACKNOWLEDGEMENTS

The technical assistance of Mrs Pirkko Erkkilä and Mrs Sirkka Soininen and the financial support of the Finnish National Research Council for the Natural Sciences are gratefully acknowledged.

References

AGRAWAL, H. C., BONE, A. H. & DAVISON, A. N. (1970) Effect of phenylalanine on protein synthesis in the developing rat brain. *Biochem. J.* **117**, 325–331

ANDREWS, T. M. & TATA, J. R. (1971) Protein synthesis by membrane-bound and free ribosomes of secretory and non-secretory tissues. *Biochem. J.* **121**, 683–694

BATTISTIN, L., GRYNBAUM, A. & LAJTHA, A. (1969) Distribution and uptake of amino acids in various regions of the cat brain *in vitro. J. Neurochem.* **16**, 1459–1468

CARVER, M. J. (1965) Influence of phenylalanine administration on the free amino acids of brain and liver in the rat. *J. Neurochem.* **12**, 45–50

DIASAMIDZE, G. A. (1970) Vliyanie nagruzki tirozinom, metioninom i lizinom na regional'noe raspredelenie fonda svobodnykh amino-kislot v golovnom mozgu belykh krys. *Vopr. Med. Khim.* **16**, 244–250

EAGLE, H., PIEZ, K. A., FLEISCHMAN, R. & OYAMA, V. I. (1959) Protein turnover in mammalian cell cultures. *J. Biol. Chem.* **234**, 592–597

FANGMAN, W. L. & NEIDHARDT, F. C. (1964) Demonstration of an altered aminoacyl ribonucleic acid synthetase in a mutant of *Escherichia coli. J. Biol. Chem.* **239**, 1839–1843

JOOSTEN, A., FAES, M. H., DULCINO, J. & DE LOECKER, W. (1966) The effect of cortisol on free amino acid levels in extra and intracellular fluids of rats. *Protides Biol. Fluids Proc. Colloq. Bruges* **14**, 701–707

LÄHDESMÄKI, P. & OJA, S. S. (1972) Effect of electrical stimulation on the influx and efflux of taurine in brain slices of newborn and adult rats. *Exp. Brain Res.* **15**, 430–438

LAJTHA, A. (1967) Transport as control mechanism of cerebral metabolite levels. *Prog. Brain Res.* **29**, 201–216

LINDROOS, O. F. C. & OJA, S. S. (1972) Hyperphenylalaninaemia and the exchange of tyrosine in adult rat brain. *Exp. Brain Res.* **14**, 48–60

LOWRY, O. H., ROSEBROUGH, N. J., FARR, A. L. & RANDALL, R. J. (1951) Protein measurement with the Folin phenol reagent. *J. Biol. Chem.* **193**, 265–275

McKEAN, C. M., BOGGS, D. E. & PETERSON, N. A. (1968) The influence of high phenylalanine and tyrosine on the concentrations of essential amino acids in brain. *J. Neurochem.* **15**, 235–241

MORGAN, H. E., EARL, D. C. N., BROADUS, A., WOLPERT, E. B., GIGER, K. E. & JEFFERSON, L. S. (1971) Regulation of protein synthesis in heart muscle. I. Effect of amino acid levels on protein synthesis. *J. Biol. Chem.* **246**, 2152–2162

OJA, S. S. (1967) Studies on protein metabolism in developing rat brain. *Ann. Acad. Sci. Fenn. Ser. A V Med.* **131**, 1–81

OJA, S. S. (1972) Incorporation of phenylalanine, tyrosine and tryptophan into protein of homogenates from developing rat brain: kinetics of incorporation and reciprocal inhibition. *J. Neurochem.* **19**, 2057–2069

PETERSON, N. A. & MCKEAN, C. M. (1969) The effects of individual amino acids on the incorporation of labelled amino acids into proteins by brain homogenates. *J. Neurochem.* **16**, 1211–1217

RAGNOTTI, G. (1971) A rapid method for the determination of ribonucleic acid in subcellular fractions. *Biochem. J.* **125**, 1057–1058

VAHVELAINEN, M.-L. & OJA, S. S. (1972) Kinetics of influx of phenylalanine, tyrosine, tryptophan, histidine and leucine into slices of brain cortex from adult and 7-day-old rats. *Brain Res.* **40**, 477–488

Discussion

Lajtha: When I mentioned (p. 277) that we found no evidence for a significant degree of compartmentation of the precursor pool for protein synthesis, I was referring to experiments in which influences on amino acid transport would not disturb the observations, for example, double-labelling experiments in which amino acids do not affect each other's transport. In short-term experiments on, for instance, the effect of phenylalanine on tyrosine incorporation, influences on amino acid transport could give a false impression that protein synthesis is altered. The major effect in your experiments could be on the initial penetration of amino acid into the preparation rather than on the incorporation into protein.

Oja: It is difficult to understand how that is possible in brain homogenates although it is most likely in brain slices.

Lajtha: Amino acids are also transported in brain particles, such as mitochondria and synaptosomes. If you incubated the homogenates with the labelled amino acid first and added the second amino acid later, transport effects might be reduced.

Oja: Tyrosine and tryptophan inhibited the incorporation of [³H]phenylalanine into polyphenylalanine and phenylalanyl-tRNA also in the subcellular preparations in which no brain particles, such as mitochondria or synaptosomes, were present.

Mandel: I am a little surprised at the straight line plots you presented. For so complex a phenomenon with at least four or five steps, inhibitory effects and different kinds of controls are possible. Surely, these K_m values are difficult to interpret, since there may be inhibitory or secondary phenomena in your system?

Oja: I am the first to admit the difficulties of interpretation; obviously, the system is complicated and consists of several steps. But if, in such a system, one step is rate-limiting, we can apply simple kinetics.

Mandel: In considering the exchange of amino acids between the different compartments or the incorporation into proteins you should take into account the fact that the half-life varies widely from protein to protein and that you are working with diverse cells. What does this oversimplification gain you?

Oja: It is difficult to design an experiment to measure true half-lives of brain proteins *in vivo* (Oja 1973*b*). The best solution would be to study the metabolism of one homogenous protein fraction (isolated from only one type of brain cells), but at present that is impossible to do. When studying a heterogenous mixture of proteins, the necessary assumptions I made are not formally correct, but what else can be done (Oja 1973*a*)? In autoradiographic studies, for example, the underlying assumptions are still less well founded: we know nothing about the specific radioactivity of precursors and about their distribution between the cells engaged in protein synthesis.

Mandel: Since specific proteins can be isolated, can't a 'clean' experiment on protein synthesis be performed?

Wurtman: Neurotubular proteins could be isolated.

Kaufman: There is an antibody for tyrosine hydroxylase. That could be used as a specific precipitating agent.

Oja: But even a single protein might originate from different types of cells in the brain. That complicates this simple-looking experimental situation (Oja 1972).

Moir: Professor Oja, the literature contains excellent data showing that, both *in vivo* and *in vitro*, aromatic amino acids interact with each other in the process of entering brain through active transport systems. However, we found (pp. 201 *et seq.*) that we could quadruple the amount of tryptophan in canine erythrocytes, (which also have an active transport system but with an alanine-preferring carrier) and increase it eightfold in brain without altering the concentration of tyrosine, that is, the amount of tyrosine available for subsequent metabolism is the same. Does competition in the dynamics of transport have any implication for the subsequent metabolism?

Oja: No.

Wurtman: Except perhaps in pathological states such as phenylketonuria, where the plasma concentration of a single amino acid (i.e. phenylalanine) is very high.

In view of the well established physiological competition among amino acids for transport into the brain, could we all agree that, in experiments on brain slices and even homogenates, it is wise to use media containing *all* the amino acids, or at least all that are known to compete for transport with the amino acid being studied? I suspect that the information that we would derive would be more physiologically relevant.

Professor Oja, your data indicate that the transport system for tryptophan is not saturated by the concentrations of tryptophan normally present in brain or plasma. Do the combined concentrations of tryptophan, tyrosine, phenylalanine and the branched chain amino acids together come nearer to saturating this particular system?

Oja: I imagine they would, but nobody has yet done such a study.

Moir: Guroff & Udenfriend (1962) showed that the uptake system of tryptophan interacts with that of tyrosine in the brain *in vivo*.

Wurtman: They gave large pharmacological doses (1–2 g) of tyrosine and tryptophan to rats.

When the plasma concentration of tryptophan increases after carbohydrate consumption and the concentrations of the other neutral amino acids decrease, the amino acids in brain exhibit parallel responses (see Fernstrom, p. 162). So I wonder whether the *total* amount of neutral amino acids in brain is constant, with only the proportions of individual amino acids changing, or whether suitable major increases in plasma amino acids could elevate *total* brain neutral amino acid concentrations.

Roberts: The total amount of amino acids in the brain may be reduced acutely after the administration of an excess of one amino acid.

Wurtman: After a loading dose of phenylalanine, which will certainly reduce the amounts of tyrosine, leucine and tryptophan, does the sum of phenylalanine plus the other neutral amino acids in the brain differ from the sum in brains of untreated animals? It is important to know whether the brain sum is more or less constant within each transport group.

Moir: Our data show that a large increase in tryptophan did not affect tyrosine.

Wurtman: But the brain concentrations of leucine, isoleucine and valine together exceed the large increase in tryptophan you mentioned.

The normal dynamic range of tryptophan concentration is considerable (100%). I wonder whether the normal increase in brain tryptophan is compensated by a corresponding decrease in one of its competitors.

Roberts: It is a mistake to draw up such a balance sheet only in terms of transport mechanisms because we have observed increased use of certain brain amino acids after a loading dose of phenylalanine which results in decreases in their concentration in the brain, entirely without regard to their position within a transport system. As you pointed out, Professor Oja, it takes large concentrations of one amino acid to inhibit the incorporation of another amino acid into protein in homogenates or into protein attached to ribosomes. However, conclusions derived from this observation neglect the fact that there are already present in these preparations basal amounts of amino acids. This phenomenon might also alter the calculation of the differences between animals of different ages, if there were different pool sizes present to start with.

Oja: I had already taken this endogenous source into account.

Gál: The half-life for the efflux of L-[^{14}C]tryptophan from rat brain is about 52–55 min (Gál *et al.* 1964). Since amino acid analogues such as *p*-chlorophenyl-alanine or α-methyltryptophan will interfere with the influx of tryptophan and 5-hydroxytryptophan, shouldn't concentration interfere with efflux of cerebral tryptophan?

Moir: Something like that must happen otherwise we could not explain the stability of the tyrosine concentration after a large tryptophan load when we know that there is interference with the dynamic parameters of uptake.

Kaufman: In phenylketonuria, where the blood concentration of phenyl-alanine goes up 25–50-fold, we are unable to detect a defect at the gross level of protein content and brain size. Only if we examine a specific protein such as myelin do we notice something wrong in the brains of phenylketonuric patients. Obviously, the brain somehow protects itself against these terrible amino acid imbalances.

Wurtman: What is known about the brain amino acid pattern in phenyl-ketonuria?

Kaufman: Nobody has detected a difference at the gross level.

References

Gál, E. M., Morgan, M., Chatterjee, S. K. & Marshall, G. D. (1964) Hydroxylation of tryptophan by brain tissue *in vivo* and related aspects of 5-hydroxytryptamine metabolism. *Biochem. Pharmacol.* **13**, 1639–1653

Guroff, G. & Udenfriend, S. (1962) Studies on aromatic amino acid uptake by rat brain. *J. Biol. Chem.* **237**, 803–806

Moir, A. T. B. (1974) Tryptophan concentration in the brain, *This Volume*, pp. 179–206

Oja, S. S. (1972) On the measurement of metabolic rates of brain proteins in *Ergebnisse der experimentellen Medizin. Vol. 10. Biochemical, Physiological and Pharmacological Aspects of Learning Processes* (Krug, M. & Winter, R., eds.), pp. 248–254, VEB Verlag Volk und Gesundheit, Berlin

Oja, S. S. (1973a) Comments on the measurement of protein synthesis in the brain. *Int. J. Neurosci.* **5**, 31–33

Oja, S. S. (1973b) Determination of transport rates in brain *in vivo* in *Research Methods in Neurochemistry*, vol. 2 (Marks, N. & Rodnight, R., eds.), pp. 183–216, Plenum Press, New York

Effects of amino acid imbalance on amino acid utilization, protein synthesis and polyribosome function in cerebral cortex

SIDNEY ROBERTS

Department of Biological Chemistry, School of Medicine and the Brain Research Institute, University of California Center for the Health Sciences, Los Angeles, California

Abstract The concentrations of free amino acids in cerebral cortex of immature and adult rats were altered acutely by intravenous administration of a loading dose of one amino acid. Within five minutes of intravenous administration of phenylalanine, leucine or valine (100 µmol/100 g body wt.), significant decreases were noted in cerebral concentrations of certain other amino acids, including aromatic compounds. The depression was progressively accentuated for 30–60 min after amino acid administration and resulted in a maximal decline of 50% in concentrations of the other amino acids. This phenomenon was accompanied by decreased uptake of amino acids from the blood into the free amino acid pool and into total proteins of cerebral cortex as well as enhanced catabolism of cerebral amino acids. However, the incorporation of amino acids administered intracisternally into cerebral protein and cerebral polyribosome formation were either unaffected or actually accentuated under these conditions. Cerebral polyribosome disruption, observed after intraperitoneal administration of massive amounts of phenylalanine or leucine, appeared to be due to the release of lysosomal hydrolytic enzymes rather than a specific effect of the resulting amino acid imbalance on protein-synthesizing mechanisms. The present results support the earlier suggestion that, although amino acid imbalance may restrict cerebral uptake of amino acids, the concurrent synthesis of cerebral proteins is not necessarily inhibited and may, in fact, be acutely stimulated.

Abnormalities of the human central nervous system in prenatal or early postnatal life have been described in connection with derangements of the metabolism of several amino acids, including aromatic, branched-chain and sulphur amino acids (Nyhan 1967; Rosenberg 1969; Gaull 1972; Wiltse & Menkes 1972). The roster of amino acid dyscrasias which are associated with brain damage has increased annually over the past decade and appears to be far from complete. In several instances, the primary defect in these diseases occurs outside the brain and involves a deficiency in amino acid transport or metabolism. Alterations in circulating concentrations of one or more amino acids and their metabolites are presumed to be directly or indirectly responsible for the resulting aberrations in brain development and function. Several observations suggest that the neurological manifestations of the aminoacidopathies derive in part from selective inhibition of the synthesis of certain brain proteins: for example, myelin proteins in human and experimental phenylketonuria and maple-syrup urine disease (Menkes 1968; Silberberg 1969; Agrawal *et al.* 1970; Shah *et al.* 1972). However, the precise relationship between alterations in extracellular or intracellular amounts of amino acids and brain metabolism has not been elucidated. To a considerable extent, this lack of knowledge may be attributed to the fact that experimentally induced variations in the concentration or distribution of one or more amino acids rapidly result in changes in cerebral metabolism of these and other amino acids as well (Roberts & Morelos 1965; Roberts 1968; Roberts *et al.* 1971). Moreover, metabolic products of amino acids appear in altered quantities and may affect the processes under investigation. These secondary metabolic alterations include not only increased production of toxic metabolites of the amino acid present in excess, but also changes in the formation or availability of important amino acid metabolites, such as the biogenic amines (Fellman 1956; Goldstein 1961; Shah *et al.* 1969; Harper *et al.* 1970; Bowden & McArthur 1972; Edwards & Blau 1972; Knott & Curzon 1972). The investigations described here are principally concerned with relatively acute alterations in amino acid and protein metabolism in the brain in response to a primary elevation in the circulating level of a single amino acid. Major hepatic conversion of the amino acid given in excess into other metabolites has thus been avoided and the results obtained may be assumed to be due principally to the ensuing amino acid imbalance. Under these conditions the major effects of the alterations in amino acid supply to the brain may include (1) decreases in cerebral uptake and concentrations of other amino acids and (2) enhanced cerebral utilization of amino acids for degradative reactions and, possibly, synthetic reactions.

METHODS

Most of the methods have been described in detail elsewhere (Roberts 1963; Roberts & Morelos 1965; Roberts 1968; Zomzely-Neurath & Roberts 1972). Male rats of an inbred Sprague–Dawley strain were obtained before weaning at 7 and 14 days of age, or were maintained on Purina laboratory chow after weaning until they were six weeks old. In some instances, the adult animals were fasted overnight (16–20 h) before the experiments. Amino acids, dissolved in saline and neutralized, were administered intraperitoneally without anaesthesia or intravenously (into the saphenous vein) under light anaesthesia with sodium pentobarbital. At various intervals (5–60 min), the rats were killed by decapitation or by exsanguination from the dorsal aorta. Both these methods gave comparable results (Roberts 1963; Zomzely et al. 1971). The cerebral cortices were dissected free of underlying tissue at 0–4 °C. Cerebral and plasma free amino acids, cerebral amino acid degradation and incorporation of amino acids into cerebral proteins in vivo were determined essentially as described previously (Roberts 1963, 1968; Roberts & Morelos 1965).

The state of aggregation of cerebral ribosomes was ascertained in the postmitochondrial supernatant fraction from cerebral cortical homogenates as well as in the polyribosomes isolated from this fraction (Zomzely et al. 1968). The tissue was homogenized in medium (1:3 w/v) containing 0.25M-sucrose, 12mM-$MgCl_2$, 100mM-KCl and 50mM-Tris-HCl at pH 7.6. Postmitochondrial supernatant fluids were layered onto a linear sucrose gradient (5 ml; 10–40%) containing buffer and salts in the same concentrations as the homogenization medium. After centrifugation at 130 000 g and 0 °C for 1.5 h in a Spinco SW 39 swinging bucket rotor, the gradients were analysed for materials which absorbed ultraviolet radiation at 254 nm. Purified polyribosomes were prepared from the postmitochondrial supernatant fractions by centrifugation for four hours at 105 000 g through 2M-sucrose which contained the same buffer and salts.

The pellet was rinsed several times with the homogenization medium and then suspended in this medium for sucrose density gradient analysis (Zomzely-Neurath & Roberts 1972). Special precautions to minimize ribonuclease action in these ribosome analyses included sterilization of media, glassware, and polyethylene ware, addition of a purified ribonuclease inhibitor protein from rat liver (Shortman 1961) to all media except the sucrose gradients, and washing of nitrocellulose tubes with diethyl pyrocarbonate, a potent protein-denaturing agent. In certain instances when the liver ribonuclease inhibitor was omitted from the media, the postmitochondrial supernatant fraction and the purified polyribosomes were assayed for ribonuclease activity by modifications (Zomzely et al. 1968; Szego et al. 1971) of the sensitive method of Barondes & Nirenberg (1962) which is based on the degradation of [^{14}C]poly(U). One modification specifically measures the activity of acid ribonuclease II, including that released from lysosomes by Triton treatment, and therefore provides an index of possible lysosomal labilization or disruption.

RESULTS AND DISCUSSION

Effect of amino acid imbalance on cerebral uptake and degradation of amino acids

The distribution of amino acids between plasma and brain is complicated by
the existence of specialized barriers which selectively restrict net uptake of
certain amino acids, as well as by the existence of competitive interactions in
membrane transport and variations in cellular metabolism of amino acids
(Chirigos *et al*. 1960; Lajtha & Toth 1961; Roberts & Morelos 1965; Blasberg &
Lajtha 1966; Roberts 1968; Scriver & Hechtman 1970). Nevertheless, primary
variations in circulating concentrations of one or more amino acids, produced
by dietary means or parenteral administration, generally result in alterations in
cerebral concentrations and use of these and other amino acids (Schwerin
et al. 1950; Kamin & Handler 1951; Tigerman & MacVicar 1951; Lajtha 1958;
Dingman & Sporn 1959; Chirigos *et al*. 1960; Lajtha & Toth 1961; Guroff &
Udenfriend 1962; Roberts *et al*. 1962; Roberts 1963, 1968; Peng *et al*. 1972).

In our earlier investigations, we found that the cerebral concentrations of
certain amino acids in adult rats rapidly and profoundly declined after intravenous administration of large doses of one other amino acid (e.g., 100 µmol/100 g
body wt.) (Roberts & Morelos 1965; Roberts 1968). Acute lowering of certain
other amino acids in the brain has also been noted after parenteral administration of the natural L-isomers of leucine (Roberts & Morelos 1965), isoleucine
(Carver 1969), phenylalanine (Carver 1965; McKean *et al*. 1968; Roberts 1968;
Lowden & LaRamée 1969; Agrawal *et al*. 1970; Wong *et al*. 1972), methionine
(Daniel & Waisman 1969), valine (Roberts 1968) and dopa (Weiss *et al*. 1971).
The investigations described below indicate that the processes in this phen-
omenon include enhanced intracerebral degradation of amino acids (see also
Roberts & Morelos 1965; Roberts 1968), as well as competition among amino
acids for brain transport mechanisms (Udenfriend 1963; Lajtha 1964).

Within five minutes of intravenous administration of phenylalanine (100
µmol/100 g body wt.) to adult male rats which had been previously fasted
overnight to reduce uncontrollable variations in endogenous amino acid
concentrations, plasma and cerebral concentrations of free phenylalanine were
roughly trebled (see for example Fig. 1). Maximal cerebral concentrations of
free phenylalanine were observed about 15 min after administration of this
amino acid while tyrosine concentration was significantly raised only after 30
min. The increase in cerebral tyrosine was due to influx of this amino acid
from plasma where elevated concentrations appeared owing to the hepatic
metabolism of phenylalanine. Cerebral concentrations of leucine and histidine
declined significantly as early as five minutes after intravenous administration

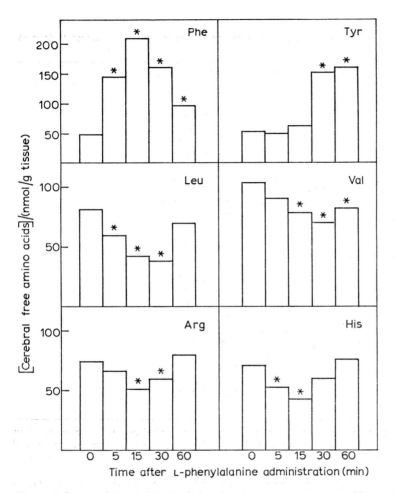

FIG. 1. Influence of elevated levels of phenylalanine on concentrations of free amino acids in rat cerebral cortex. Adult rats were fed Purina laboratory chow *ad libitum* before the experiment and were then deprived of food for 16–20 h. Phenylalanine (100 μmol/100 g body wt.) was given intravenously in 0.9% saline. The animals were killed at intervals of 5–60 min. The zero-time values were obtained from animals given 0.9% saline alone. The results represent the averages $\pm$ s.e.m. for four animals. Values which are significantly different from control values ($P \leqslant 0.01$–0.02) are marked with an asterisk (*).

of phenylalanine, but for valine, arginine and other cerebral free amino acids the decrease became significant 10–15 min later (see also Table 1) and continued for at least 30 min in most instances. Alterations in concentrations of certain cerebral amino acids were also observed after intravenous administration of equivalent amounts of L-leucine or L-valine to fasted rats (see for example

TABLE 1

Influence of elevated levels of L-phenylalanine on concentrations of free amino acids in rat cerebral cortex[a]

	Cerebral free amino acids ($\mu mol/g$ tissue)	
	Saline	L-Phenylalanine
Taurine	3.74 ± 0.09	3.73 ± 0.11
Urea	1.13 ± 0.07	1.56 ± 0.15
Aspartic acid	2.98 ± 0.11	2.97 ± 0.13
Threonine	0.603 ± 0.012	0.572 ± 0.013
Serine	1.27 ± 0.03	1.25 ± 0.03
Asparagine	0.070 ± 0.012	0.059 ± 0.002
Glutamine	3.19 ± 0.20	3.75 ± 0.23
Proline	0.059 ± 0.005	0.067 ± 0.004
Glutamic acid	9.22 ± 0.18	8.31 ± 0.25[b]
Glycine	0.682 ± 0.018	0.602 ± 0.012[b]
Alanine	0.655 ± 0.035	0.539 ± 0.021[c]
2-Aminobutyric acid	0.018 ± 0.003	0.019 ± 0.004
Valine	0.100 ± 0.006	0.080 ± 0.005[c]
Isoleucine	0.049 ± 0.005	0.029 ± 0.003[b]
Leucine	0.086 ± 0.009	0.042 ± 0.003[b]
Tyrosine	0.045 ± 0.003	0.052 ± 0.006
Phenylalanine	0.047 ± 0.002	0.159 ± 0.007[b]
β-Alanine	0.031 ± 0.003	0.032 ± 0.003
4-Aminobutyric acid	1.20 ± 0.13	1.01 ± 0.05
Ornithine	0.015 ± 0.005	0.009 ± 0.002
Lysine	0.233 ± 0.014	0.240 ± 0.005
Histidine	0.067 ± 0.003	0.041 ± 0,004[b]
Arginine	0.072 ± 0.004	0.052 ± 0.004[b]
No. of analyses	5	4

[a] Adult animals were fed Purina laboratory chow *ad libitum* before the experiment and then deprived of food for 16–20 h. The animals were killed 10 min after intravenous injection of 0.9% NaCl or L-phenylalanine (100 $\mu mol/100$ g body wt.). The results given are means ± S.E.M.
[b] $P \leqslant 0.01$–0.02 for difference between this value and the corresponding value in the saline-treated group.
[c] $P < 0.05$.

Table 2); the lowering was not restricted to those amino acids for which common carriers are thought to exist (see Blasberg & Lajtha 1966) for intravenous administration of L-leucine (100 $\mu mol/100$ g body wt.) resulted in significant declines in the cerebral concentrations of most amino acids within 10 min when the amount of leucine in the brain increased 2–3-times (Roberts 1968).

The lowering of cerebral concentrations of certain amino acids after the intravenous administration of L-phenylalanine, leucine or valine was associated with inhibition of the cerebral influx of these other amino acids. Thus, cerebral accumulation of L-[U-^{14}C]valine, administered intravenously in tracer quantities,

TABLE 2

Influence of elevated amounts of L-leucine on concentrations of free amino acids in rat cerebral cortex[a] (modified from Roberts 1968).

	Cerebral free amino acids ($\mu mol/g$ tissue)	
	Saline	L-Leucine
Taurine	5.43 ± 0.060	4.33 ± 0.06
Urea	3.66 ± 0.220	2.77 ± 0.07[b]
Aspartic acid	3.63 ± 0.040	2.76 ± 0.08[b]
Threonine	0.607 ± 0.033	0.488 ± 0.033[b]
Serine	1.39 ± 0.020	1.12 ± 0.04[b]
Asparagine and glutamine	4.74 ± 0.06	3.79 ± 0.11[b]
Proline	0.068 ± 0.002	0.054 ± 0.003[b]
Glutamic acid	11.50 ± 0.24	8.97 ± 0.22[b]
Glycine	0.699 ± 0.011	0.543 ± 0.017[b]
Alanine	0.625 ± 0.009	0.461 ± 0.011[b]
2-Aminobutyric acid	0.034 ± 0.003	0.029 ± 0.003
Valine	0.105 ± 0.003	0.064 ± 0.003[b]
Cystathionine	0.016 ± 0.001	0.012 ± 0.001[c]
Methionine	0.048 ± 0.002	0.033 ± 0.001[b]
Isoleucine	0.052 ± 0.001	0.024 ± 0.002[b]
Leucine	0.084 ± 0.002	0.208 ± 0.009[b]
Tyrosine	0.076 ± 0.004	0.041 ± 0.001[b]
Phenylalanine	0.053 ± 0.002	0.023 ± 0.001[b]
β-Alanine	0.034 ± 0.001	0.034 ± 0.004
4-Aminobutyric acid	1.58 ± 0.05	1.31 ± 0.080
Ornithine	0.013 ± 0.001	0.011
Lysine	0.235 ± 0.008	0.187 ± 0.009[b]
Histidine	0.079 ± 0.003	0.048 ± 0.004[b]
Arginine	0.092 ± 0.004	0.061 ± 0.003[b]
Total NRS[d]	41.674	32.750
No. of analyses	6	6

[a] The conditions were the same as those in Table 1, except that leucine was given instead of phenylalanine.
[b] $P \leqslant 0.01$–0.02 for difference between this value and the corresponding value in the column to the left.
[c] $P < 0.05$.
[d] Ninhydrin-reacting substances.

was depressed by the concurrent injection of L-phenylalanine (Table 3). This depression was most striking during the first 5–15 min. Radioactivity in the acid-soluble fraction of plasma was about 30% higher in the phenylalanine-injected rats than in the saline controls over this period and was localized almost entirely in valine (not shown). In contrast, the percentage of total radioactivity associated with acid-soluble material in cerebral extracts which was attributable to free valine was significantly reduced by prior administration of phenylalanine (Table 3). This finding indicated that the degradation of cerebral

TABLE 3

Influence of elevated levels of L-phenylalanine on uptake and utilization of intravenously administered L-[^{14}C]valine in rat cerebral cortex[a] (from Roberts 1968).

Time after injection (min)	Radioactivity of cerebral acid-soluble fraction (cpm/g tissue)	Radiactivity in cerebral free valine		
		(cpm/g tissue)	(cpm/μmol)	(% of total)
Saline-injected				
5	8 930	7 230	69 300	81 ± 1
15	8 380	6 200	52 350	74 ± 2
30	6 550	3 210	28 400	49 ± 2
Phenylalanine-injected				
5	4 260	2 900	27 400	68 ± 3[b]
15	5 300	3 500	33 700	66 ± 2[c]
30	5 890	2 180	26 050	37 ± 3[b]

[a] Adult animals were fed Purina laboratory chow *ad libitum* before the experiment and were then deprived of food for 16–20 h. L-Phenylalanine was given intravenously in an amount equivalent to 100 μmol/100 g body wt. L-[U-^{14}C]Valine (26 Ci/mol) was also given intravenously at the same time in an amount equivalent to 3.75 μCi/100 g body wt. The results given are means of values obtained from three animals in each instance. In addition, S.E.M. are shown for the % of total radiactivity in the column effluent from the amino acid analyser which remained in the valine peak.
[b] P<0.02 for difference between this value and the corresponding value in the saline-treated group.
[c] P<0.05.

valine was enhanced by this procedure. Similarly intravenous administration of large doses of L-leucine strikingly stimulated the intracerebral utilization of leucine (Roberts & Morelos 1965). In both situations, the major identifiable radioactive products in the effluent from the amino acid analyser included the dicarboxylic amino acids and their amides.

We have also observed striking developmental differences in cerebral degradation of certain amino acids (Roberts *et al.* 1971). For example, cerebral cortical tissue from newborn or immature (14-day-old) rats exhibited little or no capacity to degrade leucine (Table 4). In contrast, leucine was rapidly converted into other products (principally the dicarboxylic amino acids) in the mature cerebrum. This degradative ability of the adult brain may constitute a protective mechanism against elevated intracellular concentrations of amino acids. Most of the available evidence suggests that young animals are more susceptible to damage of the brain by excesses of amino acids than older individuals. Possibly, this difference in susceptibility reflects the development of enzymes catabolizing amino acids (Schimke *et al.* 1965; Nakano *et al.* 1970) as well as alterations in brain–barrier mechanisms (Lajtha 1964).

TABLE 4

Developmental alterations in conversion of L-[^{3}H]leucine into other amino acids in rat cerebral cortex[a]

Age	Time after administration (min)	Conversion[b] (%)
Newborn	5	trace
	15	2.3
14 days	5	trace
	15	trace
42 days	5	20.3
	15	41.7

[a] L-[4,5-^{3}H$_2$]Leucine (30–50 Ci/mmol) was given intracisternally to rats which had not been previously fasted. Immature animals received 25 μCi of isotope; adult rats (42 days old) received 50 μCi. The animals were killed 5 or 15 min later. Postmitochondrial supernatant fractions were prepared and analysed for content and radioactivity of free leucine and other amino acids.
[b] Radioactive products formed from L-[4,5-^{3}H$_2$]leucine include phosphoserine, taurine, glutamic acid and glutamine.

Effect of amino acid imbalance on incorporation of amino acids into cerebral proteins

Elevated quantities of one amino acid or its metabolites may alter the synthesis and degradation of proteins in the brain or, possibly, of a specific population such as myelin proteins (Roberts & Morelos 1965; Appel 1966; Agrawal et al. 1970).

The capacity of an excess of one or more amino acids to inhibit incorporation of amino acids into total protein in brain cellular and subcellular preparations *in vitro* is well documented (Zomzely et al. 1964; Rubin & Stenzel 1965; Appel 1966; Folbergrová 1966; Campagnoni & Mahler 1967; Orrego & Lipmann 1967; MacInnes & Schlesinger 1971). Acute inhibition of the incorporation of amino acids into brain proteins has also been reported after administration of loading doses of phenylalanine or other amino acids *in vivo* (Swaiman et al. 1968; Agrawal et al. 1970; Lindroos & Oja 1971; MacInnes & Schlesinger 1971). This apparent inhibition has been described for cerebral cortex as well as whole brain and seems to occur most readily in young animals. However, it is not clear that variations in precursor pool sizes were adequately controlled in these studies. Other reports suggest that ribosomal systems isolated from the brain after acute loading with phenylalanine exhibit a reduced capacity for amino acid incorporation *in vitro* (Aoki & Siegel 1970; Wong et al. 1972). Unfortunately,

these workers used impure microsomal preparations in which the contribution
of variations in endogenous amino acids or other factors induced by the treatment could not be assessed. One such factor might be the release of lysosomal
ribonucleases in response to raised cerebral concentrations of the administered
amino acid (see later).

To elucidate this problem, we investigated the incorporation of L-[^{14}C]valine
into cerebral protein in adult rats which were intravenously injected with the
radioactive amino acid at the same time as an excess of another amino acid
was administered. Incorporation was depressed over the 30 min period studied,
particularly when phenylalanine or leucine was in excess (Table 5). This
depression was not eliminated even when we allowed for the specific activity
of the precursor pool for protein synthesis in calculating the relative incorporation. However, we attached little confidence to these results in view of the fact
that competitive interactions in amino acid transport severely limited entry of
the radioactive amino acid into the brain (see Table 3). Consequently, the
appearance of the labelled amino acid in cerebral proteins might not accurately
represent the incorporation of that amino acid; that is, radioactive and non-
radioactive valine might be differentially compartmented. This view was supported by subsequent experiments, in which the radioactive amino acid was
administered intracisternally immediately after the loading dose of amino acid
had been given intravenously. We used L-[^{14}C]lysine in order to minimize the
diversion of the radioactive amino acid into pathways other than protein synthesis. We gave L-phenylalanine, L-leucine, L-valine, L-lysine, L-proline or

TABLE 5

Influence of elevated levels of L-phenylalanine on incorporation of intravenously administered
L-[^{14}C] valine into rat cerebral proteins[a]

Time after injection (min)	Radioactivity in cerebral valine (cpm/μmol)		
	Protein-bound	Free	$10^5 \times$ Relative incorporation[b]
Saline-injected			
5	215 ± 13	69 300 ± 3 570	300 ± 8
15	529 ± 32	52 350 ± 3 450	1 010 ± 40
30	1 110 ± 28	28 400 ± 1 550	3 880 ± 125
Phenylalanine-injected			
5	43 ± 3	27 400 ± 880	159 ± 22
15	148 ± 9	33 700 ± 3 290	474 ± 39
30	641 ± 23	26 050 ± 2 620	2 520 ± 235

[a] The conditions were the same as those in Table 3.
[b] Ratio of radioactivity in protein-bound valine to that in free valine in cerebral cortex.

TABLE 6

Influence of elevated levels of various amino acids on incorporation of intracisternally administered L-[^{14}C]lysine into rat cerebral proteins[a]

Substance given intravenously	Time after administration (min)	$10^6 \times$ Relative incorporation
Saline	5	312 ± 96
	10	974 ± 145
L-Leucine	5	403 ± 78
	10	1 145 ± 195
L-Phenylalanine	5	365 ± 82
	10	944 ± 80
L-Lysine	5	752 ± 105[b]
	10	2 415 ± 300[c]

[a] All animals were fed Purina laboratory chow *ad libitum* before the experiment and were then deprived of food for 16–20 h. L-Phenylalanine, L-leucine or L-lysine was given intravenously in an amount equivalent to 100 μmol/100 g body wt. L-[U-^{14}C]Lysine (4.5 Ci/mol) was given intracisternally in an amount equivalent to 5 μCi/100 g body wt. The results shown are means ± s.e.m. of values obtained from four animals in each instance.
[b] $P<0.05$ for difference between this value and the corresponding value in the saline-injected group.
[c] $P<0.02$.

L-threonine in excess (100 μmol/100 g body wt.). In most cases, the incorporation of lysine into protein was unaltered (Table 6), but when lysine itself was given in excess, the relative incorporation increased. Once again, compartmentation of radioactive and non-radioactive precursors for protein synthesis may limit interpretation of these data (see Roberts & Morelos 1965). In any event we could obtain no evidence in these experiments for an acute depression in cerebral protein synthesis in the presence of elevated circulating and cerebral concentrations of phenylalanine, leucine or other amino acids.

Effects of amino acid loading on brain polyribosomes

The investigations just described revealed that intravenous administration of a loading dose of one amino acid does not acutely depress but may even stimulate protein synthesis in the mature cerebrum (Roberts & Morelos 1965; Roberts 1968). Current studies indicate that similar phenomena occur in the cerebrum of immature rats (7–14 days old) 30–60 min after a single intravenous injection of L-phenylalanine or L-leucine (100 μmol/100 g body wt.). Contrariwise, several investigators have reported that parenteral administration of a large excess of one amino acid might disaggregate brain polyribosomes and consequently decrease brain protein synthesis especially in young animals (Aoki & Siegel 1970; MacInnes & Schlesinger 1971; Siegel *et al.* 1971; Weiss *et al.* 1971, 1972; Wong *et al.* 1972). In these latter experiments, high concentrations of amino acid were administered (e.g., 100 mg/100 g body wt.) generally intraperitoneally. Thus, the polyribosome disruption observed in brain preparations from animals given large excess of an amino acid might be an indirect effect of this treatment rather than a direct action on the protein-synthesizing machinery. Our results support this conclusion.

We found that intraperitoneal administration of L-phenylalanine (100 mg/ 100 g body wt.) to immature animals (seven days old) appeared to disaggregate cerebral polyribosomes (Fig. 2), in confirmation of the results from other laboratories. We noted a large increase in the proportion of dimeric ribosomes in purified polyribosomes isolated from cerebral cortices of these animals 30–60 min after administration of the amino acid. At the concentration of magnesium ions in the media used to isolate these polyribosomes, few monoribosomes are

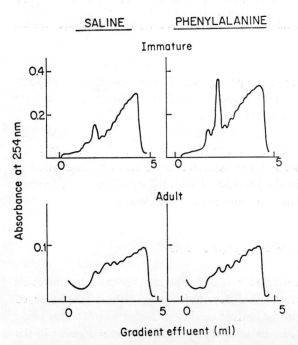

FIG. 2. Influence of phenylalanine loading on cerebral polyribosomes in immature and adult rats. Each animal was given 0.9% NaCl or 2.5% L-phenylalanine in 0.45% NaCl (4 ml/100 g body wt.) intraperitoneally. The animals were killed 60 min later and polyribosomes were isolated from the cerebri. Sucrose-density gradient analyses of the purified polyribosomes are shown. Immature rats were seven days old; adult rats were 42 days old.

normally observed. We noted no disaggregation of cerebral polyribosomes from adult rats given an excess of L-phenylalanine under comparable conditions and even found that polyribosomes isolated from adult rats given the loading dose of phenylalanine intraperitoneally may have been more highly aggregated than those obtained from control animals.

During the course of these experiments, the state of ribosomal aggregation was routinely monitored in the postmitochondrial supernatant fractions from which the purified polyribosomes were subsequently obtained by centrifugation through 2M-sucrose. Little or no indication of polyribosome disaggregation in response to intraperitoneal administration of phenylalanine was normally observed in cerebral postmitochondrial supernatant fluids from either immature or adult rats (Fig. 3). Once again, an increase in ribosome aggregation was evident in the preparation from mature cerebrum. When the cerebral postmitochondrial supernatant fractions were diluted with medium, the polyribosomes in all samples appeared to disaggregate partially (Fig. 4). However,

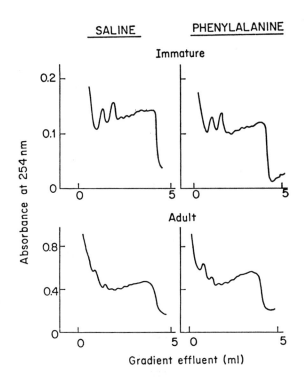

FIG. 3. Influence of phenylalanine loading on ribosomal profiles in postmitochondrial supernatant fluids prepared in the presence of liver ribonuclease inhibitor from cerebral cortices of immature and adult rats. See legend to Fig. 2 for explanations.

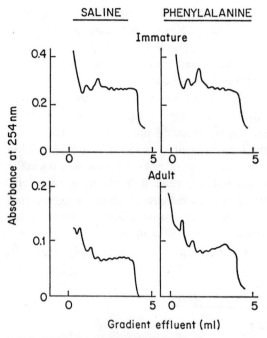

FIG. 4. Influence of phenylalanine loading on ribosomal profiles in diluted postmitochondrial supernatant fluids prepared from cerebral cortices of immature and adult rats. Before sucrose-density gradient analysis the samples were diluted 1:3 with medium free of ribonuclease inhibitor protein. See legend to Fig. 2 for other explanations.

dilution of such fractions obtained from immature rats treated with phenylalanine apparently caused more extensive polyribosome disruption than dilution of comparable samples prepared from animals given saline. This phenomenon was not noted with postmitochondrial supernatant fractions from adult rats. These results suggested that amino acid loading in the young animals led to the production or release of ribonucleases and possibly other hydrolytic enzymes from sequestered sites in the cerebrum. The reason that these enzymes apparently do not disrupt polyribosomes in the cerebral cortex *in situ* or in the undiluted postmitochondrial supernatant fraction is possibly because potent endogenous ribonuclease inhibitors are present (Roth 1956; Suzuki & Takahashi 1970). However, if the inhibitor is diluted or removed during polyribosome isolation, disruption of polyribosomes might follow in preparations from immature cerebrum. More evidence to support these conclusions came from our assays of the activity of lysosomal acid ribonuclease II in cerebral postmitochondrial supernatant fractions (Fig. 5) and in isolated cerebral polyribosomes (Fig. 6). Within 60 min of intraperitoneal administration of a loading dose of L-phenylal-

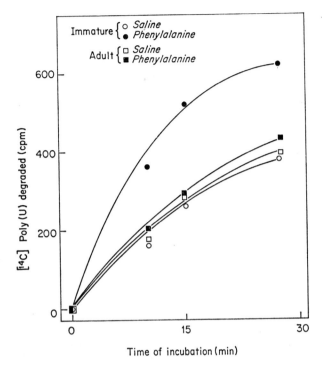

FIG. 5. Influence of phenylalanine loading on ribonuclease activity in postmitochondrial supernatant fractions obtained from cerebral cortices of immature and adult rats. Each sample for assay represents the postmitochondrial supernatant fraction from about 10 mg of tissue.

anine, ribonuclease activity was significantly enhanced in both of these ribosome-containing fractions from immature rats but not in similar preparations from adult rats. (Incidentally, significant activities of acid ribonuclease II could not be detected in cerebral polysomes from adult rats.) Endogenous ribonuclease inhibitor has been reported to increase in rat cerebrum with age (Suzuki & Takahashi 1970). This increase might help to prevent disruption of polyribosomes during isolation from the mature cerebrum.

Developmental differences in the stability of cerebral polyribosomes after administration of a loading dose of L-phenylalanine could be related to variations in the stability or concentration of lysosomes in this tissue (see, for example, Verity & Brown 1968), or to differences in the ease of penetration of the amino acid into the brain (see, for example, Weiss et al. 1971). Note, however, that cerebral polyribosomes are generally more stable in immature rats than in adult animals when subjected to identical challenges in the internal environment (Zomzely et al. 1971; Roberts 1971). Our present results do not

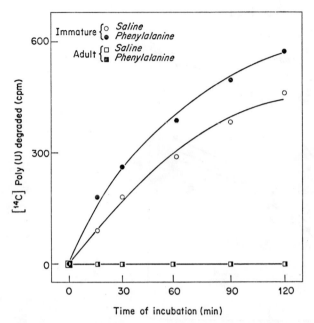

FIG. 6. Influence of phenylalanine loading on ribonuclease activity in purified polyribosomes obtained from cerebral cortices of immature and adult rats. Each sample for assay represents the polysomes isolated from about 100 mg of tissue. Carrier poly(U) was omitted in these assays.

preclude the possibility that the release of lysosomal ribonucleases in the immature cerebrum *in situ* in response to an excess of amino acid has functional significance. Thus, cerebral polyribosomes might be sensitized in these conditions, particularly if the challenge is chronic, as in the human aminoacidopathies. Alterations in the synthesis and turnover of certain cerebral proteins might ensue.

CONCLUSIONS

The major and acute consequence of intravenous administration of a moderate excess of certain amino acids (such as L-phenylalanine, L-leucine or L-valine) on cerebral metabolism appears to be a significant decline in cerebral concentrations of other amino acids. The resultant imbalance in due to enhanced intracerebral degradation of these other amino acids as well as to competition for amino acid transport systems. The turnover of total cerebral proteins is either unaffected or accelerated under these conditions. However, when the primary elevation of L-phenylalanine or L-leucine is more profound or long-continued (e.g., as a result of intraperitoneal administration of large amounts of one of these amino acids) other processes may dominate, in which case the administered amino acid or its accumulated metabolites (or both) might evoke the release of lysosomal ribonucleases and possibly other hydrolytic enzymes especially in the immature cerebrum. The release of these enzymes appears to result in the artifactual disruption of cerebral polyribosomes subsequently isolated. However, there is no evidence that polyribosomes disaggregate in the cerebrum *in situ* under these conditions. These findings do not rule out the possibility that chronic exposure to elevated circulating concentrations of certain amino acids and their metabolites (such as exist in human aminoacidopathies) leads to alterations in the synthesis and turnover of certain cerebral proteins. The nature and mechanisms underlying such possible effects remain to be elucidated.

ACKNOWLEDGEMENTS

I thank the National Institutes of Health, U. S. Public Health Service (NS–07869) and the United Cerebral Palsy Research and Educational Foundation for financial support. Mrs Beatrice S. Morelos provided expert technical assistance. I am indebted to Mrs Carole Feingold for capable bibliographic and secretarial assistance. Assays for acid ribonuclease II activity were performed by Rosemarie A. Steadman by courtesy of Dr C. M. Szego.

References

AGRAWAL, H. C., BONE, A. H. & DAVISON, A. N. (1970) Effect of phenylalanine on protein synthesis in the developing rat brain. *Biochem. J.* **117**, 325–331

AOKI, K. & SIEGEL, F. L. (1970) Hyperphenylalaninemia: disaggregation of brain polyribosomes in young rats. *Science (Wash. D.C.)* **168**, 129–130

APPEL, S. H. (1966) Inhibition of brain protein synthesis: an approach to the biochemical basis of neurological dysfunction in the amino-acidurias. *Trans. N.Y. Acad. Sci.* **29**, 63–70

BARONDES, S. H. & NIRENBERG, M. W. (1962) Fate of a synthetic polynucleotide directing cell-free protein synthesis. I. Characteristics of degradation. *Science (Wash. D.C.)* **138**, 810–817

BLASBERG, R. & LAJTHA, A. (1966) Heterogeneity of the mediated transport systems of amino acid uptake in brain. *Brain Res.* **1**, 86–104

BOWDEN, J. A. & MCARTHUR, C. L. (1972) Possible biochemical model for phenylketonuria. *Nature (Lond.)* **235**, 230

CAMPAGNONI, A. T. & MAHLER, H. R. (1967) Isolation and properties of polyribosomes from cerebral cortex. *Biochemistry* **6**, 956–967

CARVER, M. J. (1965) Influence of phenylalanine administration on the free amino acids of brain and liver in the rat. *J. Neurochem.* **12**, 45–50

CARVER, M. J. (1969) Free amino acids of fetal brain. Influence of the branched chain amino acids. *J. Neurochem.* **16**, 113–116

CHIRIGOS, M. A., GREENGARD, P. & UDENFRIEND, S. (1960) Uptake of tyrosine by rat brain *in vivo*. *J. Biol. Chem.* **235**, 2075–2079

DANIEL, R. G. & WAISMAN, H. A. (1969) The influence of excess methionine on the free amino acids of brain and liver of the weanling rat. *J. Neurochem.* **16**, 787–795

DINGMAN, W. & SPORN, M. B. (1959) The penetration of proline and proline derivatives into brain. *J. Neurochem.* **4**, 148–153

EDWARDS, D. J. & BLAU, K. (1972) Aromatic acids derived from phenylalanine in the tissues of rats with experimentally induced phenylketonuria-like characteristics. *Biochem. J.* **130**, 495–503

FELLMAN, J. H. (1956) Inhibition of DOPA decarboxylase by aromatic acids associated with phenylpyruvic oligophrenia. *Proc. Soc. Exp. Biol. Med.* **93**, 413–414

FOLBERGROVÁ, J. (1966) Incorporation of labelled amino acids into the proteins of brain cortex slices *in vitro* in the presence of other non-radioactive amino acids. *J. Neurochem.* **13**, 553–562

GAULL, G. E. (1972) in *Handbook of Neurochemistry* (Lajtha, A., ed.), vol. 7, pp. 169–190, Plenum Press, New York

GOLDSTEIN, F. B. (1961) Biochemical studies on phenylketonuria. I. Experimental hyperphenylalaninemia in the rat. *J. Biol. Chem.* **236**, 2656–2661

GUROFF, G. & UDENFRIEND, S. (1962) Studies on aromatic amino acid uptake by rat brain *in vivo*. Uptake of phenylalanine and of tryptophan; inhibition and stereoselectivity in the uptake of tyrosine by brain and muscle. *J. Biol. Chem.* **237**, 803–806

HARPER, A. E., BENEVENGA, N. J. & WOHLHEUTER, R. M. (1970) Effects of ingestion of disproportionate amounts of amino acids. *Physiol. Rev.* **50**, 428–558

KAMIN, H. & HANDLER, P. (1951) The metabolism of parenterally administered amino acids. II. Urea synthesis. *J. Biol. Chem.* **188**, 193–205

KNOTT, P. J. & CURZON, G. (1972) Free tryptophan in plasma and brain tryptophan metabolism. *Nature (Lond.)* **239**, 452–453

LAJTHA, A. (1958) Amino acid and protein metabolism of the brain. II. The uptake of L-lysine by brain and other organs of the mouse at different ages. *J. Neurochem.* **2**, 209–215

LAJTHA, A. (1964) Protein metabolism of the nervous system. *Int. Res. Neurobiol.* **6**, 1–98

LAJTHA, A. & TOTH, J. (1961) The brain barrier system. II. Uptake and transport of amino acids by the brain. *J. Neurochem.* **8**, 216–225

LINDROOS, O. F. C. & OJA, S. S. (1971) Hyperphenylalaninaemia and the exchange of tyrosine in adult rat brain. *Exp. Brain Res.* **14**, 48–60

LOWDEN, J. A. & LARAMÉE, M. A. (1969) Hyperphenylalaninemia: the effect on cerebral amino acid levels during development. *Can. J. Biochem.* **47**, 883–888

MACINNES, J. W. & SCHLESINGER, K. (1971) Effects of excess phenylalanine on *in vitro* and *in vivo* RNA and protein synthesis and polyribosome levels in brains of mice. *Brain Res.* **29**, 101–110

MCKEAN, C. M., BOGGS, D. E. & PETERSON, N. A. (1968) The influence of high phenylalanine and tyrosine on the concentrations of essential amino acids in brain. *J. Neurochem.* **15**, 235–241

MENKES, J. H. (1968) Cerebral proteolipids in phenylketonuria. *Neurology* **18**, 1003–1008

NAKANO, K., KISHI, T., KURITA, N. & ASHIDA, K. (1970) Effect of dietary amino acids on amino acid-catabolizing enzymes in rat liver. *J. Nutr.* **100**, 827–836

NYHAN, W. L. (ed.) (1967) *Amino Acid Metabolism and Genetic Variation*, McGraw-Hill, New York

ORREGO, F. & LIPMANN, F. (1967) Protein synthesis in brain slices. Effects of electrical stimulation and acidic amino acids. *J. Biol. Chem.* **242**, 665–671

PENG, Y., TEWS, J. K. & HARPER, A. E. (1972) Amino acid imbalance, protein intake, and changes in rat brain and plasma amino acids. *Am. J. Physiol.* **222**, 314–321

ROBERTS, S. (1963) Regulation of cerebral metabolism of amino acids. II. Influence of phenylalanine deficiency on free and protein-bound amino acids in rat cerebral cortex: relationship to plasma levels. *J. Neurochem.* **10**, 931–940

ROBERTS, S. (1968) Influence of elevated circulating level of amino acids on cerebral concentrations and utilization of amino acids. *Prog. Brain Res.* **29**, 235–243

ROBERTS, S. (1971) in *Handbook of Neurochemistry* (Lajtha, A., ed.), vol. 5, pp. 1–48, Plenum Press, New York

ROBERTS, S. & MORELOS, B. S. (1965) Regulation of cerebral metabolism of amino acids. IV. Influence of amino acid levels on leucine uptake, utilization and incorporation into protein *in vivo*. *J. Neurochem.* **12**, 373–387

ROBERTS, S., SETO, K. & HANKING, B. H. (1962) Regulation of cerebral metabolism of amino acids. I. Influence of phenylalanine deficiency on oxidative utilization *in vitro*. *J. Neurochem.* **9**, 493–501

ROBERTS, S., ZOMZELY, C. E. & BONDY, S. C. (1971) in *Cellular Aspects of Neural Growth and Differentiation*. (UCLA Forum in Medical Sciences No. 14) (Pease, D. C., ed.), ch. 20, pp. 447–471, University of California Press, Los Angeles

ROSENBERG, L. E. (1969) in *Duncan's Diseases of Metabolism* (Bondy, P. K., ed.), 6th edn., pp. 23–45, Saunders, Philadelphia

ROTH, J. S. (1965) Ribonuclease V. Studies on the properties and distribution of ribonuclease inhibitor in the rat. *Biochim. Biophys. Acta* **21**, 34–43

RUBIN, A. L. & STENZEL, K. H. (1965) *In vitro* synthesis of brain protein. *Proc. Natl. Acad. Sci. U.S.A.* **53**, 963–968

SCHIMKE, R. T., SWEENEY, E. W. & BERLIN, C. M. (1965) The roles of synthesis and degradation in the control of rat liver tryptophan pyrrolase. *J. Biol. Chem.* **240**, 322–331

SCHWERIN, P., BESSMAN, S. P. & WAELSCH, H. (1950) The uptake of glutamic acid and glutamine by brain and other tissues of the rat and mouse. *J. Biol. Chem.* **184**, 37–44

SCRIVER, C. R. & HECHTMAN, P. (1970) Human genetics of membrane transport with emphasis on amino acids. *Adv. Hum. Genet.* **1**, 211–274

SHAH, S. N., PETERSON, N. A. & MCKEAN, C. M. (1969) Inhibition of sterol synthesis *in vitro* by metabolites of phenylalanine. *Biochim. Biophys. Acta* **187**, 236–242

SHAH, S. N., PETERSON, N. A. & MCKEAN, C. M. (1972) Lipid composition of human cerebral white matter and myelin in phenylketonuria. *J. Neurochem.* **19**, 2369–2376

SHORTMAN, K. (1961) Studies on cellular inhibitors of ribonuclease. I. The assay of the ribonuclease-inhibitor system, and the purification of the inhibitor from rat liver. *Biochim. Biophys. Acta* **51**, 37–49

SIEGEL, F. L., AOKI, K. & COLWELL, R. E. (1971) Polyribosome disaggregation and cell-free protein synthesis in preparations from cerebral cortex of hyperphenylalaninemic rats. *J. Neurochem.* **18**, 537–547

SILBERBERG, D. H. (1969) Maple syrup urine disease metabolites studied in cerebellum cultures. *J. Neurochem.* **16**, 1141–1146

SUZUKI, Y. & TAKAHASHI, Y. (1970) Developmental and regional variations in ribonuclease inhibitor activity in brain. *J. Neurochem.* **17**, 1521–1524

SWAIMAN, K. F., HOSFIELD, W. B. & LEMIEUX, B. (1968) Elevated plasma phenylalanine concentration and lysine incorporation into ribosomal protein of developing brain. *J. Neurochem.* **15**, 687–690

SZEGO, C. M., SEELER, B. J., STEADMAN, R. A., HILL, D. F., KIMURA, A. K. & ROBERTS, J. A. (1971) The lysosomal membrane complex. Focal point of primary steroid hormone action. *Biochem. J.* **123**, 523–538

TIGERMAN, H. & MACVICAR, R. (1951) Glutamine, glutamic acid, ammonia administration, and tissue glutamine. *J. Biol. Chem.* **189**, 793–799

UDENFRIEND, S. (1963) Factors in amino acid metabolism which can influence the central nervous system. *Am. J. Clin. Nutr.* **12**, 287–290

VERITY, M. A. & BROWN, W. J. (1968) Structure-linked activity of lysosomal enzymes in the developing mouse brain. *J. Neurochem.* **15**, 69–80

WEISS, B. F., MUNRO, H. N. & WURTMAN, R. J. (1971) L-Dopa: disaggregation of brain polysomes and elevation of brain tryptophan. *Science (Wash. D.C.)* **173**, 833–835

WEISS, B. F., MUNRO, H. N., ORDONEZ, L. A. & WURTMAN, R. J. (1972) Dopamine: mediator
of brain polysome disaggregation after L-dopa. *Science (Wash. D.C.)* **177**, 613–616
WILTSE, H. E. & MENKES, J. H. (1972) in *Handbook of Neurochemistry* (Lajtha, A., ed.),
vol. 7, pp. 143–167, Plenum Press, New York
WONG, P. W. K., FRESCO, R. & JUSTICE, P. (1972) The effect of maternal amino acid imbalance
on fetal cerebral polyribosomes. *Metab. (Clin. Exp.)* **21**, 875–881
ZOMZELY, C. E., ROBERTS, S. & RAPAPORT, D. (1964) Regulation of cerebral metabolism of
amino acids. III. Characteristics of amino acid incorporation into protein of microsomal
and ribosomal preparations of rat cerebral cortex *J. Neurochem.* **11**, 567–582
ZOMZELY, C. E., ROBERTS, S., GRUBER, C. P. & BROWN, D. M. (1968) Cerebral protein synthesis.
II. Instability of cerebral messenger ribonucleic acid-ribosome complexes. *J. Biol. Chem.*
243, 5396–5409
ZOMZELY, C. E., ROBERTS, S., PEACHE, S. & BROWN, D. M. (1971) Cerebral protein synthesis.
III. Developmental alterations in the stability of cerebral messenger ribonucleic acid-
ribosome complexes. *J. Biol. Chem.* **246**, 2097–2103
ZOMZELY-NEURATH, C. E. & ROBERTS, S. (1972) in *Research Methods in Neurochemistry*
(Marks, N. & Rodnight, R., eds.), vol. 1, pp. 95–137, Plenum Press, New York

Discussion

Wurtman: Isn't incubation of brain slices with single amino acids analogous
to giving the intact animal a loading dose of an amino acid? If we investigated
further, perhaps we might find that ribonuclease was activated and polysomes
disaggregated in the brain slices in those conditions.

Aoki & Siegel (1970) reported that they could protect the young rat against
the polysome disaggregation produced by phenylalanine by giving tryptophan
at the same time. Have you confirmed this? Might tryptophan suppress the
ribonuclease activation?

Roberts: We have not tried to reverse this effect with tryptophan; obviously
a number of questions remain to be answered.

Wurtman: In maturation, is it possible that something happens to the
ribonuclease-activating mechanism that renders it no longer susceptible to
activation by amino acid imbalance?

Roberts: Lysosomes appear to be more unstable in the brain of young
animals than in adults.

Kaufman: Are lysosomes generally more unstable in all tissues of young
animals or just in their brains?

Roberts: I believe the developmental decrease in stability is a general pattern.

Kaufman: Have you added phenylalanine to lysosomes *in vitro* to see if the
effect of phenylalanine on lysosomal ribonuclease can be observed?

Roberts: No. Not yet.

Kaufman: Must phenylalanine be hydroxylated before it has this effect on lysosomes? I ask because peroxide might be formed as a side-product during phenylalanine catabolism and the lysosomal membrane is known to be unstable to peroxide. This possibility could be investigated in rats treated with *p*-chlorophenylalanine which inhibits phenylalanine hydroxylase.

Roberts: The chloro-compound might itself disaggregate polysomes. The work of Drs Wurtman and Munro hints that certain types of receptors are necessary for the disaggregation phenomenon.

Gál: When I add 5-hydroxytryptamine in concentrations comparable with those *in vivo* (1–10 μM) to an *in vitro* preparation of pure brain ribosomes, I do not see any disaggregation, but I have not investigated many amino acids (see pp. 343–354).

I was particularly interested by your data in Fig. 1; they seem to bear out the ideas which Dr Moir and I proposed. It seems that after intravenous injection of phenylalanine, the concentration of the free amino acids returned to normal after 60 min.

Munro: You inject [^{14}C]lysine intracisternally. Lysine has many fewer catabolic pathways than other amino acids. Is lysine metabolized at all by brain?

Roberts: The utilization or transformation of lysine was minimal in these acute experiments. Apparently, pathways other than protein synthesis were not extensively involved.

Munro: At what point does the ribonuclease activity fall during growth of the rat? In other words, what level of enzyme activity distinguishes a mature rat from a young rat? Is the low amount of ribonuclease you detected in these rats of different ages due to the lack of penetration of phenylalanine into the tissue of older rats or are you implying that there is only a little ribonuclease in the brain at *that* stage?

Roberts: The total ribonuclease activity in the brain seems to decline progressively during development until adulthood, but it is difficult to compare different strains of animals. Our 42-day-old animals weigh about 200 g and are mature sexually.

Munro: When we inject dopa into larger animals, a dose ten times that required for the infant is needed to produce disaggregation. Have the polysomes in dopa-treated infants acquired more ribonuclease than in the adults as a result of damage to the lysosomal compartment which may be stronger in the adult?

Roberts: Your results may be due to a combination of factors. The amino acid presumably does not enter the brain as readily in the older animals. Secondly, it is more effectively catabolized. Thirdly, the lysosomes become less fragile.

Wurtman: In the youngest animals, the dose of phenylalanine needed to disaggregate the polysomes is already considerably greater than the dose of dopa. The disaggregation induced by phenylalanine is caused by the amino acid *per se*. It can be blocked by the simultaneous administration of tryptophan but not by decarboxylase inhibition. In contrast, the disaggregation produced by dopa or 5-hydroxytryptophan depends entirely on their decarboxylation to neurotransmitter monoamines.

Roberts: The question of the nature of receptors is fascinating. Activation of these receptors may result in lysosomal release of ribonuclease. How is this accomplished? Possible mechanisms include those that require adenylate cyclase receptors, because cyclic AMP appears to affect lysosomal stability.

Wurtman: Can the polysome disaggregation in liver produced by amino acid imbalance also be related to activation of ribonuclease?

Munro: With many liver preparations care has been taken to exclude breakdown by the addition of the liver cell sap which has a ribonuclease inhibitor in it. The patterns obtained in the presence of sap still show changes due to treatment of the rat.

Roberts: That may be, but when the polysomes are isolated the inhibitor is apparently removed and some of the ribonuclease is still attached to the polysome.

Munro: Polysomes are often harvested directly from postmitochondrial supernatant liquid. In disaggregating brain polysomes, phenylalanine operates by a different mechanism from dopa for we have evidence that production of dopamine from dopa and 5-hydroxytryptophan from tryptophan is needed to cause disaggregation. Secondly, receptors for these amines are necessary.

Roberts: I did not mean this to be a unitarian hypothesis. Of course, other factors may contribute.

Weiner: Ignarro *et al.* (1973) have shown that catecholamines stabilize lysosomes. You could be witnessing the stabilization of lysosomes which may then be collected by centrifugation in the ribosome fraction. Lysis at this time would lead to ribonuclease activation and consequent polyribosome disaggregation.

Mandel: Your results might be due to the presence of more ribonuclease or to a decrease in the amount of its natural inhibitor, of which there are several in nuclei and cytoplasm.

Roberts: We add an inhibitor to all our preparations.

Mandel: Even so, the activity could still increase owing to a decline in the inhibition of a specific ribonuclease.

Roberts: That is possible. The amount of endogenous ribonuclease inhibitor in brain rises with age. The assay we use dissociates ribonuclease from the inhibitor and so we should be measuring all the ribonuclease that is present in such circumstances. We still don't know how variations in ribonuclease inhibitor affect polysome disruption caused by an excess of amino acids.

Barondes: How did you measure the ribonuclease?

Roberts: We used your method (Barondes & Nirenberg 1962). In one series of experiments, we measured the activity under the conditions to which the polysomes were exposed, namely in a Tris–sucrose buffer at pH 7.6. In other studies, the lysosomal ribonuclease was released by a preincubation of the material with Triton, which disrupts lysosomes and then the ribonuclease activity was measured at the acid pH optimal for the lysosomal enzyme.

Barondes: That might all be *latent* enzyme contained in lysosomes. It might be completely inactive in the cell.

Roberts: The latent ribonuclease is that which is sequestered within the lysosome. As far as I know, there is little non-lysosomal ribonuclease in brain.

Barondes: But being sequestered, it does not have access to messenger RNA.

Roberts: No, not until it is released, which presumably happens in the presence of an excess of amino acids.

Barondes: You can measure what is not in the lysosomes but in the 'soluble fraction' (e.g. the supernatant of centrifugation at 100 000 g for one hour) without Triton treatment.

Roberts: That would also include what was there initially as well as what was released. I agree, this should be done. We are anxious to try to detect how much ribonuclease has been released from the lysosomes and whether there is evidence for lysosomal disruption.

Lajtha: Dr Roberts, compartmentation of the amino acids might account for some of your results. The enhanced incorporation of lysine (Table 6) indicates to me that with the increased concentration of lysine the exchange between the lysine precursor pool and the added [^{14}C]lysine is greater. Perhaps, at low concentrations lysine does not penetrate fast enough to keep the internal specific activities on a par with the external ones. The initially greater variations in Fig. 1 could also be interpreted as a slow equilibrium: once equilibrium is attained, the effects observed are less.

Roberts: Long-term infusion experiments are necessary to settle this point.

Wurtman: What are the other explanations for this apparent increase after administration of non-radioactive lysine?

Roberts: There are only two explanations. First, the radioactive lysine is preferentially used and the presence of the carrier lysine injected at the same time facilitates the use of that external pool for protein synthesis. That is the prevailing view in the literature. The other plausible explanation is that the turnover of protein is actually increased in these conditions.

Wurtman: Is the effect of lysine on the incorporation of phenylalanine into protein a special property of basic amino acids?

Roberts: I have no data to exclude that.

Fernstrom: For how many other amino acids is this effect observed?

Roberts: An excess of leucine disaggregates polysomes but we have not studied the effect of leucine on ribonuclease activity yet. Several other amino acids have also been reported to cause disaggregation of brain polysomes (see, for example, MacInnes & Schlesinger 1971).

Gessa: Is there any age difference in the disaggregating response to electro-convulsive shock?

Roberts: I don't know. Vesco & Giuditta (1968) found no differences in ribonuclease activity in rabbit cerebral cortex after electroconvulsive shock.

Wurtman: What is the current hypothesis for the mechanism by which lack of tryptophan causes the hepatic polysomes to disaggregate?

Munro: Tryptophan is the least abundant amino acid. If the concentration of other amino acids is drastically reduced, then these acids become limiting for protein synthesis.

Wurtman: Is this because of an initiation factor?

Munro: No, it is probably because the rate of insertion determines the rate at which the ribosomes travel along and elongate the peptide chain.

Roberts: The explanation I proposed was not intended to account for all types of polysomal disaggregation observed in the presence of altered concentrations of amino acids. Of course, other mechanisms will obtain in some of these circumstances.

Munro: We always seem to produce unitary hypotheses for events in which nature has cleverly coordinated a whole series of mechanisms. For example, the effect of tryptophan on liver polysomes is independent of messenger supplies since large doses of actinomycin D fail to suppress the response (Fleck *et al.* 1965). Nevertheless, several people have demonstrated that RNA synthesis and ribosomal maturation in the nucleus are affected by the supply of amino acid, specifically tryptophan (Henderson 1970). This implies that two events coordinate but are not necessarily interdependent.

Similarly, the charging of tRNA and the stability of the elongation factor, (EFI) complex are related to the relative amounts of EFI and of aminoacyl-tRNA; EFI is less stable when less aminoacyl-tRNA is present and *vice versa*. Thus, we have to disentangle innumerable interlocking controls, some of which are dependent, some of which are parallel.

Barondes: It may be relevant that cycloheximide, which blocks protein synthesis, stabilizes polysomes.

Munro: Cycloheximide suppresses transport of the peptide chain on the ribosome as it elongates. At adequate doses, puromycin disaggregates poly-somes, because the mechanism is different; removal of the peptidyl chain is involved. Godin (1967) found that the amino acid pools of the body enlarge after administration of puromycin, presumably because protein synthesis no longer removes supplies. This can be reversed by inhibiting the aminotransferases with hydrazine; the free amino acid pools increase and protein synthesis in liver slices is augmented (Amenta & Johnson 1963). Thus, we have a dynamic equilibrium in which the inputs and outputs to the pool are rapid. By reducing one of these outlets the size of the pool can be increased and so the amount passing through the other areas is augmented.

Barondes: Puromycin and cycloheximide act in opposite ways on polysomes. Puromycin disaggregates polysomes for the specific reason that peptidyl puromycin comes off the polysome which then falls apart. The condition that you are discussing, where aminoacyl-tRNA might be limiting, is different. I do not think there is evidence that this causes the disaggregation of polysomes.

Fernstrom: Suppose tryptophyl-tRNA was lacking and that a tryptophan residue was required at the end of the peptide chain. As the ribosome came to the end of the messenger, it would stop at that point and wait for a tryptophyl-tRNA. Other ribosomes would have to queue up behind. So, in some situations where protein synthesis is slower, polysomes might aggregate more than dis-aggregate.

Barondes: That might be more analogous to the effect of cycloheximide.

Munro: About one in every 100 residues is tryptophan and the ribosome covers about 30 codons. By chance, the last two or three ribosomes at the end of the chain for a mixture of proteins with random tryptophan distribution might be able to skip off while the remainder are still waiting for the insertion tryptophyl-tRNA.

References

AMENTA, J. S. & JOHNSON, E. H. (1963) The effects of hydrazine upon the metabolism of amino acids in the rat liver. *Lab. Invest.* **12**, 921–928

AOKI, K. & SIEGEL, F. L. (1970) Hyperphenylalaninemia: disaggregation of brain polysomes in young rats. *Science (Wash. D.C.)* **168**, 129–130

BARONDES, S. H. & NIRENBERG, M. W. (1962) Fate of a synthetic polynucleotide directing cell-free protein synthesis. I. Characteristics of degradation. *Science (Wash. D.C.)* **138**, 810–817

FLECK, A., SHEPHERD, J. & MUNRO, H. N. (1965) Protein synthesis in rat liver: influence of amino acids in diet on microsomes and polysomes. *Science (Wash. D.C.)* **150**, 628–629

GÁL, E. M. (1974) Synthetic *p*-halogenophenylalanines and protein synthesis in the brain, *This Volume*, pp. 343–354

GODIN, C. (1967) Effect of puromycin on the metabolism of phenylalanine in rats. *Can. J. Biochem.* **45**, 1961–1964

HENDERSON, A. R. (1970) The effect of feeding with a tryptophan-free amino acid mixture on rat liver magnesium ion-activated deoxyribonucleic acid-dependent ribonucleic acid polymerase. *Biochem. J.* **120**, 205–214

IGNARRO, L. J., KRASSIKOFF, N. & SLYWKA, J. (1973) Release of enzymes from a rat liver lysosome fraction: inhibition by catecholamines and cyclic 3′,5′-adenosine monophosphate, stimulation by cholinergic agents and cyclic 3′,5′-guanosine monophosphate. *J. Pharmacol. Exp. Ther.* **183**, 86–99

MACINNES, J. W. & SCHLESINGER, K. (1971) Effects of excess phenylalanine on *in vitro* and *in vivo* RNA and protein synthesis and polyribosome levels in brains of mice. *Brain Res.* **29**, 101–110

MUNRO, H. N. (1974) Control of plasma amino acid concentrations, *This Volume*, pp. 5–18

VESCO, C. & GIUDITTA, A. (1968) Disaggregation of brain polysomes induced by electro-convulsive treatment. *J. Neurochem.* **15**, 81–85

L-Dopa, polysomal aggregation and cerebral synthesis of protein

BETTE F. WEISS, L. E. ROEL, H. N. MUNRO and R. J. WURTMAN

Department of Nutrition and Food Science, Massachusetts Institute of Technology

Abstract Administration of L-dopa or 5-hydroxytryptophan to rats profoundly disaggregates whole-brain polyribosomes and concurrently suppresses synthesis of protein in the brain *in vivo*. These effects are mediated by the corresponding amines, dopamine and 5-hydroxytryptamine. Answers to several important questions raised by these results are presently being sought.

Aromatic amino acids apparently suppress synthesis of protein in brain by at least two mechanisms. In one, the naturally occurring amino acids act as such, possibly by activating lysosomal enzymes or competing with tryptophan uptake. Seemingly, the hydroxylated precursors of the monoamines act on specific 'receptors'.

High doses of phenylalanine are known to disaggregate polysomes in the brain and at the same time reduce the free tryptophan content of the brain. This disaggregation can be prevented by the administration of tryptophan with the phenylalanine (Aoki & Siegel 1970). We thought that L-dopa might also disaggregate brain polysomes, perhaps by competing with tryptophan for transport, and thereby suppress protein synthesis. We found that exogenous L-dopa transiently disaggregated polysomes in the brains of rats of various ages (Weiss *et al.* 1971). The effect lasted as long as dopa could be detected in the brains. The minimum effective dose increased with the age of the animals; in adult rats, the standard dose is 300–500 mg/kg. This dose is about five times that given to adult humans with Parkinson's disease.

L-Dopa, far from decreasing the amount of tryptophan in the brain like phenylalanine, increased it (Weiss *et al.* 1971). Clearly, dopa acts by a different mechanism. We then showed that dopamine is an essential intermediate in the disaggregation (Weiss *et al.* 1972), and suggested that its site of action was not the synapses but the receptors present in most brain cells. Subsequently, we have found that L-5-hydroxytryptophan also disaggregates polysomes through the intermediacy of its metabolite, 5-hydroxytryptamine (Weiss *et al.* 1973).

The dopa-induced disaggregation of brain polysomes is associated with a substantial suppression of cerebral protein synthesis, including the synthesis of neurotubular proteins which are precipitable with vinblastine (L. Roel, unpublished observations). The relationships, if any, between these effects of L-dopa and its therapeutic or toxic actions in human subjects await clarification.

Fig. 1 shows the marked disaggregation of polysomes after the intraperitoneal

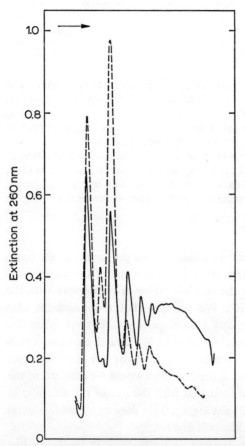

FIG. 1. Effect of L-dopa on brain polysome profiles in 50 g male rats. Whole brains were taken 60 min after the intraperitoneal administration of L-dopa (500 mg/kg) in 0.05M-HCl, or the diluent alone. Polysomes were isolated with discontinuous sucrose gradients, and samples were spun on continuous 10–40% sucrose density gradients for 70 min at 105 000 *g*. With a flow cell, the RNA concentrations in continuous portions of the gradient were then measured by reading absorbance at 260 nm (Weiss *et al.* 1971, 1972). Solid line, control; dashed line, L-dopa-treated animal. The arrow indicates top of gradient.

administration of L-dopa to rats. Doses of 50–300 mg/kg disaggregated polysomes in 7–9-day-old rats 40–60 min after dopa injection. This effect was
manifest for 20 min and persisted for about two hours.

We investigated whether the changes dopa brought about in polysome aggregation coincided with *in vivo* synthesis of protein in the brain (Weiss *et al.*
1974). To do this, we injected four male 50 g rats intraperitoneally with L-dopa
(500 mg/kg) or its solvent (hydrochloric acid). After 45 min, they received
either [U-^{14}C]leucine or [U-^{14}C]lysine and were killed 7, 15 or 30 min later.
Their brains were immediately assayed for radioactive proteins. In each experiment, we found a significant decrease in the incorporation of amino acids into
protein in the dopa-treated rats. Fig. 2 shows the results for injection of [U-^{14}C]-
lysine. There is a good temporal correlation between the polysome profile and
the suppression of incorporation of lysine into the cerebral protein. The results
for leucine are similar. We have also found that four hours after administration
of dopa, when polysome profiles are back to normal, the rate at which radio-

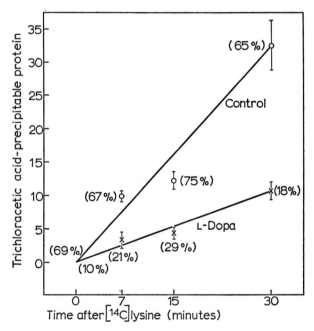

FIG. 2. Correlation of cerebral disaggregation of polysomes in rats with decreased *in vivo*
protein synthesis after L-dopa administration. Male rats (50 g) received 1.5 μCi [U-^{14}C]lysine
intracisternally 45 min after L-dopa (500 mg/kg) in 0.05M-HCl (or the diluent alone) had been
injected intraperitoneally. They were killed after 7, 15 or 30 min. The percentage of homogenate counts found in the protein fraction precipitable with trichloroacetic acid was compared
in controls and L-dopa-treated rats as was the percentage of rRNA present as polysomes.

active lysine is incorporated into protein is 30–40% greater than in control animals (L. Roel, unpublished observations).

We decided to investigate the metabolic fate of dopa in the brain in order to try to discover the mechanism of its action. Dopa can be decarboxylated to dopamine which can be β-hydroxylated to noradrenaline (Carlsson *et al.* 1958; Chalmers *et al.* 1971); these two monoamines can then be destroyed by mono-amine oxidase (see Fig. 3). Alternatively, dopa (and its catecholamine deriva-tives) can be *O*-methylated to form 4-hydroxy-3-methoxyphenylalanine (or the corresponding 3-methoxy-amines). This process, catalysed by catechol methyl-transferase (EC 2.1.1.6) (Axelrod & Tomchick 1958; Bartholini & Pletscher 1968; Wurtman *et al.* 1970*a*), depletes *S*-adenosylmethionine in the brain (Wurtman *et al.* 1970*b*) (see Fig. 3). We set out to reproduce each of the steps in dopa metabolism (Fig. 3) to see whether polysomes were disaggregated.

D-Dopa is taken up by the brain. Its *O*-methylation, like that of L-dopa, requires *S*-adenosylmethionine (Wurtman *et al.* 1970*b*) and leads to significant quantities of 4-hydroxy-3-methoxyphenylalanine in brain. D-Dopa is not decarboxylated in brain and thus does not elevate catecholamine concentra-tions. We found that 60 min after administration of either D-dopa or 4-hydroxy-3-methoxyphenylalanine (500 mg/kg) to 50 g male rats in quantities sufficient to produce changes that should parallel the effects on brain methylation produced by L-dopa there was no effect on polysome aggregation (see Table 1). There-fore, we thought we could rule out the depletion of *S*-adenosylmethionine or the formation of the methoxy compound as mechanisms by which L-dopa works. If we administer a decarboxylase inhibitor, RO4-4602, in doses large

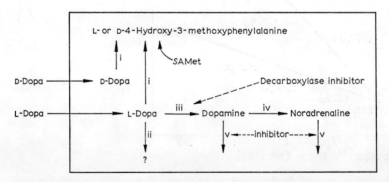

FIG. 3. Metabolism of L- and D-dopa in rat brain (after Weiss *et al.* 1972). Enzymes: i, catechol methyltransferase; ii, dopa aminotransferase; iii, aromatic L-amino acid decarboxylase; iv, dopamine hydroxylase; v, monoamine oxidase. *S*AMet is *S*-adenosylmethionine. The decarboxylase inhibitor was RO4–4602 and the monoamine oxidase inhibitor used was pheniprazine.

TABLE 1. Effects of L-dopa and related drugs on brain polysome profiles and dopa metabolism of 50 g male rats (Weiss et al. 1972). Significance of differences was evaluated by student's t-test. Abbreviations: i.p., intraperitoneal; i.c., intracisternal.

Treatment	Route of administration	Dose (mg/kg)	Time after injection (min)	Polysomes (percentage of profile)	Dopa (μg/g)	3-O-Methyl-dopa (μg/g)	S-Adenosyl-methionine (percentage of control)*	Dopamine (ng/g)	Noradrenaline (ng/g)
Control	i.p.	0	60	65	<0.1	<0.1			
L-Dopa	i.p.	500	40		18.7[a]	0.5[a]			
3-O-Methyldopa	i.p.	500	60	43[a]	6.8[a]	1.2[b]			
			90	60	0.3[c]	4.6[c]			
			120	65	0.3[a]	11.8[a]			
Control	i.p.	0	60	60	<0.1	<0.1	100	505	300
L-Dopa	i.p.	500	60		9.5[a]	3.4[a]	32[a]	2 230[b]	330
D-Dopa	i.p.	500	60	61	10.7[a]	2.9[a]	27[a]	640	310
RO4-4602	i.p.	800	90	65	2.9[c]	0.1	45[b]	346[b]	188[a]
RO4-4602 plus L-dopa	i.p.	800, 500	90, 60	60	40.2[a]	4.3[a]	39[b]	298[b]	200[a]
Control	i.p.	0	60	73	<0.1	<0.1		508	305
L-Dopa	i.p.	100	60	75	2.3[a]	<0.1		780[c]	348
Pheniprazine	i.p.	10	180	62	<0.1	<0.1		779	398[c]
Pheniprazine plus L-dopa	i.p.	10, 100	180, 60	32[a]	1.2[a]	<0.1		2 890[a]	618[b]
Pheniprazine plus L-dopa	i.p.	10, 500	180, 60	41[b]	17.0[a]	<0.1		30 000[a]	801[a]
Control	i.c.	0 (μg/rat)	45	60	<0.1	<0.1	100	470	289
Dopamine	i.c.	100 (μg/rat)	15	59	<0.1	<0.1	92	6 020[c]	308
			45	58	<0.1	<0.1	86	903	212[c]
Noradrenaline	i.c.	100 (μg/rat)	15	57	<0.1	<0.1	94	555	20 800[a]
			45	54	<0.1	<0.1	82	593	10 200[a]

* Control brains contained an average of 13.4 μg S-adenosylmethionine/g.   [a] $P<0.001$ differs from control group.   [b] $P<0.01$ differs from control group.   [c] $P<0.05$ differs from control group.

enough for it to enter the brain, we completely block the effect of dopa on polysome disaggregation. These observations suggested that the catecholamines, dopamine and/or noradrenaline, mediate the action of dopa on brain polysomes.

In order to potentiate the elevation of concentrations of these cerebral catecholamines after injection of L-dopa, we pretreated the animals with a monoamine oxidase inhibitor, pheniprazine. The effect of this inhibition was, indeed, to potentiate the disaggregation of the polysomes (see Table 1).

We also examined the effect of 5-hydroxytryptophan. Like L-dopa, 500 mg/kg of this amino acid, injected intraperitoneally, significantly disaggregated brain polysomes after one hour in 50 g male rats (see Fig. 4) (40% compared with 71% in controls). When the decarboxylase inhibitor, RO4-4602, is given before 5-hydroxytryptophan or L-dopa, the disaggregation is blocked. Thus, polysome aggregation by the hydroxylated amino acids that are precursors for monoamine neurotransmitters appears to be mediated by the corresponding biogenic amines.

Treatment of the rats with pargyline before giving them lower doses of 5-

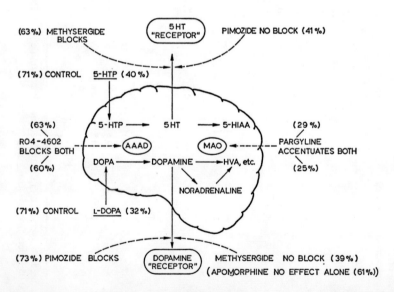

FIG. 4. Synthesis of biogenic amines from L-dopa and L-5-hydroxytryptophan (5HTP) in rat brains: effects of pretreatment with related drugs on brain polysomes: 5HT, 5-hydroxytryptamine; AAAD, aromatic L-amino acid decarboxylase; MAO, monoamine oxidase; 5-HIAA, 5-hydroxyindoleacetic acid; HVA, homovanillic acid. RO4–4602 inhibits AAAD. Pargyline inhibits MAO. Pimozide blocks dopamine receptors. Apomorphine stimulates dopamine receptors. Methysergide blocks 5-hydroxytryptamine receptors. The percentage of total rRNA present as polysomes 60 min after L-dopa or L-5-HTP administration is indicated beside each drug. Apomorphine was given alone, 60 min before autopsy.

hydroxytryptophan potentiates the polysomal disaggregation, just as it potentiates the effect of L-dopa (Fig. 4). This confirms the hypothesis that 5-hydroxytryptophan acts through its monoamine product 5-hydroxytryptamine. On the basis of pharmacological evidence, we suspect that both dopamine and 5-hydroxytryptamine influence polysome aggregation by acting on pharmacologically-specific receptors. Methysergide and cyproheptadine block receptors for 5-hydroxytryptamine and block the effect of 5-hydroxytryptophan on polysome disaggregation, whereas pimozide (a specific blocking agent for dopamine receptors) does not. In contrast, pimozide fully blocks the effect of dopa on polysome disaggregation whereas methysergide and cyproheptadine do not (Fig. 4).

Where are these pharmacologically distinct receptors? Are these compounds being taken up within monoaminergic neurons and there converted into their corresponding amines, which are then released synaptically? We doubt this for several reasons. First, we think there are not enough monoaminergic neurons in the brain to produce a reduction in polysome disaggregation from 70% to 40–30%. This great reduction requires the majority of brain cells, most of which are glia, to respond to L-dopa or 5-hydroxytryptophan. Histochemical fluorescence studies do not reveal enough synapses in which the presynaptic terminals are dopaminergic, noradrenergic and 5-hydroxytryptaminergic neurons to explain this amount of polysome disaggregation. Moreover, intracisternal injections of dopamine or noradrenaline in doses that are high enough to raise brain dopamine concentrations to what they would be after L-dopa have no effect on polysome disaggregation; such material is known to be concentrated selectively within monoaminergic neurons. Similarly, large doses of apomorphine have no effect. Pretreatment with 2-(2,4,5-trihydroxyphenyl)ethylamine (6-hydroxydopamine) in regimens that destroy most catecholaminergic terminals in brain does not prevent disaggregation after injection of L-dopa. Hence we suspect that the dopamine or 5-hydroxytryptamine formed after dopa or 5-hydroxytryptophan is working intracellularly to modify polysome aggregation. That is, dopa and 5-hydroxytryptophan, as aromatic amino acids, are taken up by most brain cells—neurons and glia—and are decarboxylated to form monoamines. These monoamines then presumably interact with specific pharmacologically defined intracellular receptors to affect polysome function and protein synthesis. We do not know why such intracellular receptors should exist. From an evolutionary standpoint, how did they come to develop within cells that have never contained dopamine or 5-hydroxytryptamine?

It is interesting to compare the polysomal responses to dopa and 5-hydroxytryptophan with the responses to phenylalanine. Phenylalanine will also cause polysome disaggregation and suppress protein synthesis but only at very high doses (1 g/kg) in young animals and barely in adult animals. Its effects on polysome aggregation are not blocked if the decarboxylation is suppressed. Its effect is apparently mediated by phenylalanine itself and not by its decarboxylated product.

At least two mechanisms exist, apparently, by which aromatic amino acids can suppress brain protein synthesis. The naturally occurring circulating amino acids do so as such, perhaps by activating lysosomal enzymes, or competing with tryptophan uptake, or by any of the previously mentioned theories. The hydroxylated precursors of the monoamines seem to act by entirely different mechanisms which involve specific 'receptors'.

References

AOKI, K. & SIEGEL, F. L. (1970) Hyperphenylalaninemia: disaggregation of brain polysomes in young rats. *Science (Wash. D.C.)* **168**, 129–130

AXELROD, J. A. & TOMCHICK, R. (1958) Enzymatic *O*-methylation of epinephrine and other catechols. *J. Biol. Chem.* **233**, 702–705

BARTHOLINI, G. & PLETSCHER, A. (1968) Cerebral accumulation and metabolism of ^{14}C-dopa after selective inhibition of peripheral decarboxylase. *J. Pharmacol. Exp. Ther.* **161**, 14–20

CARLSSON, A., LINDQVIST, M., MAGNUSSON, T. & WALDBECK, B. (1958) On the presence of 3-hydroxytryptamine in brain. *Science (Wash. D.C.)* **127**, 471

CHALMERS, J. P., BALDESSARINI, R. J. & WURTMAN, R. J. (1971) Effects of L-dopa on norepinephrine metabolism in the brain. *Proc. Natl. Acad. Sci. U.S.A.* **68**, 662–666

WEISS, B. F., MUNRO, H. N. & WURTMAN, R. J. (1971) L-Dopa: disaggregation of brain polysomes and elevation of brain tryptophan. *Science (Wash. D.C.)* **173**, 833–835

WEISS, B. F., MUNRO, H. N., ORDONEZ, L. A. & WURTMAN, R. J. (1972) Dopamine: mediator of brain polysome disaggregation after L-dopa. *Science (Wash. D.C.)* **177**, 613–166

WEISS, B. F., WURTMAN, R. J. & MUNRO, H. N. (1973) Disaggregation of brain polysomes by L-5-hydroxytryptophan: mediation by serotonin. *Life Sci.*, **13**, 411–416

WEISS, B. F., ROEL, L. E., MUNRO, H. N. & WURTMAN, R. J. (1974) The effect of L-dopa on brain polysomes and protein synthesis; probable mediation by intracellular dopamine. *Adv. Neurol.* **5**, in press

WURTMAN, R. J., CHOU, C. & ROSE, C. (1970a) The fate of [^{14}C]dihydroxyphenylalanine ([^{14}C]dopa) in the whole mouse. *J. Pharmacol. Exp. Ther.* **174**, 351–356

WURTMAN, R. J., ROSE, C. M., MATTHYSSE, S., STEPHENSON, J. & BALDESSARINI, R. (1970b) L-Dihydroxyphenylalanine: effect on *S*-adenosylmethionine in brain. *Science (Wash. D.C.)* **169**, 395–397

Discussion

Sandler: Have you investigated *m*-tyrosine? That is readily decarboxylated to *m*-tyramine.

Wurtman: No, we haven't.

Carlsson: I question your assumption that dopa decarboxylase is so wide-spread. Present evidence indicates two locations for it—monoaminergic neurons and capillaries—between which one should be able to distinguish. For example, one could selectively block the capillary decarboxylase and the other peripheral decarboxylases with MK 486. Alternatively, one could remove the neuronal decarboxylase by axotomy thereby allowing the fibres to degenerate. Using both methods, one could rid the rat brain of all the decarboxylase.

Wurtman: We have not done axotomy; however, we have used 6-hydroxy-dopamine both intracisternally and intraventricularly. I don't believe that blocking the capillary decarboxylase will make any difference. We do intend to perform the MK 486 experiment. Both L- and D-dopa markedly deplete cerebral S-adenosylmethionine, which fact suggests that both enter all the brain cells in large amounts. Don't histochemical fluorescence studies show that decarboxylase is present in most brain cells?

Carlsson: No, the green fluorescence in the brain is due to dopa. A simple experiment is to cut the spinal cord, whereupon the decarboxylase activity drops although it does not disappear. If then MK 486 is given, virtually all activity ceases (Andén *et al.* 1972). The inhibitor does not enter the brain. Thus, we have evidence of the two locations.

Munro: Are you implying that dopamine is produced in the capillary walls and then filters through into the cell population of the brain?

Carlsson: Yes, to some extent perhaps. But the dopamine formed in the monoaminergic neurons is probably more important functionally.

Munro: But dopa not dopamine is 3-*O*-methylated in the brain cells.

Carlsson: I am not saying that all the dopa is decarboxylated in the capillaries. Some must escape and enter the parenchyma and there be decarboxylated or *O*-methylated. In addition, the *O*-methylated metabolite formed in the liver easily enters the brain.

Wurtman: Dr Carlsson, what fraction of all the cells in the brain, including glia, receive monoaminergic terminals? For the polysomal effect of dopa to be mediated by dopamine release at synapses, about 80% of the brain cells would have to receive such synapses.

Carlsson: Probably a very small fraction. But if there was a chain reaction only a few cells need receive an innervation.

Munro: That is excluded by the experiments with 6-hydroxydopamine.

Carlsson: Even so, you probably had about 10% of catecholamine neurons left.

Wurtman: We would all be delighted if it turned out that polysome disaggregation was an effect of dopamine acting at synapses. We could then postulate that catecholamines normally controlled brain protein synthesis—a hypothesis more easy to reconcile with prevailing views than one based on an intracellular effect.

Carlsson: A possibility you have not allowed for, even though unlikely, is that we are seeing the consequences of the tremendous cardiovascular effect of such doses of dopa.

Wurtman: Do you see the same results with 5-hydroxytryptophan in doses of 300 mg/kg?

Carlsson: Certainly, we see strong peripheral actions at this dose.

Gessa: How do you explain the fact that intracisternal dopamine and noradrenaline had no effect on polysomes?

Wurtman: These compounds do not seem to enter glia or non-monoaminergic neurons; the only cells that concentrate these compounds well are those that normally contain them. Injected dopa, in contrast, is probably taken up within most brain cells and converted into monoamines. When dopamine or noradrenaline are injected intracisternally they are rapidly concentrated from the cerebrospinal fluid by monoaminergic terminals near the surface of the brain. Hence the period of time during which cells deep within the brain would be exposed to injected catecholamine is relatively short.

Roberts: It is unnecessary to invoke different explanations for the action of different effectors of polysome disaggregation or even to invoke the presence of specific receptors, although I hope they exist. I say unnecessary because possibly we are simply dealing with the effect of substances which are capable of labilizing lysosomes. If this is general, then all these substances that are effective might be good lysosome labilizers and the others not.

Wurtman: For that, you must postulate that dopa does not labilize lysosomes whereas phenylalanine does.

Roberts: Some of its abnormal by-products, which could be formed as a result of high concentrations of the amino acid and the inhibition of amino acid decarboxylation, could have this result.

Gessa: If there are no specific receptors, how can one explain the prevention of this disaggregation by pimozide and methysergide?

Wurtman: The blocking agents have no effect on polysome aggregation by themselves, they simply protect against the effects of dopa and 5-hydroxytryptophan.

Is there evidence that monoamines can compete intracellularly with polyamines?

Mandel: Nobody has tried that.

Reference

ANDÉN, N.-E., ENGEL, Y. & RUBENSON, A. (1972) Central decarboxylation and uptake of L-dopa. *Naunyn-Schmiedeberg's Arch. Pharmacol.* **273**, 11–26

α-Methyldopa, an unnatural aromatic amino acid[*]

L. MAÎTRE, P. R. HEDWALL and P. C. WALDMEIER

Research Department, Pharmaceuticals Division, CIBA-GEIGY Limited, Basle

Abstract   The mechanism of the antihypertensive action of α-methyldopa is still unknown. The drug was originally designed as a decarboxylase inhibitor but its antihypertensive effect could not be explained on this basis. α-Methyldopa is metabolized to α-methyldopamine and thence to α-methylnoradrenaline. On this basis, the 'false transmitter' hypothesis was put forward, in support of which sound evidence was adduced but only from experiments on peripheral neurons. This hypothesis had to explain the previous finding that the major site of attack was in the central nervous system. Recent results suggest that α-methylated amine metabolites mediate the antihypertensive action of the drug in the central nervous system. Effects on non-catecholaminergic neurons have also been demonstrated but their significance is still unknown.

Elsewhere in this volume, Sourkes describes some biochemical effects of the unnatural aromatic amino acid, α-methyltryptophan and how it interferes with the metabolism of its natural analogue, tryptophan. 3,4-Dihydroxy-α-methyl-phenylalanine (α-methyldopa) is a similar unnatural aromatic amino acid which is widely used as an antihypertensive drug (for several comprehensive reviews see Holtz & Palm 1966; Muscholl 1966, 1972; Stone & Porter 1967; Kopin 1968a, b; Henning 1969; Thoenen 1969). Here, we shall summarize the current views on the antihypertensive effect of α-methyldopa and report some relevant new findings. We will also describe the several ways in which α-methyldopa interacts with natural amino acids (besides dopa) and related amines.

The mechanism underlying the antihypertensive action of α-methyldopa is still not clearly understood. Its inhibition of dopa decarboxylase was soon established (Dengler & Reichel 1958); originally it was synthesized as a possible inhibitor of the enzyme on the assumption that such inhibitors would possess

[*]   This contribution was received after the Symposium.

antihypertensive properties (see Sourkes & Rodriguez 1967). At the same time, α-methyldopa was shown to inhibit the formation of 5-hydroxytryptamine from 5-hydroxytryptophan (Westermann et al. 1958). This was the first evidence of an interaction between α-methyldopa and 5-hydroxytryptamine.

After Oates et al. (1960) demonstrated that the compound inhibits decarboxylase in humans and reduces high blood pressure, it soon became apparent that the degree of decarboxylase inhibition did not correlate with the antihypertensive properties. The discovery of α-methyldopamine and α-methylnoradrenaline in the brain and heart of mice treated with α-methyldopa (Carlsson & Lindqvist 1962) was important because not only did it show that the metabolism of α-methyldopa in the body follows a similar path to that of its corresponding endogenous amino acid dopa but also it represented the first step toward the theory of 'false transmitters'. According to this theory, the unnatural metabolites, α-methyldopamine and α-methylnoradrenaline, replace the natural amines, dopamine and noradrenaline, in their subcellular stores. Upon sympathetic nerve stimulation, the unnatural (false) amines are released in place of the natural ones and function as transmitters. The weaker sympathomimetic activity of the false amine impairs sympathetic transmission, thereby reducing sympathetic drive with a consequent decrease in blood pressure.

The false transmitter hypothesis rests primarily on two experimental findings: first, sympathetic nervous stimulation in rabbits treated with α-methyldopa releases α-methylnoradrenaline from the heart and, secondly, α-methyldopa is antihypertensive only when the amino acid undergoes decarboxylation. Against the hypothesis is the fact that impaired sympathetic transmission in animals treated with α-methyldopa can be demonstrated conclusively in some, but not all, animals, only at low stimulation frequencies and after a particular schedule of treatment. The responsiveness of different test organs to adrenergic blockade by a false transmitter varies widely. In many circumstances, no impairment of sympathetic transmission by α-methyldopa can be detected. This lack of effect implies that the false transmitters are not as false as this elegant hypothesis suggests.

The notion that such a mechanism could largely explain the antihypertensive effect of α-methyldopa is based on observations made in peripheral tissues. Indirect evidence (Henning 1969) shows that the site of the antihypertensive effect of α-methyldopa is at least partly located within the central nervous system, a view which has been substantiated by Day et al. (1973) and by Finch & Haeusler (1973). All conclude that the production of α-methylnoradrenaline in the central nervous system is an important if not essential mediator of the antihypertensive effect of α-methyidopa.

The theory of false transmitters can only be accepted if it can account for the events within the central neurons. Substitution of a false for a natural transmitter and release of a false transmitter are less easy to demonstrate in the central nervous system than in peripheral tissues.

After treating renal hypertensive rats with α-methyldopa, we have measured the blood pressure and cerebral concentrations of α-methyldopa, α-methyldopamine, α-methylnoradrenaline, dopamine and noradrenaline either at different times after treatment with a fixed dose of α-methyldopa or at a standard time after treatment with variable doses. The antihypertensive effect paralleled the increase in concentration of α-methyldopamine and the decrease in that of dopamine. When we studied dose–response relations, we also found a close correlation between the reduction in blood pressure and the rise in α-methyldopamine or the fall in dopamine. In both series of experiments, the amounts of α-methyldopa, α-methylnoradrenaline and noradrenaline were unrelated to the antihypertensive effects of α-methyldopa (Waldmeier *et al.* 1973; Waldmeier, Maître & Hedwall, unpublished data, 1973). A good correlation does not, of course, necessarily mean a causal relationship, and we cannot infer from these results that the antihypertensive efficacy of α-methyldopa is determined by the concentration of α-methyldopamine or the decrease in the dopamine concentration in the brain. Nevertheless, they provide good evidence that only a minor portion of the α-methylnoradrenaline present in the brain induces the decrease in blood pressure if it is involved at all. In that case, this small portion must consist of freshly synthesized metabolite and therefore depend directly on the availability of its precursor, α-methyldopamine.

So far, most studies on α-methyldopa have been concerned with its action on its endogenous analogues, tyrosine and dopa, and their physiologically-active metabolites. Its effects on the transport of other amino acids are, unfortunately, poorly documented. Young & Edwards (1964) reported that in rats it caused specific reversible aminoaciduria involving histidine, taurine, serine, alanine and glutamic acid. Milne *et al.* (1960) believe that histidine and the neutral amino acids are actively reabsorbed from the nephron by a common transport mechanism. Young & Edwards (1964) proposed that α-methyldopa competitively inhibits this transport mechanism and later (1966) showed that it exerted a more general inhibitory action on the transport of other amino acids in the jejunal loop of the rat: it inhibited the transport of histidine, glutamic acid, glycine, lysine, taurine and phenylalanine but not that of methionine or proline. α-Methyldopa inhibits uptake of dopa in the isolated guinea-pig vas deferens but leaves tyrosine uptake unaltered (Thoa *et al.* 1972). Thus, there seems to be conclusive evidence that α-methyldopa interferes with the transport of at least some other amino acids, though the significance of these findings is still obscure.

Bhargava (1967) found that the acetylcholine content of the brain, heart and ileum of rats treated with α-methyldopa was almost twice that of control animals. Antonaccio & Cote (1974) have suggested that the antihypertensive effect of α-methyldopa in the spontaneously hypertensive rat is potentiated by centrally active anticholinergic drugs. The significance of these results is more readily understood in view of the findings of Corrodi *et al.* (1972) who showed that anticholinergics decrease dopamine turnover in the neostriatum and limbic forebrain. Furthermore, we found a good correlation between the antihypertensive effect of α-methyldopa and the decrease in the dopamine content of the brains of renal hypertensive rats (Waldmeier *et al.* 1973; Waldmeier, Maître & Hedwall, unpublished data, 1973). Thus, if the ratio of dopamine to α-methyldopamine is assumed to be somehow related to the antihypertensive action of α-methyldopa, high acetylcholine concentrations and blockade of the cholinergic receptors might potentiate this effect by further decreasing the amount of dopamine.

Many reports indicate that the drug exerts an influence on the 5-hydroxy-tryptaminergic system; the reduction it causes in the concentration of 5-hydroxytryptamine in various tissues (Smith 1960; Porter *et al.* 1960; Hess *et al.* 1961) can be ascribed to its decarboxylase-inhibiting properties (Sourkes 1954). The concentrations of the amine in animal brains recover much faster than those of noradrenaline; recovery time roughly coincides with the duration of decarboxylase inhibition (Hess *et al.* 1961).

The isomer 2,3-dihydroxy-α-methylphenylalanine can induce equivalent degrees of decarboxylase inhibition *in vivo* without causing any significant decrease in noradrenaline or 5-hydroxytryptamine concentrations (Porter 1961, quoted in Hess *et al.* 1961). Furthermore, Drain *et al.* (1962) were unable to demonstrate any effect of the potent decarboxylase inhibitor, NSD 1023, on cerebral 5-hydroxytryptamine. Thus, it seems unlikely that the lowering of the amine concentration by α-methyldopa can be accounted for by decarboxylase inhibition.

Another mechanism could be the displacement of 5-hydroxytryptamine by the drug or its metabolites. This seems improbable, however, because of the decreases observed in the acidic metabolites of the amine in tissues (Sharman & Smith 1962) and in urinary 5-hydroxyindoleacetic acid (Oates *et al.* 1960).

A third possible mechanism, suggested by the observation (Burkard *et al.* 1964) that α-methyldopa markedly inhibited tryptophan and phenylalanine hydroxylation in the rat liver *in vitro* and *in vivo*, is that the inhibition of trypto-phan hydroxylase is responsible for the decrease in the concentration of

5-hydroxytryptamine. This explanation does not conflict with the available evidence.

We still do not understand the pharmacological and clinical significance of the reduction in 5-hydroxytryptamine induced by α-methyldopa, and in particular we are not clear whether there is any connection between this effect and the antihypertensive action of the drug.

As a part of a more general study of the mechanism of the antihypertensive effect of α-methyldopa, we measured the amounts of 5-hydroxytryptamine and 5-hydroxyindoleacetic acid in the brains of renal hypertensive rats treated with the drug. Blood pressure was recorded before and after administration. We compared the variation of the concentration of the two substances with the course of the antihypertensive effect (Fig. 1). Both varied similarly, so much so that, if the concentration of 5-hydroxytryptamine or 5-hydroxyindoleacetic acid is plotted against the corresponding decreases in blood pressure in the individual animals, highly significant correlations are found for both the amine (Fig. 2) and the acid (Fig. 3). We do not know whether the depletion of 5-hydroxytryptamine and its metabolite are responsible for the antihypertensive effect of α-methyldopa. Most likely, the falls are the consequence of the

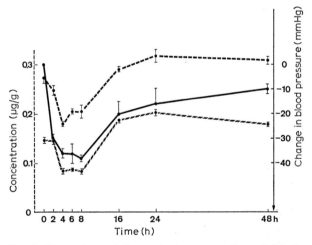

FIG. 1. Comparison of the cerebral concentrations of 5-hydroxytryptamine (▬ ▬ ▬) and 5-hydroxyindoleacetic acid (▰▰▰▰) with the decrease in blood pressure (▬▬▬▬) after administration of α-methyldopa.

Groups of four male renal hypertensive Goldblatt rats weighing 240–290 g were treated with α-methyldopa (300 mg/kg subcutaneously). Blood pressure was recorded plethysmographically (Wilson & Byrom 1939). The values given are differences (in mmHg) in relation to initial preinjection values. The mean blood pressure immediately before treatment was 200 ± 2 mmHg ($n = 36$). 5-Hydroxytryptamine and 5-hydroxyindoleacetic acid were estimated fluorometrically (Curzon & Green 1970).

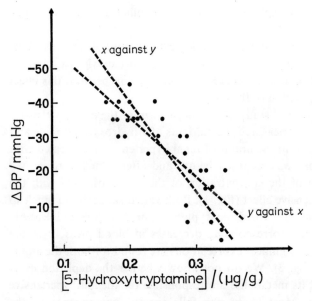

FIG. 2. Correlation between the concentration of 5-hydroxytryptamine and decrease in blood pressure (ΔBP) in individual animals. The regression of both x on y and of y on x was calculated. The angle between the two regression lines gives an impression of the degree of correlation ($r = 0.8045$, $P < 0.001$).

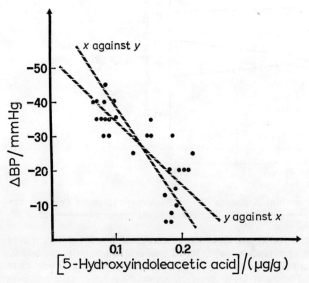

FIG. 3. Correlation between concentration of 5-hydroxyindoleacetic acid and decrease in blood pressure (ΔBP) in individual animals ($r = 0.7933$, $P < 0.001$) (see Fig. 2).

inhibition of tryptophan hydroxylase. Possibly, α-methyldopa competes with tryptophan for the uptake sites in the brain. Finally, from the foregoing results, we must consider the possibility that the antihypertensive action of α-methyldopa is not exclusively a consequence of the interference of the drug with catecholamine metabolism.

ACKNOWLEDGEMENT

The skilful technical assistance of Mrs G. Schlicht, Miss A. M. Buchle and Mr K. Stöcklin is gratefully acknowledged.

References

ANTONACCIO, M. J. & COTE, D. (1974) Evidence for an adrenergic–cholinergic interaction in the regulation of blood pressure in the spontaneous hypertensive rat. *Eur. J. Pharmacol.*, in press

BHARGAVA, V. (1967) Effect of α-methyl-Dopa on acetylcholine content of rat brain, heart and intestine. *Nature (Lond.)* **215**, 202

BURKARD, W. P., GEY, K. F. & PLETSCHER, A. (1964) Inhibition of the hydroxylation of tryptophan and phenylalanine by α-methyldopa and similar compounds. *Life Sci.* 27–33

CARLSSON, A. & LINDQVIST, M. (1962) *In vivo* decarboxylation of α-methyl-Dopa and α-methyl-metatyrosine. *Acta Physiol. Scand.* **54**, 87–94

CORRODI, H., FUXE, K. & LIDBRINK, P. (1972) Interaction between cholinergic and catecholaminergic neurons in rat brain. *Brain Res.* **43**, 397–416

CURZON, G. & GREEN, A. R. (1970) Rapid method for the determination of 5-hydroxytryptamine and 5-hydroxyindoleacetic acid in small regions of rat brain. *Br. J. Pharmacol.* **39**, 653–655

DAY, M. D., ROACH, A. G. & WHITING, R. L. (1973) The mechanism of the antihypertensive action of α-methyldopa in hypertensive rats. *Eur. J. Pharmacol.* **21**, 271–280

DENGLER, H. & REICHEL, G. (1958) Hemmung der Decarboxylase durch α-Methyldopa *in vivo*. *Naunyn-Schmiedeberg's Arch. Pharmakol. Exp. Pathol.* **234**, 275–281

DRAIN, D. J., HORLINGTON, M., LAZARE, R. & POULTER, G. A. (1962) The effect of α-methyl-Dopa and some other decarboxylase inhibitors on brain 5-hydroxytryptamine. *Life Sci.* 93–97

FINCH, L. & HAEUSLER, G. (1973) Further evidence for a central hypotensive action of α-methyldopa in both the rat and cat. *Br. J. Pharmacol.* **47**, 217–228

HENNING, M. (1969) Studies on the mode of action of α-methyldopa. *Acta Physiol. Scand.* *(Suppl.)* 322

HESS, S. M., CONNAMACHER, R. H., OZAKI, M. & UDENFRIEND, S. (1961) The effects of α-methyl-Dopa and α-methyl-metatyrosine on the metabolism of norepinephrine and serotonin *in vivo*. *J. Pharmacol. Exp. Ther.* **134**, 129–138

HOLTZ, P. & PALM, D. (1966) Brenzkatechinamine und andere sympathicomimetische Amine. Biosynthese und Inaktivierung. Freisetzung und Wirkung. *Ergeb. Physiol. Biol. Chem. Exp. Pharmakol.* **58**, 1–580

KOPIN, I. J. (1968a) False adrenergic transmitters. *Annu. Rev. Pharmacol.* **8**, 377–394

KOPIN, I. J. (1968*b*) The influence of false adrenergic transmitters on adrenergic neurotransmission in *Adrenergic Neurotransmission*, pp. 95–104, Churchill, London

MILNE, M. D., CRAWFORD, M. A., GIRAO, V. B. & LOUGHBRIDGE, L. W. (1960) The metabolic disorder in Hartnup disease. *Q. J. Med. N. S.* **29**, 407–421

MUSCHOLL, E. (1966) Autonomic nervous system: newer mechanisms of adrenergic blockade. *Annu. Rev. Pharmacol.* **6**, 107–128

MUSCHOLL, E. (1972) In *Handbook of Experimental Pharmacology*, vol. 33, pp. 618–660, Springer, Berlin & New York

OATES, J. A., GILLESPIE, L. G., UDENFRIEND, S. & SJOERDSMA, A. (1960) Decarboxylase inhibition and blood pressure reduction by α-methyl-3,4-dihydroxy-DL-phenylalanine. *Science (Wash. D.C.)* **131**, 1890–1891

PORTER, C. C., TOTARO, J. A. & LEIBY, C. M. (1960) Effect of decarboxylase inhibitors on serotonin and catecholamine levels in mice. *Pharmacologist* **2**, 81

SHARMAN, D. F. & SMITH, S. E. (1962) The effect of α-methyldopa on the metabolism of 5-hydroxytryptamine in rat brain. *J. Neurochem.* **9**, 403–406

SMITH, S. E. (1960) The pharmacological actions of 3,4-dihydroxyphenyl-α-methylalanine (α-methyldopa), an inhibitor of 5-hydroxytryptophan decarboxylase. *Br. J. Pharmacol.* **15**, 319–327

SOURKES, T. L. (1954) Inhibition of dihydroxyphenylalanine decarboxylase by derivatives of phenylalanine. *Arch. Biochem. Biophys.* **51**, 444–456

SOURKES, T. L. & RODRIGUEZ, H. R. (1967) α-Methyldopa and other decarboxylase inhibitors in *Antihypertensive Agents*, vol. 7, pp. 151–189, Academic Press, London & New York

STONE, C. A. & PORTER, C. C. (1967) Biochemistry and pharmacology of methyldopa and some related structures. *Adv. Drug Res.* **4**, 71–93

THOA, N. B., JOHNSON, D. G. & KOPIN, I. J. (1972) Inhibition of norepinephrine biosynthesis by α-methyl amino acids in the guinea pig vas deferens. *J. Pharmacol. Exp. Ther.* **180**, 71–77

THOENEN, H. (1969) Bildung und funktionelle Bedeutung adrenerger Ersatztransmitter in *Experimentelle Medizin, Pathologie und Klinik*, vol. 27, pp. 1–85, Springer, Berlin & New York

WALDMEIER, P. C., MAÎTRE, L. & HEDWALL, P. R. (1973) Studies on α-methyldopa: relationship between metabolism, biogenic amines and antihypertensive effect. *Naunyn-Schmiedeberg's Arch. Pharmakol.* **277** *(Suppl.)*, R 86

WESTERMANN, E., BALZER, H. & KNELL, J. (1958) Hemmung der Serotoninbildung durch α-Methyl-Dopa. *Naunyn-Schmiedeberg's Arch. Pharmakol. Exp. Pathol.* **234**, 194–205

WILSON, C. & BYROM, F. B. (1939) Renal changes in malignant hypertension. *Lancet* **1**, 136–139

YOUNG, J. A. & EDWARDS, K. D. G. (1964) Studies on the absorption, metabolism and excretion of methyldopa and other catechols and their influence on amino acid transport in rats. *J. Pharmacol. Exp. Ther.* **145**, 102–112

YOUNG, J. A. & EDWARDS, K. D. G. (1966) Competition for transport between methyldopa and other amino acids in rat gut loops. *Am. J. Physiol.* **210**, 1130–1136

Synthetic *p*-halogenophenylalanines and protein synthesis in the brain

E. MARTIN GÁL

Neurochemical Research Laboratory, Deparment of Psychiatry, University of Iowa, Iowa

Abstract   Peripheral administration of *p*-halogenophenylalanines produces two simultaneous mechanisms of interference with the activity of cerebral and hepatic hydroxylases. One effect is a transient competitive inhibition of the hydroxylases by *p*-halogenophenylalanines (or α-methyl analogues) and their metabolites. This inhibition gradually disappears over two and a half days. The other effect, lasting about a week, is a selective and irreversible inactivation of hepatic phenylalanine 4-hydroxylase and cerebral tryptophan 5-hydroxylase. We demonstrated that *p*-chlorophenyl[^{14}C]alanine was incorporated into protein and proposed that the selective inactivation of the two hydroxylases was due to substitution of *p*-chlorophenylalanine into the primary structure near or at the active centre of these enzymes. I shall discuss data from our studies *in vitro* and *in vivo* showing (1) incorporation of *p*-chlorophenyl[^{14}C]alanine into tryptophan 5-hydroxylase with and without added inhibitors of protein synthesis, (2) prevention of enzymic inactivation by *p*-chlorophenylalanine in the brains of animals pretreated by intracerebral injection of cycloheximide, (3) that purified brain ribosomes, unlike those from *E.coli*, use both codons (UUU and UUC) for reading off *p*-chlorophenylalanyl-tRNA and (4) that the mediation by ribosomes from the brain of rats treated with *p*-chlorophenylalanine of incorporation of [^{14}C]tyrosine, phenylalanine or tryptophan is not impaired.

Also I shall present evidence that administration of *p*-chlorophenylalanine does not lead to polysomal disaggregation in the brain of either young or foetal rats unless the animals were pretreated with an inhibitor of monoamine oxidase. I shall propose a hypothesis to explain this transient polysomal disaggregation *in vivo*.

Synthetic analogues of natural amino acids have been frequently employed in the study of the binding sites of native enzymes or of substitution for the natural compound under conditions of growth. Most act as competitive inhibitors of the enzyme in question. Structural modifications of phenylalanine, for example, have included changes in the ring and side chain and substitution on the

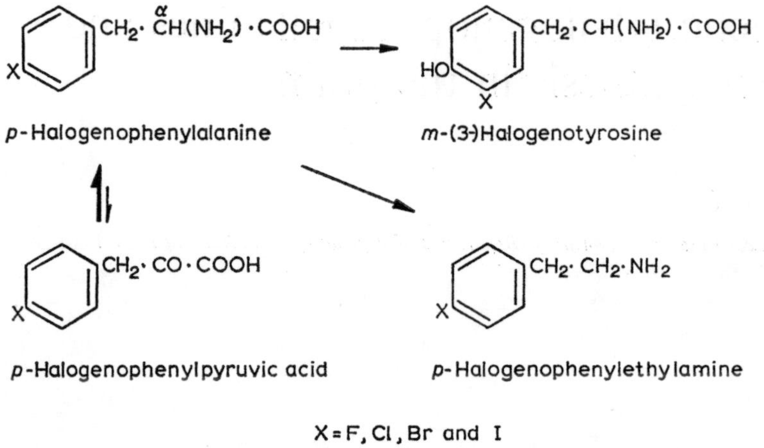

FIG. 1. *p*-Halogenophenylalanines and their possible metabolites. The chloro-substituted compounds are the most stable. Little *p*-halogenophenylethylamine is normally formed.

phenyl ring (see Fig. 1). One of the most representative and effective substituents is a halogen in the 4-(*para*-) position of the phenyl ring. This particular analogue is incorporated into bacterial protein (Shive & Skinner 1963; Fowden 1972).

p-Chlorophenylalanine has been reported to deplete specifically cerebral 5-hydroxytryptamine in rats (Koe & Weissman 1966). At the same time, it was noted that after its administration *p*-fluorophenyl[^{14}C]alanine was recovered from isolated serum and tissue protein (Dolan & Godin 1966; Westhead & Boyer 1961). Soon it was demonstrated that the depletion of 5-hydroxytryptamine was the result of a selective and irreversible inactivation of cerebral tryptophan 5-hydroxylase (Jequier *et al.* 1967), the rate-limiting enzyme in the synthesis of the amine.

As the mechanism of this irreversible inactivation was not clear, we undertook the study of this problem using *p*-chlorophenyl[^{14}C]alanine. Our decision to use this compound was predicated upon two considerations: (*a*) *p*-chlorophenylalanine was the most effective of the *p*-halogenophenylalanines in producing enzymic inactivation and (*b*) *p*-fluorophenylalanine was unsuitable for quantitative studies since 50% was hydroxylated to tyrosine both *in vivo* (Gál & Millard 1971) and *in vitro* (Kaufman 1961).

We have reported the incorporation of *p*-chlorophenyl[^{14}C]alanine into a variety of enzymes (Gál *et al.* 1970). Such general incorporation was inconsistent with the selective and irreversible inactivation of tryptophan and phenylalanine hydroxylases: we postulated that *p*-chlorophenylalanine was incorporated either into the catalytically active site thereby causing inactivation or into inactive sites in enzymes like tyrosine hydroxylase and muscle aldolase leaving both unaffected (Gál & Millard 1971). Inactivation of the enzyme by *p*-chloro-phenylalanine was accompanied by changes in the higher order of structures, according to optical rotary dispersion measurements and electrophoretic patterns of pure hepatic phenylalanine 4-hydroxylase (Gál 1972).

Experimental evidence indicated only one possible way in which *p*-chloro-phenylalanine could selectively inactivate tryptophan 5-hydroxylase, that is by participation in the cerebral synthesis of protein. I shall discuss certain aspects of such participation in this paper.

EFFECT OF *p*-CHLOROPHENYLALANINE ON TRYPTOPHAN 5-HYDROXYLASE *IN VIVO*

Intraperitoneal administration of *p*-chlorophenyl[2-^{14}C]alanine (300 mg/kg) to Sprague–Dawley rats resulted in two simultaneous interferences with the hydroxylation of phenylalanine and tryptophan. In one mechanism, the onset of competitive inhibition by the *p*-chlorophenylalanine is rapid both *in vitro* and *in vivo*. This feature is not unexpected since we had previously shown that the natural amino acid, L-phenylalanine (0.1 mmol/l), produced 50% inhibition of cerebral tryptophan 5-hydroxylase *in vitro* (Gál *et al.* 1966). Incubation of *p*-chlorophenyl[^{14}C]alanine for 30 min with purified tryptophan 5-hydroxylase (Gál & Millard 1971) and with tyrosine hydroxylase resulted in 80% inhibition of both enzymes. The competitive inhibition of these enzymes in brain is readily apparent; the concentrations of both 5-hydroxytryptamine and nor-adrenaline are reduced (Koe & Weissman 1966). However, the reduction of cerebral noradrenaline persists for no longer than 48 h. Simultaneously, a selective and irreversible inactivation of phenylalanine and tryptophan 5-hydroxylase gradually sets in. The progress of the bimodal effect of *p*-chloro-phenylalanine on this latter enzyme, in which lies our immediate interest, is illustrated in Table 1, along with the change of total radioactivity in the brain (about 90% of the total radioactivity corresponded to the *p*-chlorophenyl-alanine). The enzyme activity representing the combined effects of competitive inhibition and irreversible inactivation is expressed by the values obtained from the hydroxylase assay of supernatant fraction (30 000 *g*) of brain stem (Gál *et al.* 1970). The extent of irreversible inactivation was assessed from samples of the purified enzyme (see Gál & Millard 1971). This enzyme was fortified with the diffusible stimulating factor (C-1) (Gál & Roggeveen 1973). The results imply that during the initial 6–12 h competitive inhibition by *p*-chlorophenyl-alanine dominates. The onset of the irreversible inactivation is somewhat delayed. The calculated half-life, $T_{1/2}$, of inactivation is about 18 h, a value

TABLE 1

The activity of cerebral tryptophan 5-hydroxylase and radioactivity after injection of DL-p-chlorophenyl[2-^{14}C]alanine ethyl ester hydrochloride.[a]

Time (h)	Enzyme activity[b]		$10^4 \times$ Radioactivity in brain (d.p.m.)
	30 000 **g** supernatant (%)	Purified enzyme (%)	
1	95	100	—
2	92	—	2.6
4	63	—	2.4
6	49	80	2.3
12	26	60	2.1
24	11	38	1.4
48	7	16	0.7
72	14	10	0.5

[a] Rats (300 g) were intraperitoneally injected with 135 mg of the hydrochloride of the ester (containing 7.6 $\times$ 10^6 d.p.m.).
[b] Compared to control enzyme.

somewhat inconsistent with the value of 2.5–3 days for the half-life of the enzyme obtained by us and others (Meek & Neff 1972). At present we are unable to explain this difference. A re-examination of this problem must await isolation of pure tryptophan 5-hydroxylase. The return of the enzyme activity and the concomitant rise in concentration of 5-hydroxytryptamine lasts till the 10th day when its normal level is reached.

IN VIVO INCORPORATION OF p-CHLOROPHENYL[2-^{14}C]ALANINE IN BRAIN PROTEIN WITH TRYPTOPHAN 5-HYDROXYLASE ACTIVITY

As mentioned earlier, our experiments *in vivo* disclosed that, after injection of p-chlorophenyl[^{14}C]alanine, enzyme proteins became labelled. After hydrolysis of the purified tryptophan 5-hydroxylase, the radioactive amino acid was identified by chromatography as p-chlorophenyl[^{14}C]alanine (Gál et al. 1970).

The results of injecting two groups of animals with labelled p-chlorophenyl-alanine and L-tyrosine are given in Table 2. Aliquot portions (1 mg protein) of the purified enzyme were tested for activity while the rest was either counted for total activity or analysed chromatographically after acid hydrolysis. Our data indicate a reasonably good incorporation of p-chlorophenyl[^{14}C]alanine compared with [^{14}C]tyrosine. Even higher specific activities were obtained for the enzyme protein from brain stem of rats treated with p-chlorophenyl[^{14}C]alanine when the enzyme was purified with ammonium sulphate as well as alcohol (Gál

TABLE 2

Incorporation of [^{14}C]tyrosine, *p*-chlorophenyl[^{14}C]alanine in partially purified cerebral tryptophan 5-hydroxylase

Enzyme	Amino acid	Enzyme protein radioactivity	Acid hydrolysate		Enzyme activity (%)
		Incorpora-tion (d.p.m./mg)	Tyrosine (d.p.m./mg)	p-Chlorophenyl-alanine (d.p.m./mg)	
Brain trypto-phan 5-hydr-oxylase	L-[U-^{14}C]Tyrosine	77	62	—	100
	DL-*p*-Chlorophenyl-[2-^{14}C]alanine^a	27	—	21	51

Two animals for each group were injected with amino acid (1.6 nmol; 90 μCi/kg intraperi-toneally) and killed six hours later. Tryptophan 5-hydroxylase was prepared as described in the text.
^{a} The 2-position is labelled α in Fig. 1.

& Millard 1971). Incidentally, when labelled *p*-chlorophenylalanine was added to brain homogenates of control rats in zero-time experiments, no radioactivity was detected in the purified protein. Of course, we cannot yet estimate how much of the label found in the purified tryptophan 5-hydroxylase is associated with the pure enzyme *per se*. However, it is to be noted that similar incorpora-tion of the *p*-chlorophenylalanine was also evident in purified tyrosine hydroxyl-ase but without concomitant inactivation of this enzyme (Gál *et al.* 1970).

We undertook subsequent experiments to prove that the *p*-chlorophenyl-alanine was not adsorbed onto the enzymes but actively incorporated during protein synthesis. In order to verify the active incorporation of the *p*-chloro-phenylalanine, we used cycloheximide and puromycin, which are potent inhibitors at two different levels of protein synthesis. We discerned no radio-activity (see Table 3) in the brain protein of animals treated with cycloheximide for the four hours preceding their death. We removed the brains and analysed them for total RNA, soluble RNA and protein by modification of an existing procedure (Nemeth & de la Haba 1962). Cycloheximide inhibits incorporation of *p*-chlorophenylalanine and tyrosine in similar ways. (The intraperitoneal injection of cycloheximide was aimed at producing simultaneous inhibition of hepatic protein synthesis.)

These experiments substantially confirm that *p*-chlorophenyl[^{14}C]alanine is not adsorbed onto proteins but is actively incorporated through *p*-chloro-phenyl[^{14}C]alanyl-tRNA and the 80 S ribosomal system as a function of protein synthesis.

TABLE 3

Effect of cycloheximide on the incorporation of p-chlorophenyl[2-^{14}C]alanine and [^{14}C]-tyrosine into brain protein

Compounds injected	Brain activities			
	Amino acid[a] (d.p.m./mg)	Aminoacyl-RNA (total)[b] (d.p.m./mg RNA)	Aminoacyl-RNA (soluble)[c] (d.p.m./mg RNA)	Protein (d.p.m./mg)
Tyrosine	4 300	910	—	98
Tyrosine + cycloheximide	11 700	280	—	7
p-Chlorophenylalanine	26 250	300	21	16
p-Chlorophenylalanine + cycloheximide	20 000	100	1	1

Cycloheximide (2.0 mg), [^{14}C]tyrosine (2 μCi) and p-chlorophenyl[^{14}C]alanine (1.8 μCi) were injected intraperitoneally at 0, 1, 2 and 3 h; cycloheximide (400 μg) was also injected into each cortical hemisphere at 0 h. Animals were killed after four hours.
[a] Free amino acid (cold $HClO_4$ extract).
[b] Amino acid esterified to total RNA (hot $HClO_4$ extract).
[c] Amino acid esterified to soluble RNA (hot $HClO_4$ extract).

Peterkofsky & Tomkins (1967) suggested that inhibitors of protein synthesis would interfere with the catabolism of enzyme protein. Therefore, we studied the disappearance of labelled protein from brains of rats given [^{14}C]tyrosine intraperitoneally 24 h before injection of cycloheximide (2 mg/kg). We pursued these studies for several days and at no time could we discern any difference in radioactivity of the purified protein containing hydroxylase activity from control or cycloheximide-treated animals.

We were interested in proving that inactivation of tryptophan hydroxylase by p-chlorophenylalanine was due to incorporation of the chloride into the primary structure at or near the active enzymic centre. If this were so, might not the inhibitors of protein synthesis quantitatively interfere with inactivation of tryptophan 5-hydroxylase? To investigate this, we injected rats intracerebrally with cycloheximide four hours before intraperitoneal administration of p-chlorophenylalanine (Table 4). Hydroxylase assays were performed with partially purified enzyme samples obtained from pooled brain stems. The results show that a single injection of cycloheximide diminished the inactivation of tryptophan 5-hydroxylase by p-chlorophenylalanine over a three-day period, thereby lending support to our belief that p-chlorophenylalanine is inhibitory by virtue of its incorporation into the peptidic sequence at or near the active centre of the affected enzymes.

TABLE 4

Effect of cycloheximide on inhibition of tryptophan 5-hydroxylase by *p*-chlorophenylalanine[a]

| Treatment | Tryptophan 5-hydroxylase (nmol mg^{-1} h^{-1}) | | | |
	24 h	Activity (%)	72 h	Activity (%)
Control	3.14	100	2.84	100
Control + cycloheximide	2.60	83	2.00	70
p-Chlorophenylalanine ethyl ester hydrochloride	0.99	32	0.34	12
p-Chlorophenylalanine ethyl ester hydrochloride + cycloheximide	1.79	57	0.87	31

[a] *p*-Chlorophenylalanine (300 mg/kg) was administered intraperitoneally four hours after intracerebral injection of cycloheximide (400 µg into each hemisphere). The enzyme was purified from pooled subcortical areas of four animals for each group; the assays were made in duplicate.

BRAIN RIBOSOMES: ACTIVATION AND EFFECT OF *p*-CHLOROPHENYL-ALANINE

Our experiments *in vivo* (see Table 3) demonstrated, albeit indirectly, the ability of *p*-chlorophenylalanine to serve as substrate for aminoacyl-tRNA synthetase. Furthermore, several reports concur that *p*-fluorophenylalanine is activated almost as well as phenylalanine with synthetase preparation from *E. coli* (Conway *et al.* 1962). Competition studies with cell-free preparations of rabbit reticulocytes demonstrated that the active synthetase showed less affinity for *p*-fluorophenylalanine than for phenylalanine (Arnstein & Richmond 1964). This seems to agree well with our observation from estimating aminoacyl-RNA from liver that the formation of phenylalanyl-tRNA has outstripped that of *p*-chlorophenylalanyl-tRNA by a ratio of about 3:1. We were inclined to interpret this as an inability of *p*-chlorophenylalanine to charge tRNA species containing GAA anticodons in a fashion similar to the cell-free system from *E.coli*. In this microbial system, in spite of the identical initial rate of transfer of phenylalanine and *p*-fluorophenylalanine to tRNA, only phenylalanine was accessible to both UUU and UUC codons. There was convincing evidence that a decrease in incorporation of *p*-fluorophenyl[14C]alanine into protein was functionally related to the increase in UC copolymers in the incubation system (Dunn & Leach 1967). Intrigued by this, we prepared a cell-free–protein-synthesizing system from rat brain. Ribosomes and the 'pH 5' enzyme were prepared from the postmitochondrial supernatant fractions (Zomzely-Neurath 1972). The ribosomal system was preincubated to remove endogenous 'messenger' (Gelboin 1964) before the addition of the synthetic polymer 14C-labelled amino acids and extra phosphoenolpyruvate and pyruvate kinase. At this time the ribosomal system was further incubated for a varying length of time. A considerable excess of synthetic RNA (100 µg) was added since the requirement of cell-free ribosomal system of mammalian origin is several times greater than that of *E.coli* for optimal stimulation. At the end of the incubation, the samples were chilled and 20% perchloric acid ($HClO_4$) was added. The precipitated protein was collected and purified (Campbell *et al.* 1966). The dry protein was digested in NCS at 60 °C and subsequently counted by liquid scintillation.

At the same substrate concentrations, the incorporation of L-phenylalanine was about ten times greater than that of *p*-chlorophenylalanine (see Table 5). Additionally, unlike the cell-free system from *E.coli*, both amino acids respond to poly(U) or to the heteropolymer, poly(UC), to almost the same extent. It is apparent, therefore, that in rat brain as in rat liver (Gál *et al.* 1973) the degenerate codons are operant for halogenophenylalanines.

So far our discussion has centred on how *p*-halogenophenylalanines have participated in protein synthesis rather than how they affect protein synthesis. The concentration of *p*-chlorophenylalanine added to rat brain ribosomal system *in vitro* had to be raised to 0.5 mol/l before incorporation of phenylalanine or tyrosine was inhibited. However, this inhibition was not specific: 3-fluoro- or 3-chloro-tyrosine (see Fig. 1) and *p*-chlorophenyl-α-methylalanine were equally as inhibitory. At this time, we have no information about whether

TABLE 5

Incorporation of L-phenylalanine and L-*p*-chlorophenylalanine by a ribosomal system from the rat brain

Conditions	Activity in ribosomal protein (d.p.m./mg)	Incorporation (%)
L-Phenylalanine (containing 0.75 µCi L-phenyl[^{14}C]alanine)		
Complete system	3 000	100
Complete system + poly(U)	4 560	152
Complete system + poly(UC) (1:1)	3 830	128
L-*p*-Chlorophenylalanine (containing 0.15 µCi L-*p*-chlorophenyl[^{14}C]alanine)		
Complete system	316	100
Complete system + poly(U)	380	120
Complete system + poly(UC) (1:1)	456	144

Ribosomal system consisted of 1.0M-Tris-acetate, pH 7.8, 5mM-magnesium acetate, 60mM-potassium chloride, 6mM-2-mercaptoethanol, 2.0mM-ATP, 0.1mM-GTP, 0.1mM-^{12}C-containing amino acid mixture (with or without L-*p*-chlorophenylalanine), 20mM-potassium phosphoenolpyruvate, 100 µg pyruvate kinase, 1 mg ribosomal protein, 5.0 mg 'pH 5 enzyme' protein; total volume 1 ml. After preincubation for 10 min to remove endogenous 'messenger' 100 µg of poly(U) or poly(UC) (1:1), the ^{14}C-labelled amino acids were added and the system was incubated for another 45 min at 37 °C.

these compounds like *p*-chlorophenylalanine can be activated at concentrations of about 0.1 mmol/l to serve as substrates for tRNA synthesis.

To assess whether *p*-chlorophenylalanine would affect the ability of cerebral ribosomes to catalyse protein synthesis, we killed animals one day after administering *p*-chlorophenylalanine (300 mg/kg) to them, a time when tryptophan 5-hydroxylase began to be inactivated. To localize some of the sites of possible interference in protein synthesis, we prepared the ribosomal pellet as well as the pH 5 enzymes from brains of control and *p*-chlorophenylalanine-treated animals. The pH enzymes (pH 5-E$_c$ and pH 5-E$_{pcp}$) from these two sources were used with ribosomes from the brains of control animals while the ribosomes from brains of animals injected with *p*-chlorophenylalanine were used with pH 5 enzymes from the control brains only. Using these ribosomal systems, we studied the incorporation of labelled L-phenylalanine, L-tryptophan and L-tyrosine (see Table 6).

Administration of *p*-chlorophenylalanine at concentrations which inactivate phenylalanine and tryptophan hydroxylases did not affect either the ability of ribosomes to catalyse protein synthesis nor did it interfere with the aminoacyl-synthetase activity of the pH 5 preparation.

It has been reported that hyperphenylalaninaemia in seven-day-old (Siegel *et al.* 1971) and in 14-day-old mice (MacInnes & Schlesinger 1971) led to the disaggregation of brain polysomes into monosomes and disomes with a concomitant impairment of protein synthesis *in vitro*. Administration of L-dopa (Weiss *et al.* 1971) and L-5-hydroxytryptophan produced disaggregation of cerebral polysomes from young rats (90–100 g).

Because of our preoccupation with the overall effects of *p*-chlorophenyl-alanine in the brain, it seemed logical to evaluate the conditions which could

TABLE 6

Effect of administration of *p*-chlorophenylalanine on amino acid incorporation by rat brain ribosomal system[a]

Conditions Ribosomes pH 5 enzyme	Activity in ribosomal protein (d.p.m./mg)[b]		
	L-*Tryptophan*	L-*Tyrosine*	L-*Phenylalanine*
R$_c$ + pH 5-E$_c$	2 660 ± 60	1 780 ± 40	1 490 ± 150
R$_c$ + pH 5-E$_{pcp}$	2 580 ± 110	1 490 ± 60	1 530 ± 190
R$_{pcp}$ + pH 5-E$_c$	2 640 ± 110	1 610 ± 70	1 470 ± 100

[a] The system is described in Table 5 except that it contained 1 mg ribosomal (R) and 2 mg pH 5 enzyme protein and no polymers were added. Radioactivity of the amino acids corresponded to 0.5 µCi. *p*-Chlorophenylalanine (300 mg/kg) was intraperitoneally injected into rats (90–100 g) 24 h before their death.
[b] S.E. from duplicate determinations.

influence disaggregation of cerebral polysomes in rats injected with this compound. Accordingly, we gave p-chlorophenylalanine (300 mg/kg) to groups of young rats with or without prior intraperitoneal administration of a monoamine oxidase inhibitor (pargyline, 10 mg/kg). The animals were killed 1, 3 and 24 h after injection of the amino acid and the polysomal profiles from their continuous density gradient were recorded (Weiss *et al.* 1971). We discovered that administration of p-chlorophenylalanine (300 mg/kg) did not lead to polysomal disaggregation at any time but if a monoamine oxidase inhibitor was given three hours before the injection marked disaggregation ensued. This phenomenon was transient; we could detect it 3 or 24 h after injection of p-chlorophenyl-alanine (Fig. 2). Indeed, no traces of cerebral p-chlorophenylethylamine could be detected (see Fig. 1) (Gál *et al.* 1970) after injection of the chloro-alanine unless the rats were pretreated with pargyline (Edwards & Blau 1972).

These and others' results clearly indicate that many agents elicit polysomal disaggregation in an unspecific fashion. The disaggregation *in vivo* lasts a short time and seems to be related in most instances to a rapid rise in concentration of cytoplasmic monoamines. For the present, we guardedly conclude that the

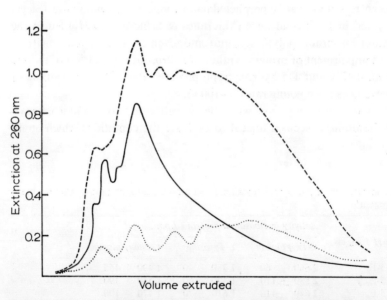

Fɪɢ. 2. Polysomal profiles from brains of rats treated with pargyline and p-chlorophenyl-alanine. Rats were injected with the latter (300 mg/kg) and again (150 mg/kg) 20 h later followed three hours later by pargyline (10 mg/kg). Ultraviolet extinction at 260 nm: – – – – –, three hours after pargyline injection; . . . ., 24 h after pargyline injection; ————, profile obtained from rats given pargyline (50 mg/kg) and p-chlorophenylalanine (300 mg/kg) two hours later and then killed one hour after that.

excessive accumulation of an amine is a requirement for polysomal disaggregation (Weiss *et al.* 1971). However, we believe that during the rapid cytoplasmic increase of an amine the lysosomes rather than the polysomes become the primary target. The changes caused in the lysosomal system will result in the release of ribonuclease and phosphates. Polyribosomal instability demonstrated in the presence of ribonuclease *in vitro* (Zomzely *et al.* 1968) supports this notion. Rupture of lysosomes in addition to ribonuclease release which affects tRNA will also produce a transient intracellular change in pH which might facilitate dissociation of the ribosomal polyamines, spermine and spermidine (Siekevitz & Pallade 1962), from the ribosomal structure. Apparently, the polyamines stabilize the polysome structure by preventing dissociation into subparticles even in absence of magnesium and might increase protein synthesis (Nathans & Lipmann 1961).

It is known that disaggregated polysomal systems barely support protein synthesis *in vitro*. Therefore, according to any theory we developed which explained disaggregation as the result of a direct effect of the amine on ribosomal structure, we ought to be able to produce disaggregation and inhibition of protein synthesis with 10µm-5-hydroxytryptamine or dopamine *in vitro*. Various tests (Campbell *et al.* 1966) to assess the response of the protein-synthesizing system to the addition of neurotransmitters *in vitro* suggest that concentrations greater than 0.5–1.0 mmol/l were required to produce 40–50% inhibition. It is evident from our preliminary experiments that during disaggregation the spermine and spermidine content of the polysomes decreased from 0.1–0.2 µg/mg ribosomal protein to undetectable amounts. Ribosomal polyamines were determined by the butanol-extraction method (Raina 1963) followed by the modified procedure with thin layer chromatography of Hammond & Herbst (1968).

We realize that the complex biochemical system for protein synthesis depends on many factors. One factor is the nature of the existing amino acid pool (Appel 1966; Munro 1970). An excess of one amino acid could interfere with the synthesis of aminoacyl-tRNA. The presence of various aminoacyl-tRNA in amounts less than those necessary for ribosomal integrity could result in disaggregation (MacInnes & Schlesinger 1971).

In view of the above, we advance our hypothesis for the amine-catalysed and polysomally-mediated ribosomal disaggregation as one possible mechanism arising during a particular set of circumstances rather than one to explain everything.

In our opinion, this hypothesis warrants the further research upon which we are now embarked.

ACKNOWLEDGEMENTS

I gratefully acknowledge the technical assistance of P. A. Christiansen and C. Malmanger.

References

APPEL, S. H. (1966). *Trans. N.Y. Acad. Sci.* **29**, 63
ARNSTEIN, H. R. V. & RICHMOND, M. H. (1964). *Biochem. J.* **91**, 340
CAMPBELL, M. K., MAHLER, H. R., MOORE, W. J. & TEWARI, S. (1966). *Biochemistry* **5**, 1174
CONWAY, R. W., LANSFORD, E. M. & SHIVE, W. (1962). *J. Biol. Chem.* **237**, 2850
DOLAN, G. & GODIN, C. (1966). *Biochemistry* **5**, 922–925
DUNN, T. F. & LEACH, F. R. (1967). *J. Biol. Chem.* **242**, 2693
EDWARDS, D. J. & BLAU, K. (1972). *J. Neurochem.* **19**, 1829
FOWDEN, L. (1972) in *Carbon-Fluorine Compounds. Chemistry, biochemistry and biological activities (Ciba Found. Symp. 2, new series)* pp. 141–159, Associated Scientific Publishers, Amsterdam
GÁL, E. M. (1972) in *Studies of Neurotransmitters at the Synaptic Level*, pp. 149–163, Raven Press, New York
GÁL, E. M. & MILLARD, S. A. (1971). *Biochem. Biophys. Acta* **227**, 32
GÁL, E. M. & ROGGEVEEN, A. E. (1973) *Science (Wash. D.C.)* **179**, 809
GÁL, E. M., ARMSTRONG, J. C. & GINSBERG, B. (1966). *J. Neurochem.* **12**, 643
GÁL, E. M., ROGGEVEEN, A. E. & MILLARD, S. A. (1970). *J. Neurochem.* **17**, 1221
GELBOIN, H. V. (1964). *Biochim. Biophys. Acta* **91**, 130
HAMMOND, J. E. & HERBST, W. J. (1968). *Anal. Biol.* **22**, 474
JEQUIER, E., LOVENBERG, W. & SJOERDSMA, A. (1967). *Mol. Pharmacol.* **3**, 274
KAUFMAN, S. (1961). *Biochim. Biophys. Acta* **51**, 619
KOE, B. K. & WEISSMAN, A. (1966). *J. Pharmacol. Exp. Ther.* **154**, 499
MACINNES, J. W. & SCHLESINGER, K. (1971). *Brain Res.* **29**, 101
MEEK, J. L. & NEFF, N. H. (1972). *J. Neurochem.* **19**, 1519–1525
MUNRO, H. N. (1970) in *Mammalian Protein Synthesis*, Academic Press, New York
NATHANS, D. & LIPMANN, F. (1961). *Proc. Natl. Acad. Sci. U.S.A.* **48**, 2171
NEMETH, A. M. & DE LA HABA, G. (1962). *J. Biol. Chem.* **237**, 1190
PETERKOFSKY, B. & TOMKINS, G. M. (1967). *J. Mol. Biol.* **30**, 49
RAINA, A. (1963). *Acta Physiol. Scand.* **60** (Suppl. 218), 7
SHIVE, W. & SKINNER, C. G. (1963) in *Metabolic Inhibitors*, pp. 1–73, Academic Press, New York
SIEGEL, F. L., AOKI, K. & COLWELL, R. E. (1971). *J. Neurochem.* **18**, 537
SIEKEVITZ, P. & PALLADE, G. E. (1962). *J. Cell Biol.* **13**, 217
WEISS, B. F., MUNRO, H. N. & WURTMAN, R. J. (1971). *Science (Wash. D.C.)* **173**, 833
WESTHEAD, E. W. & BOYER, P. D. (1961). *Biochim. Biophys. Acta* **54**, 145–156
ZOMZELY, C. E., ROBERTS, S., GRUBER, C. P. & BROWN, D. M. (1968). *J. Biol. Chem.* **243**, 5396
ZOMZELY-NEURATH, E. C. (1972) in *Protein Synthesis in Nonbacterial Systems*, vol. 2, p. 147, Marcel Dekker, New York

Discussion

Wurtman: You need larger doses than 300 mg/kg of phenylalanine to disaggregate polysomes.

Roberts: The doses should be about 0.75–1.00 g/kg.

Gál: Even with the ethyl ester, the animals will not tolerate doses of *p*-chlorophenylalanine higher than 300–400 mg/kg. For comparison, we keep the amount of phenylalanine the same.

Wurtman: Isn't there evidence that other amino acids which are not substrates for aromatic amino acid decarboxylase can disaggregate polysomes *in vitro*?

Gál: I don't know. I implied that disaggregation is not specific.

Carlsson: Kuhar *et al.* (1971) showed that midbrain raphe lesions which destroy 5-hydroxytryptamine cell bodies led to the slow disappearance of tryptophan hydroxylase in the brain over about eight days.

Gál: Meek & Neff (1972) have transected the spinal cords of rats and estimated that the enzyme activity is constant in the proximal part of the cord but disappears in three days in the distal part. They also used *p*-chlorophenylalanine and from the return of enzyme activity found a half-life of 2.5–3 days. If we calculate the reappearance of enzyme activity from the second day after injection of *p*-chlorophenylalanine, we obtain the same value as they did.

I don't believe that the rate of disappearance in the distal part is the same thing as *de novo* synthesis because, for one thing, the extent of proteolysis could be greater there than in the proximal part. Also, I don't believe that you can measure the half-life of an enzyme solely from its disappearance; you have to measure it from its synthesis.

Carlsson: Saying that the enzyme survives in the distal part after the axons have been cut means that the enzyme has a long lifespan. It is hard to understand how in three days all the enzyme is lost just by blocking the synthesis of new enzyme molecules. Andén & Modigh (1972) have shown that *p*-chlorophenyl-alanine still actively inhibits synthesis of 5-hydroxytryptamine after cutting the spinal cord. Under these conditions, new enzyme molecules are probably not synthesized and thus *p*-chlorophenylalanine appears to act by some other mechanism.

Gál: This depends on the time after its administration. In the first 48 h after transection, there might be a competitive inhibition of the enzyme in the cell. In spite of transection, *p*-chlorophenylalanine is incorporated into the enzyme (Harvey & Gál 1974) in the septum. Whether this is genuine incorporation into tryptophan 5-hydroxylase we are still unable to tell because of the low yield of the enzyme isolated from the septum.

Knapp & Mandell (1972) puzzled us by saying that only particulate enzyme and no soluble enzyme is present. We found that septal tryptophan 5-hydroxylase can be recovered as soluble enzyme if the septum is homogenized in Tris-acetate buffer (Harvey & Gál 1974). It was interesting that medial forebrain bundle lesions, even after 10 days, had no effect on the specific activity of the septal tryptophan 5-hydroxylase compared with that of the septi from sham-operated rats. Incidentally, our assays for individual 5-hydroxyindoles and tryptophan hydroxylase activity are sensitive enough to measure a single septum or a single rat pineal.

Kaufman: The discrepancy in the half-life seems to depend on whether it is estimated after transection or after administration of *p*-chlorophenylalanine. Could this compound labilize the enzyme? In *E. coli*, when an analogue of an amino acid is incorporated into a protein, the resultant abnormal protein may be digested at a faster rate than the normal protein. Couldn't that happen in this case?

You were surprised that *p*-chlorophenylalanine was incorporated into tyrosine hydroxylase and yet did not inhibit it. I assume that you were surprised because you expected tyrosine and phenylalanine hydroxylases to behave similarly. I hope I didn't emphasize too strongly the notion that there is a family of pterin-dependent hydroxylases. Thus, although the antibodies to phenylalanine hydroxylase cross-react with (but do not precipitate) adrenal and brain tyrosine hydroxylases, antibodies to bovine adrenal tyrosine hydroxylase do not cross-react with rat liver phenylalanine hydroxylase. They are separate enzymes. I am not at all surprised that they behave differently after incorporation of *p*-chlorophenylalanine.

Is it possible that the incorporation of *p*-chlorophenylalanine into phenylalanine hydroxylase has nothing to do with the inhibition of the activity, that is, that these are two parallel events? *p*-Chlorophenylalanine is an uncoupler of the reaction of phenylalanine hydroxylase and during its hydroxylation one of the products is presumably hydrogen peroxide. Could the irreversible inhibition of phenylalanine hydroxylase be mediated by hydrogen peroxide?

Gál: Why should an inhibition mediated by hydrogen peroxide be unique for phenylalanine hydroxylase? We racked our brains for a plausible explanation for the selective inactivation of this enzyme by *p*-chlorophenylalanine since none of its possible metabolites inactivates the enzyme. Tryptophan and phenylalanine hydroxylases are mechanistically similar to tyrosine hydroxylase but the latter was not inactivated while tryptophan and phenylalanine hydroxylases were. Pure phenylalanine 4-hydroxylase contained *p*-chlorophenylalanine which had been incorporated by protein synthesis. ORD studies revealed that the presence of the *p*-chloro compound in the enzyme molecule also causes changes in the secondary structure of this enzyme (Gál 1972). We are pursuing similar studies with tryptophan 5-hydroxylase.

Kaufman: That fits in well with the idea that the incorporation of an analogue of an amino acid, such as *p*-chlorophenylalanine, leads to a more labile enzyme.

Gál: Unfortunately, it does not explain to me why tyrosine hydroxylase is totally unaffected by *p*-chlorophenylalanine.

Kaufman: Is *p*-chlorophenylalanine incorporated into peptide linkages?

Gál: We are still investigating this. The fact that cycloheximide inhibits its incorporation suggests that *p*-chlorophenylalanine joins the peptide sequence through *p*-chlorophenylalanyl-tRNA.

Kaufman: That doesn't prove it has been incorporated into phenylalanine hydroxylase.

Gál: It is hard to envisage any explanation in which the chloro compound is not in the primary structure. We can recover *p*-chlorophenyl[^{14}C]alanine from hydrolysates or peptides of pure phenylalanine hydroxylase purified either by your method or by our method from livers of rats treated with the amino acid.

Kaufman: Tentative evidence points to epoxides as intermediates in some enzyme-catalysed hydroxylations (Jerina *et al.* 1968). With an abnormal substrate like *p*-chlorophenylalanine, the steady-state concentration of the epoxide intermediate on the enzyme surface is possibly much higher than normal, so that the epoxide has a chance to react with the protein to make a covalent (but not a peptidic) bond.

Weiner: To say that *p*-chlorophenyl[^{14}C]alanine is incorporated into both tyrosine hydroxylase and tryptophan hydroxylase means that you must have isolated pure enzymes.

Gál: *p*-Chlorophenylalanine inactivates phenylalanine 4-hydroxylase *in vivo* and its presence in the enzyme is verified by our amino acid analyses. Analogously, it also inactivated tryptophan 5-hydroxylase purified 6–8 times from rat brain and that enzyme contained *p*-chlorophenyl[^{14}C]alanine.

Weiner: But that enzyme is still very impure. The radioactivity could simply be due to material adsorbed onto the tyrosine hydroxylase or its fragments.

Gál: Precisely; crude tyrosine hydroxylase, which appears to have incorporated *p*-chlorophenyl[^{14}C]alanine, was not inactivated. Since one enzyme contains the chlorophenylalanine and is inactivated whereas the other enzyme also contains it but it is not affected, I surmise that incorporation of *p*-chlorophenylalanine is connected with inactivation of the enzyme but only if it is incorporated at or near the active centre of the enzyme.

Weiner: Alternatively, it has not been incorporated into tyrosine hydroxylase. Surely you find radioactivity in almost every protein fraction of the brain you recover?

Gál: Yes, but consider another enzyme, muscle aldolase, obtained from *p*-chlorophenylalanine-treated rats. The amino acid is present in the crystalline enzyme yet the enzyme activity is unimpaired.

Weiner: That doesn't prove anything about tyrosine hydroxylase.

Gál: But it proves to me that *p*-chlorophenylalanine is in the aldolase without, however, affecting the active centre of the enzyme.

Wurtman: What is known about the amino acid sequences of the active sites of phenylalanine 4-hydroxylase and tryptophan 5-hydroxylase?

Gál: To my knowledge, nothing as yet.

Wurtman: Have you examined hepatic proteins that are synthesized in large amounts, such as albumin or ferritin?

Gál: Dolan & Godin (1966), using *p*-fluorophenylalanine, studied pancreatic protein and albumin from which they recovered it. Also, Westhead & Boyer (1961) have recovered *p*-fluorophenyl[^{14}C]alanine from glyceraldehyde dehydrogenase. The reason we didn't use the *p*-fluoro analogue was that a considerable amount was converted into tyrosine.

Wurtman: How do spermine and spermidine become undetectable (p. 353)? Is a lysosomal enzyme released?

Gál: As Dr Roberts suggests, lysosomal ribonuclease might act on the ribosomes thereby causing disaggregation. Equally, lysosomal phosphates might mobilize the polyamines from ribosomal structure with the same result. If the cells survive they will return to normal after one or two hours.

Wurtman: The intracellular formation of 5-hydroxytryptamine or dopamine through the lysosomes transiently decreases the concentration of spermine and spermidine. What is the mechanism that relates the high intracellular concentration of amine to polyamines?

Gál: Administration of α-methyldopa had no effect on ribosomal integrity; whether we can get disaggregation with α-methyltryptophan remains to be seen. As you have shown, high doses of intraperitoneally injected 5-hydroxytryptophan and dopa, both with and without pretreatment with pargyline, induced ribosomal disaggregation. However, *p*-fluorophenylalanine and phenylalanine require pargyline pretreatment otherwise there is no disaggregation.

Weiner: What happens with pargyline alone? If the rapidly increasing amounts of monoamines cause disaggregation, why is phenylalanine required?

Gál: The amino acid concentration required to produce phenylethylamine may be excessive as Dr Axelrod showed (see p. 57).

Weiner: But the amount of phenylethylamine produced would be minute.

Gál: Yes. That's exactly why there is no disaggregation with *p*-chlorophenylalanine or phenylalanine unless we pretreat with pargyline. Using gas chromatography, Edwards & Blau (1973) have confirmed this by studying cerebral concentrations of amines. They also showed that pargyline treatment of rats was necessary before *p*-chlorophenylalanine will rapidly yield appreciable amounts of *p*-chlorophenylethylamine (0.2–0.4 μg/g in the rat brain) (see p. 352).

Gessa: If *p*-chlorophenylalanine acts through the formation of a monoamine, would you not expect the same effect with *p*-chloroamphetamine?

Gál: I assume one would require doses larger than 20 mg/kg for ribosomal disaggregation. The animals would not survive when given higher doses of *p*-chloroamphetamine combined with pargyline.

Gessa: If the concentrations of *p*-chlorophenylethylamine and *p*-chloro-amphetamines are comparable, wouldn't you expect the same degree of dis-aggregation?

Gál: I certainly would, but I have not done this experiment. When we measured the amount of *p*-chloroamphetamine in the rat brain after a dose of 10 mg/kg, we were surprised at the effect of only 15–20 ng/g of this compound; but in order to have massive amounts of the amine, we would have to inject high doses of *p*-chloroamphetamine with pargyline.

References

ANDÉN, N.-E. & MODIGH, K. (1972) Effects of *p*-chlorophenylalanine and a monoamine oxidase inhibitor on the 5-hydroxytryptamine in the spinal cord after transection. *J. Neural Transm.* **33**, 211–22

DOLAN, G. & GODIN, C. (1966) *In vivo* formation of tyrosine from *p*-fluorophenylalanine. *Biochemistry* **5**, 922–925

EDWARDS, D. J. & BLAU, K. (1973) Phenylethylamines in brain and liver of rats with experimentally induced phenylketonuria-like characteristics. *Biochem. J.* **132**, 95–100

GÁL, E. M. (1972) in *Studies in Neurotransmitters at the Synaptic Level*, pp. 149–163, Raven Press, New York

HARVEY, J. A. & GÁL, E. M. (1974) *Science (Wash. D.C.)*, in press

JERINA, D. M., DALY, J. W., WITKOP, B., ZALTZMAN-NIRENBERG, P. & UDENFRIEND, S. (1968) The role of arene oxide–oxepin systems in the metabolism of aromatic substrates. III. Formation of 1,2-naphthalene oxide from naphthalene by liver microsomes. *J. Am. Chem. Soc.* **90**, 6525–6527

KNAPP, S. & MANDELL, A. J. (1972) *p*-Chlorophenylalanine—its three phase sequence of interactions with the two forms of brain tryptophan hydroxylase. *Life Sci.* **16**, 761–771

KUHAR, M. J., ROTH, R. H. & AGHAJANIAN, G. K. (1971) Selective reduction of tryptophan hydroxylase activity in rat forebrain after midbrain raphe lesions. *Brain Res.* **35**, 167–176

MEEK, J. L. & NEFF, N. H. (1972) Tryptophan 5-hydroxylase: approximation of half-life and rate of axonal transport. *J. Neurochem.* **19**, 1519–1525

WESTHEAD, E. W. & BOYER, P. D. (1961) The incorporation of *p*-fluorophenylalanine into some rabbit enzymes and other proteins. *Biochim. Biophys. Acta* **54**, 145–156

Effects of α-methyltryptophan on tryptophan, 5-hydroxytryptamine and protein metabolism in the brain

THEODORE L. SOURKES

Departments of Psychiatry and Biochemistry, Faculty of Medicine, McGill University, Montreal

Abstract α-Methyltryptophan acts on the regulatory mechanisms of tryptophan metabolism in the central nervous system and peripheral tissues. The administration of α-methyltryptophan leads to the formation of 5-hydroxy-α-
methyltryptamine in the brain and spinal cord in excess of the normal concentration of 5-hydroxytryptamine; these high levels persist for many days. Explanations of the concomitant decrease of cerebral 5-hydroxytryptamine at the
same time include (i) actions affecting tryptophan hydroxylase such as inhibition
of the enzyme by α-methyltryptophan; (ii) substitution of α-methyltryptophan
for the natural substrate; (iii) reduction in the concentrations of tryptophan in
blood and brain stemming from a reduction in free tryptophan of the liver, and
this in turn from accelerated catabolism of tryptophan caused by activation of
hepatic oxygenase stimulated by α-methyltryptophan. Interference by α-methyltryptophan with tryptophan transport into nervous tissue (cerebral cortical
slices) is probably unimportant.

   Injection of α-methyltryptophan also increases the rates of gluconeogenesis,
urea formation and oxidation of many amino acids *in vivo*. The concentrations
of alanine, cysteine, methionine, proline, lysine and ornithine increase while that
of glycine decreases; the amounts of phenylalanine and tyrosine are unchanged.
The activity of hepatic tyrosine aminotransferase is increased for 1–2 days.

   α-Methyltryptophan inhibits incorporation of amino acids and aminoacyl-
tRNA into protein in liver and in brain but has no such effect *in vitro*. The overall
rate of amino acid activation is not changed.

We have extensively explored the biochemical actions of α-methyltryptophan
for many years. We began our work after observing that administration of this
compound to rats stimulated the activity of tryptophan oxygenase (EC 1.13.1.12)
(Sourkes & Townsend 1955; Sankoff & Sourkes 1962), as does the injection of
tryptophan (Knox 1951) (Table 1). Civen & Knox (1960) showed that this
happens in adrenalectomized animals as well as in intact rats. However,
tryptophan and its α-methyl analogue differ in one important aspect: the increase

TABLE 1

Effect of loading rats with tryptophan or α-methyltryptophan on the metabolism of tryptophan

Function measured	Amino acid loaded		
	None	Tryptophan	α-Methyltryptophan
A. Hepatic oxygenase activity (μmol kynurenine formed h^{-1} [g dry wt. rat-liver supernatant fraction]$^{-1}$)[a]			
Expt. 1	4.0	21.3	47.2
Expt. 2	22.3	44.8	62.5
B. Rate of oxidation of [^{14}C]tryptophan *in vivo* (% administered label appearing as expired $^{14}CO_2$ in 6 h)			
Intact rats			
Unlabelled amino acid[b] given:			
1 h before [β-^{14}C]tryptophan[c]	6.8	35.0	13.0
14 h before [β-^{14}C]tryptophan[c]	6.8	10.8	26.5
at same time as [2-^{14}C]tryptophan[d]	7.6	42	9
24 h before [2-^{14}C]tryptophan[d]	7.6	7	20
Adrenalectomized rats			
Unlabelled amino acid given:			
at same time as [2-^{14}C]tryptophan[d]	7.8	34	9
24 h before [2-^{14}C]tryptophan[d]	7.8	18	11
Unlabelled amino acid[e] given:			
16 h before L-[*carboxy*-^{14}C]tryptophan	8.4		17.9
C. Recovery of ^{14}C-containing metabolites from urine of intact rats in 7 h[b]			
Unlabelled amino acid given:			
1 h before [β-^{14}C]tryptophan[c]	2.0	6.2	4.4
14 h before [β-^{14}C]tryptophan[c]	2.0	1.8	7.2

[a] Enzyme prepared four hours after subcutaneous injection of DL-amino acid (0.5 mmol; Sourkes & Townsend 1955).

[b] Unlabelled L-tryptophan or DL-α-methyltryptophan, 1 mmol/kg body weight.

[c] See structure below for position of label (Moran & Sourkes 1963); about 0.2 mg/rat DL-[^{14}C]tryptophan.

[d] See structure below for position of label (Madras & Sourkes 1968); about 0.2 mg/rat DL-[^{14}C]tryptophan.

[e] (–)-α-Methyltryptophan (1.25 mg/kg). Respiratory $^{14}CO_2$ collected for four hours (Sourkes *et al.* 1970a).

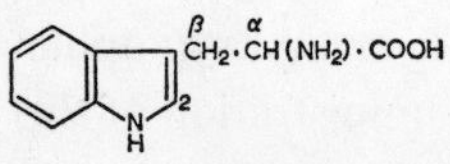

in oxygenase activity after a tryptophan load lasts for about 12 h but the effect of α-methyltryptophan persists for many days (Sankoff & Sourkes 1962). One action of α-methyltryptophan seems to be the inhibition of the degradation of oxygenase (Civen & Knox 1960; Schimke *et al.* 1965; Sourkes & Townsend 1955).

The effects of α-methyltryptophan on oxygenase activity in the liver, the only mammalian organ in which the enzyme is readily detectable, are set out in Table 1. With higher activity a more vigorous rate of degradation of free tryptophan in the liver can be expected. Some years ago, we found evidence of this: the excretion of kynurenic and xanthurenic acids (4-hydroxyquinoline-2-carboxylic and 4,8-dihydroxyquinoline-2-carboxylic acids, respectively) in the urine increased for at least two days after the injection of a large dose of α-methyltryptophan (Madras & Sourkes 1965). As α-methyltryptophan is not a substrate for tryptophan oxygenase, these acidic metabolites must come from the breakdown of tryptophan. Increased excretion of the quinaldic (quinoline-2-carboxylic) acids signifies either a metabolic block resulting from pyridoxine deficiency, or an elevated rate of tryptophan catabolism at least as far as kynurenine and 3-hydroxykynurenine; some of these amino acids are funnelled off into a terminal pathway initiated by their transamination. The latter explanation seems likely, for there is an actual decrease in the concentration of tryptophan in the tissues (Sourkes et al. 1970a) (see Table 2). These effects resemble those seen in animals which have eaten tryptophan-deficient diets (Eber & Lembeck 1958; Gál & Drewes 1962; Culley et al. 1963; Korbitz et al. 1963). We are currently attempting to determine the effect of α-methyltryptophan not only on the plasma concentrations of tryptophan but on the ratio of free to protein-bound amino acid.

TABLE 2

Effect of α-methyltryptophan on the concentration of tryptophan in rat tissues

	Concentration[a] of tryptophan in:			
	Liver	Brain	Blood	Serum
Series	1	1	1	2
No. of determinations (control/ α-methyltryptophan)	5/5	4/4	8/8	10/8
Saline (controls)	2.76 ± 0.23	0.31 ± 0.06	1.01 ± 0.12	2.60 ± 0.19
Injected with α-methyltryptophan	1.52 ± 0.10	0.12 ± 0.03	0.30 ± 0.03	3.19 ± 0.21

Control rats (140–160 g) received 0.9% NaCl (10 ml/kg body weight). In series 1 the rats were injected with DL-α-methyltryptophan (100 mg/kg in saline intraperitoneally) and were killed after 16 h for the determination of tryptophan by a method based on the specific action of partially purified tryptophan synthase (from E. coli) on L-tryptophan. α-Methyltryptophan is neither substrate for nor inhibitor of this enzyme. We determined the indole product as before (Sourkes et al. 1970a). In series 2, the animals (number in parentheses) were killed 48 h (4), 96 h (2) or 120 h (2) after injection of α-methyltryptophan. Ten controls were used in parallel. The method of Denckla & Dewey (1967) was used to measure total tryptophan and tryptophan-like substances (including D- and L-α-methyltryptophan and tryptamine). Data are from Sourkes & Linda Vadnais (unpublished results, 1973).
[a] Concentrations in mg/100 ml or mg/100 g, mean ± S.E. Note that the control concentration of tryptophan in blood in series 1 is equivalent to 0.05 mmol/l.

We can also measure oxygenase activity in the intact animal. Moran & I (1963) showed that when the hepatic oxygenase is stimulated by the administration of unlabelled tryptophan, the increased activity serves to oxidize a tracer dose of labelled amino acid at a greater rate than in controls; this effect is no longer evident by about 16 h after administration of the load. Injection of α-methyltryptophan takes longer to accelerate tryptophan catabolism (Table 1B, intact rats) but its action is still evident a week after the load has been given. The injection of α-methyltryptophan (or tryptophan) is a stress which causes the release of an excess of adrenal cortical hormone, and glucocorticoids themselves increase the activity of tryptophan oxygenase (Knox 1966). Hence, we had to test how α-methyltryptophan affected tryptophan catabolism in adrenalectomized rats (Madras & Sourkes 1968). In these animals, the effect of the α-methyl analogue is qualitatively the same as in intact rats, but it is not so pronounced or long-lasting (Table 1B).

The dynamic factors in the processes described so far seem to be the increased concentrations of circulating tryptophan stemming from the administered load of the amino acid and the increased activity of the hepatic oxygenase. The large quantity of enzyme soon reduces the excess of the amino acid and, when tissue levels of tryptophan have returned to normal, the oxygenase activity does so also. On the other hand, if α-methyltryptophan is given, the circulating concentration of tryptophan decreases, although this can be recognized only if a specific method is used that will detect tryptophan but not α-methyl analogue (Table 2, Series 1). Using the method of Denckla & Dewey (1967) (in which tryptophan and α-methyltryptophan and their derivatives which can form norharman-like substances [β-carbolines] are reactants), we found no apparent change in the concentration of detected substances after administration of α-methyltryptophan, at least between two and five days, in serum (Table 2, Series 2). However, the percentage of free tryptophan (or what is measured as tryptophan) increases significantly from $14.1 \pm 1.5\%$ ($n=10$) in control rats to $27.4 \pm 2.4\%$ ($n=4$) after two days in those receiving DL-α-methyltryptophan. This measurement was made at 37 °C with the sera under 5% CO_2. Simultaneously with these changes in circulating tryptophan, the activity of the hepatic oxygenase rises, and remains high for many days. These results indicate that it is possible to have an increased rate of catabolism of tryptophan with either an elevated or depressed level of circulating tryptophan but only when the activity of the oxygenase is increased.

Does tryptophan concentration in no way determine the flux of tryptophan through the liver? If it does, is it the total concentration in the plasma or only that of free tryptophan? Evidence from at least three laboratories (Kim & Miller 1969; Powanda & Wannemacher 1970; Yuwiler 1973) suggests that the rate of

tryptophan degradation through the oxygenase-initiated pathway depends entirely on tryptophan concentrations. Knox (1966) and Curzon (1971) emphasize the
role of oxygenase in the regulation. Because regulation of this process is so
important for the delivery of tryptophan to the brain for formation of 5-hydroxy-
tryptamine, besides all its other functions, it is important to obtain an answer
to this question.

We chose to study the rate of degradation of labelled tryptophan in rats in
which both oxygenase activity and tryptophan concentrations were altered.
One important aim was to establish the effects of high tryptophan concentrations in animals whose oxygenase activity had already been increased by the
administration of α-methyltryptophan. If oxygenase levels determine the regulation of the tryptophan flux, then the tryptophan load should have no additional
effect on the rate of catabolism of a tracer dose of the amino acid. If the concentration of tryptophan in the plasma is the primary regulatory factor, then no
matter whether oxygenase activity is elevated, an increase in the concentration
of tryptophan in the plasma should result in an increased rate of catabolism of
the labelled material. The results of these experiments are shown in Fig. 1.
Treatment of rats with α-methyltryptophan alone leads to a considerable
increase in the rate of conversion of labelled tryptophan into $^{14}CO_2$. Thus, the

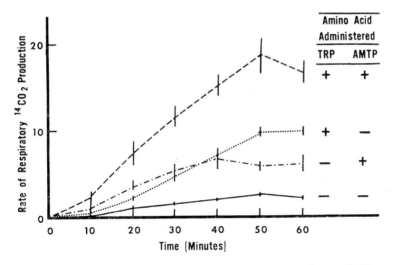

FIG. 1. Effect of altering oxygenase and tryptophan concentrations on $^{14}CO_2$ production in
the rat. Rate of $^{14}CO_2$ production (as percentage of administered ^{14}C released in expired
gases per h) is plotted against time (in min) after injection of [2-^{14}C]tryptophan. Values are
means ± s.e. Each point represents six adrenalectomized rats. The DL-α-methyltryptophan
(AMTP) (100 mg/kg) was given 65 h before the radioactive tryptophan (TRP) (S.N. Young &
Sourkes, unpublished results, 1972).

rise in oxygenase activity increases the rate of catabolism of the tryptophan even when serum tryptophan has been reduced below normal levels. This type of result has been obtained in both intact and adrenalectomized rats. However, when rats treated with α-methyltryptophan 65 h before administration of the labelled tryptophan (to ensure that the oxygenase activity is elevated but that there is no excess of circulating α-methyltryptophan) were given a tryptophan load, there was a striking increase in the rate of $^{14}CO_2$ production over the initial period of the experiment (i.e. the first hour, as shown in Fig. 1), over and above that with either α-methyltryptophan or tryptophan alone. This clearly indicates that both factors are important in determining the flux of tryptophan and this experiment shows one way in which α-methyltryptophan has been used to answer a crucial question about the regulation of tryptophan metabolism.

We were interested in the effect of α-methyltryptophan on other amino acids besides tryptophan. Sankoff & I (1962) had observed the temporary loss of weight in rats after the injection of this compound, which was later associated with an increased rate of protein catabolism as indicated by a large increase in the excretion of urea (Oravec & Sourkes 1969) and by enhanced rates of oxidation of (labelled) alanine, glutamate, leucine, isoleucine, histidine and threonine to carbon dioxide (Oravec & Sourkes 1969) in rats that had received α-methyl-tryptophan 16 h before. In some cases, the stimulated rate of oxidation depended entirely upon the presence of the adrenal glands; this was evident with alanine, glutamate and isoleucine. α-Methyltryptophan-stimulated oxidation of leucine and threonine was only partially dependent upon the adrenals. Only with histidine, among the amino acids tested, was the rate of catabolism increased to the same extent by α-methyltryptophan in both intact and adrenalectomized rats (Oravec & Sourkes 1969). At the same time, some of the carbon skeletons arising from catabolism of the amino acids could be detected in increased concentrations in hepatic glycogen (Oravec & Sourkes 1969). These data are summarized in Table 3.

Oravec has shown that α-methyltryptophan immediately affects the rate of catabolism of tryptophan. Rats were given the tracer dose of tryptophan and exactly 30 min later, when the oxidation of this material was already proceeding, α-methyltryptophan was injected (100 mg/kg). Expired $^{14}CO_2$ was sampled every five minutes for an hour. The increased catabolism was detectable at once and became statistically significant after 20 min. In a somewhat similar experiment, we used uniformly labelled leucine: α-methyltryptophan was injected and [U-^{14}C]leucine was given at intervals of 0–5 h afterwards. Catabolism was not stimulated until the fifth hour after injection of α-methyltryptophan. Before that, there was even some inhibition of the oxidation.

TABLE 3

Rate of oxidation of ^{14}C-labelled amino acids to $^{14}CO_2$ in rats[a]

Amino acid injected	Intact rats		Adrenalectomized rats	
	Control	α-Methyltryptophan	Control	α-Methyltryptophan
Leu	9 ± 0.9	24 ± 1.2	15 ± 0.5	20 ± 0.9
Ala	34 ± 1.1	47 ± 1.2	35 ± 0.3	40 ± 1.1
Glu	51 ± 1.0	56 ± 1.0	55 ± 0.8	55 ± 1.0
Thr	4 ± 0.3	15 ± 1.0	6 ± 0.7	11 ± 0.6
Ile	11 ± 1.0	26 ± 1.0	15 ± 0.8	19 ± 0.8
His	8 ± 0.5	15 ± 0.5	4 ± 0.2	7 ± 0.6
Number of rats per group	3	4	4	5

[a] Percentage of injected radioactivity recovered as $^{14}CO_2$ in expired gases in two hours; mean ± S.E.

Besides tryptophan oxygenase, other enzymes were measured by Oravec (1969) (see Table 4). The activity of tyrosine aminotransferase underwent a large increase after the administration of α-methyltryptophan but this lasted only 1–2 days. The activity of ornithine carbamoyltransferase showed a mean increase but this was not statistically significant. Ornithine aminotransferase, serine dehydratase and histidine ammonia-lyase did not change significantly.

TABLE 4

Effect of α-methyltryptophan on some hepatic enzymes *in vivo*

Enzyme	EC Number	Temperature of assay (°C)	Activities	
			Control	α-Methyl-tryptophan[a]
Tryptophan oxygenase	1.13.1.12	37	0.1	1.1
Aspartate aminotransferase	2.6.1.1	22	96	94
Tyrosine aminotransferase	2.6.1.5	37	1.7	7.3
Ornithine aminotransferase	2.6.1.13	37	4.3	3.8
Ornithine carbamoyltransferase	2.1.3.3	37	75	101
Serine dehydratase	4.2.1.13	22	14.0	15.4
Histidine ammonia-lyase	4.3.1.3	22	6.1	6.5

[a] DL-α-Methyltryptophan (100 mg/kg), suspended in saline, was injected intraperitoneally into rats weighing about 150 g; controls received saline alone. Animals were fasted for 16 h and then killed for assay of the hepatic enzymes. Enzymic activities are expressed as nmol product min^{-1} (mg protein)$^{-1}$. Each mean is based upon data from 3–7 rats. Oxygenase results were estimated from previous data (Sourkes & Townsend 1955; Sankoff & Sourkes 1962); all the other data are abstracted from Oravec (1969). In this series, the administration of α-methyltryptophan to the rats evoked significant increases ($P<0.01$) only in the cases of tryptophan oxygenase and tyrosine aminotransferase.

TABLE 5

Effect of α-methyltryptophan on the free amino acid content of rat liver[a]

Amino acid	Amount of amino acid (μmol/g liver)	
	Controls	α-Methyltryptophan
Lys + Orn	0.48 ± 0.05	0.89 ± 0.13[b]
His + its methyl analogues	0.34 ± 0.03	0.36 ± 0.04
Pro	0.09 ± 0	0.14 ± 0.01[c]
Gly	1.67 ± 0.14	1.37 ± 0.11[b]
Ala	0.31 ± 0.02	0.99 ± 0.16[c]
Cys	0.05 ± 0	0.12 ± 0.01[c]
Val	0.14 ± 0	0.19 ± 0.03
Met	0.06 ± 0	0.10 ± 0.01[b]
Leu	0.14 ± 0.01	0.15 ± 0.01
Ile	0.09 ± 0	0.09 ± 0.011
Tyr	0.05 ± 0	0.04 ± 0
Phe	0.05 ± 0	0.05 ± 0

[a] The pool size of the tabulated free amino acids of liver was determined with a Beckman Automatic Amino Acid Analyzer, Model 120B; column length was 60 cm × 0.9 cm. Rats received DL-α-methyltryptophan (100 mg/kg) and were killed 16 h later. Values are means ± S.E. Standard errors are reported as 0 if less than 0.005.
[b] $P<0.05$.
[c] $P<0.01$.

The size of the hepatic pool of some amino acids was determined in α-methyltryptophan-treated rats; the results are listed in Table 5. Of those amino acids whose concentration was determined, proline, alanine, cysteine, methionine and lysine plus ornithine became significantly more plentiful in the livers of rats that had received α-methyltryptophan (100 mg/kg) 16 h earlier. The greatest increase was observed in the free alanine pool: this may simply reflect increased transamination, for example, of tyrosine (Table 5) with pyruvate as the amino group acceptor. The amounts of proline, cysteine and methionine also rose; glycine significantly decreased in concentration.

Because of the increased rate of catabolism of tryptophan under the influence of α-methyltryptophan, along with the depressed levels of this amino acid in the liver, protein synthesis should be affected. In normal animals, tryptophan is the least abundant amino acid in the free amino acid pool and in proteins formed from that pool (Munro 1968), so the concentration of tryptophan could be rate-limiting for protein synthesis and thus determine the state of polyribosomal aggregation in the liver of the intact animal. Again, α-methyltryptophan treatment and its consequences should be analogous to tryptophan deficiency. The classic result of nutritional tryptophan deficiency is, of course, loss of body weight in mature animals (Osborne & Mendel 1914). After receiving α-methyltryptophan, rats lose weight temporarily but within a few days resume a normal rate of growth (Sankoff & Sourkes 1962). When rats are force-fed a tryptophan-free diet, their livers lose the ability to synthesize protein (Sidransky *et al.* 1967; Pronczuk *et al.* 1970), the hepatic polyribosomes disaggregate (Fleck *et al.* 1965; Wunner *et al.* 1966), and the concentration of tryptophyl-tRNA is reduced (Allen *et al.* 1969). Thus, the availability of tryptophan in the tissues plays a special role in the regulation of protein synthesis in the rat liver.

Administration of α-methyltryptophan to rats decreased the ability of the postmitochondrial fraction of liver and brain to incorporate leucine and leucyl-tRNA into proteins (Table 6). Moreover, it causes the disaggregation of hepatic polyribosomes, with a much higher proportion of mono-, di- and tri-ribosomes than that seen in preparations from liver of normal animals. This result is analogous to that obtained with rats force-fed a tryptophan-free diet (Sidransky *et al.* 1968) and can be reversed if tryptophan is given to the α-methyltryptophan-treated animals shortly before killing them to prepare the amino acid-incorporating system from their livers. However, the addition of L-tryptophan to that system *in vitro* is without effect (Oravec & Sourkes 1970).

Oravec (1969; and unpublished results) has argued that a decrease in trypto-phyl-tRNA synthetase activity could deplete the tryptophyl-tRNA at the ribosomal binding site, which might then account for the disaggregation of the polyribosomes and the diminished rate of incorporation of [^{14}C]leucine and

TABLE 6

Effect of α-methyltryptophan on the incorporation of the radioactive label from [^{14}C]leucine or [^{14}C]leucyl-tRNA into protein *in vitro*

					Rate of protein synthesis	
Series	Organ	Substrate	α-Methyl-tryptophan	No. of Expts.	Incorporation into protein (d.p.m./sample)	Percentage of control
A	Liver	Leucine	—	6	22 400 ± 1 600	
			+	6	12 300 ± 1 700^a	55
	Brain	Leucine	—	6	17 700 ± 450	
			+	7	14 300 ± 220^a	81
B	Liver	Leucine	—	4	13 700 ± 940	
			+	4	10 100 ± 680^b	74
	Liver	Leucyl-tRNA	—	4	25 200 ± 850	
			+	4	21 900 ± 460^b	86

Male rats received an intraperitoneal injection of DL-α-methyltryptophan (100 mg/kg) and, after fasting for 16 h (overnight), were killed. Organs were prepared for measurement of the rate of incorporation of labelled amino acid into protein according to the method of Sidransky *et al.* (1967). Values are means ± S.E. (from Oravec & Sourkes 1970).
^{a} Difference from control mean is significant, $P < 0.01$.
^{b} Difference from control mean is significant, $P < 0.02$.

[^{14}C]leucyl-tRNA into proteins. Using a preparation of the cell sap from rat liver and measuring amino acid activation by ATP–[^{32}P]pyrophosphate exchange, he has been unable to detect an effect of administration of α-methyltryptophan on the overall rate (complete amino acid mixture available in the mixture), nor did there seem to be any activation of α-methyltryptophan itself by enzymic preparations from normal rat liver. Moreover, when labelled tryptophan was used in a system containing aminoacyl-tRNA synthetase, there was no significant effect on synthetase activity.

Thus, the manner in which α-methyltryptophan inhibits protein synthesis is not clear, although it probably reduces the concentration of free tryptophan in the tissues (Sourkes et al. 1970a). Studies by Allen et al. (1969) on the regulation of ribosomal activity indicate the consequences of tryptophan deficiency in the rat. It is still possible that α-methyltryptamine or a related amine metabolite has some depolymerizing action, as Wurtman has claimed for dopamine and 5-hydroxytryptamine.

Let us now consider the entry of tryptophan from plasma into the brain. The transport of this amino acid has only recently received much attention but we now know that tryptophan is actively transported in cerebral cortex (Barbosa et al. 1968, 1970; Kiely & Sourkes 1972) and in synaptosomes (Grahame-Smith & Parfitt 1970). α-Methyltryptophan can interfere to some extent with the uptake of L-tryptophan into cerebral cortex slices if the incubation time is prolonged but not at the shorter times (e.g. 15 min), within which the rate of uptake is linear (Kiely & Sourkes 1972) (Table 7). When low concentrations of

TABLE 7

Effect of α-methyltryptophan on transport of tryptophan into slices of rat cerebral cortex

Experimental set	Concentration of amino acids (mM)		Incubation time (min)	Ratio[a]	
	L-Tryptophan	DL-α-Methyl-tryptophan		Control	+α-Methyl-tryptophan
1	1.0	1.0	60	7.06 ± 0.05	6.00 ± 0.18
2	1.0	1.0	15	3.25 ± 0.14	4.05 ± 0.31
3a	0.20	1.0	15	7.06 ± 0.34	5.45 ± 0.73
3b	0.20	0.10	15	7.06 ± 0.34	6.84 ± 0.23
4a	0.05	0.10	15	8 79 ± 0.94	7.62 ± 0.91
4b	0.05	0.04	15	8.79 ± 0.94	7.39 ± 0.97

The slices were incubated at 37 °C under an atmosphere of pure oxygen as described before (Kiely & Sourkes 1972).
[a] Ratio refers to (c.p.m./ml cell water)/(c.p.m./ml medium). Ratios are given as means ± s.e., and are based on four determinations, except for the control values in sets 1 and 2, where $n = 3$. The effect is statistically significant ($P<0.01$) only in the case of set 1.

tryptophan are used to mimic those one might observe in the circulating plasma, α-methyltryptophan may cause a little inhibition, but this is variable and not statistically significant.

Green & Curzon (1968) reported a decrease in the concentration of cerebral 5-hydroxytryptamine of rats given α-methyltryptophan. After confirming this observation (Sourkes *et al.* 1970*b*), we then found that a certain time after the administration of α-methyltryptophan, 5-hydroxy-α-methyltryptamine appeared in the brain. We established this by extracting the amines from the brain using Amberlite ion-exchange columns before solvent extraction (n-butanol) at pH 10, transfer to aqueous medium with acid and, finally, paper chromatography (for details, see Roberge *et al.* 1971, 1972). Within a few hours of administration of α-methyltryptophan, most of the 5-hydroxytryptamine-like material in the brain is 5-hydroxy-α-methyltryptamine (Table 8). Moreover, for many days after injection of α-methyltryptophan the total concentration of the two amines in the brain and spinal cord (Table 9) is above normal value for 5-hydroxytryptamine alone. This may occur earlier in the tryptophan-deficient rat. Although the concentrations of 5-hydroxytryptamine and 5-hydroxyindoleacetic acid in the brains of deficient rats are only about half the values found in rats nutritionally favoured with tryptophan, 16 h after the administration of α-methyltryptophan the latter group has a low concentration of 5-hydroxytryptamine-like substances in the brain whereas the deficient animals show a significant increase (Table 9). This may represent an occupation of normal storage sites for 5-hydroxytryptamine by 5-hydroxy-α-methyltryptamine.

The 5-hydroxy-α-methyltryptamine is formed by hydroxylation of α-methyl-

TABLE 8

Formation of 5-hydroxy-α-methyltryptamine (5HMT) from α-methyltryptophan in rat brain

Time after administration (h)[a]	*[5-Hydroxytryptamine]/[5HMT]*	*[5HMT], as percentage of total 'apparent 5-hydroxytryptamine'*
2	1.64	38
6	0.44	69
16	0.24	81
48	0.14	88
96	0.33	75
240	—	0

[a] Male rats weighing about 150 g were injected with DL-α-methyltryptophan (100 mg/kg, i.p.) and at the times indicated were killed and their brains were removed for extraction of amines. 5-Hydroxytryptamine and its α-methyl analogue were separated by paper chromatography (see Roberge *et al.* 1972). Half of the stated dose was used in the six hour experiment.

TABLE 9

Effect of α-methyltryptophan on 5-hydroxytryptamine metabolism in the central nervous system of the rat

Series	Diet	Region	α-Methyl-tryptophan	Duration of action (h)	5-Hydroxytryptamine		5-Hydroxyindoleacetic acid	
					No. of rats (n)	Amount (µg/g)	No. of rats (n)	Amount (µg/g)
1	Purina chow	Brain	−		10	0.69 ± 0.02	10	0.22 ± 0.01
			+	6	13	0.54 ± 0.02	13	0.13 ± 0.01
2	Purified ingredients including:	Brain						
	0.23% tryptophan		−		8	0.40 ± 0.01	8	0.37 ± 0.02
			+	16	8	0.25 ± 0.02	4	0.14 ± 0.02
	0.05% tryptophan		−		8	0.21 ± 0.02	7	0.18 ± 0.03
			+	16	7	0.30 ± 0.02	7	0.07 ± 0.01
3	Purina chow	Brain	−		3	0.54 ± 0.11	3	0.44 ± 0.11
			+	24	3	0.52 ± 0.03	3	0.29 ± 0.01
		Spinal cord	−		3	0.26 ± 0.01	3	0.62 ± 0.03
			+	24	3	0.74 ± 0.04	3	0.38 ± 0.06
		Brain	−		3	0.38 ± 0.07	3	0.44 ± 0.15
			+	48	3	0.67 ± 0.01	3	0.27 ± 0.02
		Spinal cord	−		3	0.28 ± 0.01	3	0.64 ± 0.09
			+	48	3	0.68 ± 0.04	3	0.38 ± 0.02

Male rats were injected with DL-α-methyltryptophan, 50 mg/kg (Expts. 1 and 2) or 100 mg/kg (Expt. 3) i.p., or with 0.9% NaCl, 10 ml/kg (controls). After 6 or 16 h, as indicated, they were killed, and the amount of 5-hydroxytryptamine and 5-hydroxyindoleacetic acid in the brain (less cerebellum) and spinal cord determined. Animals received the diet of purified ingredients, with two different amounts of L-tryptophan added, for about one week before α-methyltryptophan was injected. A content of 0.23% was injected. A content of 0.23% tryptophan permitted good growth, but 0.5% led to negative nitrogen balance and loss of weight. Concentrations of metabolites in brain are expressed as µg/g wet wt. tissue, mean ± S.E. Data for series 1 are taken from Sourkes et al. (1970a); series 2 are based on unpublished experiments of Roberge & Sourkes, 1970; series 3 are unpublished data of Sourkes & Linda Vadnais, 1973.

tryptophan with tryptophan 5-hydroxylase, followed by the decarboxylation of the product by aromatic amino acid decarboxylase. In addition to its role as substrate for tryptophan 5-hydroxylase, α-methyltryptophan is an inhibitor of this enzyme (E. M. Gál & P. A. Christiansen, unpublished results, 1973). Apparently, all the biological and biochemical activity of DL-α-methyltryptophan resides in the laevorotatory isomer (Sourkes 1971a,b), small amounts of which have been available for some of these tests.

Note that the use of α-methyltryptophan generally reduces the amount of 5-hydroxyindoleacetic acid; it never increases it (see Table 9). It is well known that compounds like 5-hydroxy-α-methyltryptamine, bearing the isopropylamine side chain, are not substrates for monoamine oxidase. Hence, the 5-hydroxy-indoleacetic acid that is measured is derived from 5-hydroxytryptamine only and represents in some way the metabolism of that substance. Despite the erratic changes in concentration of 5-hydroxytryptamine-like substances in the brain during the first 24 h after injection of α-methyltryptophan, the concentration of 5-hydroxyindoleacetic acid declines steadily to a low level and remains there for over a week (Roberge et al. 1972). More recent data (Table 9) indicate that this decline is true for the spinal cord as well, at least up to 48 h. Thus, 5-hydroxyindoleacetic acid may here reflect the concentration of 5-hydroxytryptamine.

Interestingly, when we made these measurements on the small intestine of rats treated with α-methyltryptophan, the concentration of 5-hydroxytrypt-amine decreased sharply and remained low for many days.

The fact that the α-methyl-amine is formed in situ from the injected amino acid raises the possibility that it can act as a false transmitter. It has some inhibitory action on monoamine oxidase and may affect 5-hydroxytrypt-amine receptor sites. Even if the action of this α-methyl-amine on receptors is weaker than the natural transmitter, the amine might inhibit the re-uptake of 5-hydroxytryptamine by nerve terminals and thereby augment its effect. More-over, the presence of the amine in the brain for as long as eight days after the injection of α-methyltryptophan signifies that long-lasting effects of this amino acid on the central nervous system may derive from some action of its metabolite in neuronal systems. α-Methyltryptophan may, therefore, prove useful as an agent for lowering the concentration of 5-hydroxytryptamine and as a precursor of a false neurotransmitter. Compounds with interesting properties of this kind can materially aid research on the functions of cerebral 5-hydroxytrypt-amine in relation to physiological functions of sleep and temperature regula-tion, as well as to neurological diseases and mental depression.

Research on α-methyltryptophan has two aims: the intrinsic interest that attaches to a biologically active analogue of an indispensable amino acid in

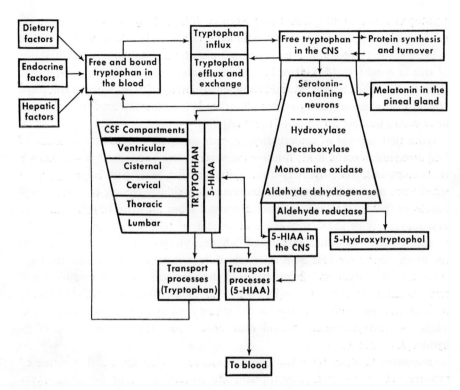

FIG. 2. Regulation of tryptophan concentration of brain (for nomenclature see Editors' note, p. IX).

mammalian nutrition and its use as a tool in elucidating mechanisms of regulation of tryptophan metabolism peripherally and centrally (Fig. 2). Some of these known or putative regulators have already been discussed: tryptophan concentrations in plasma and the ratio of free to bound tryptophan there; hepatic oxygenase activity; transport by energy-requiring processes across cell membranes, in particular those of the cerebral cortex; concentration of tryptophan in the brain. The use of α-methyltryptophan has aided the clarification of some of these steps. We are concerned also with the question of the relation between tryptophan and 5-hydroxytryptamine concentrations in the brain.

If the concentration of tryptophan is rate-limiting in the brain for formation of 5-hydroxytryptamine, then under some circumstances this concentration could be used as a measure of the steady-state amount of the latter. Previously, the concentration of 5-hydroxyindoleacetic acid has often been taken as an index of the amount of the amine and some of the data presented in this paper have favoured this view. Thus, we might expect to have a cascade of correlated concentrations, in series: plasma tryptophan (free or total); tryptophan, 5-hydroxytryptamine and 5-hydroxyindoleacetic acid in the central nervous system; and, finally, of the acid alone in the cerebrospinal fluid. In man, the closest access we have to process of formation and catabolism of 5-hydroxytryptamine in brain during life is through the concentrations of tryptophan and the indole acid in the cerebrospinal fluid. Recent work in my laboratory (Garelis & Sourkes 1973; Sourkes 1973) and elsewhere has demonstrated that the acid is produced at all levels of the central nervous system and some of it penetrates into the cerebrospinal fluid. Moreover, it is not difficult to show that tryptophan is also present in that fluid in readily detectable amounts. Most of this tryptophan is free, for albumin to which it can bind is present in cerebrospinal fluid at very low concentrations (about 0.1 g/l compared with about 45 g/l in serum) (see Table 10). It is apparent that the downward gradient is sharper for the acid than for tryptophan, even though the acid apparently enters the cerebrospinal fluid at all levels. This could mean that proportionately more tryptophan moves from the spinal cord into the subarachnoid space or that transport of the amino acid from this space into the blood is slower. The correlation between these concentrations (S. Young, S. Lal, E. Garelis, P. Molina-Negro & Sourkes, unpublished data) is not statistically significant. Of course, these measurements of the cerebrospinal fluid deal with spent metabolites as far as neural metabolism is concerned so that perhaps no correlation need have been expected in the first place, but we are still seeking a correlation with plasma tryptophan values. In any case, we are also concerned with the normal concentrations of these substances so that our tabulated values may more readily serve in the detection of disturbances of tryptophan metabolism resulting from pathological changes in the central nervous system.

TABLE 10

Concentration (in ng/ml) of tryptophan and 5-hydroxyindoleacetic acid in some cerebrospinal fluid compartments

Compartment	n	Tryptophan concentration	n	5-Hydroxyindoleacetic acid concentration
Ventricular	8	860 ± 68[a]	20	105 ± 49
Cisternal			2	85,86
Mixed[b]	7	603 ± 77	18	37 ± 14
Lumbar	34	531 ± 35	17	29 ± 10

The mean for the acid in ventricular cerebrospinal fluid and one cisternal value are taken from Moir et al. (1970), and the other data are from this laboratory. The tabulated values for the acid have been summarized by Garelis & Sourkes (1973); those for tryptophan are by Young, Lal, Garelis, Molina-Negro & Sourkes (unpublished).

[a] Mean ± S.E.

[b] Cerebrospinal fluid drawn after the injection of air for pneumoencephalography.

ACKNOWLEDGEMENT

Current research is supported by a grant of the Medical Research Council (Canada).

References

ALLEN, R. E., RAINES, P. L. & REGEN, D. M. (1969) Regulatory significance of transfer RNA charging levels. I. Measurements of charging levels in livers of chow-fed rats, fasting rats, and rats fed balanced or imbalanced mixtures of amino acids. *Biochim. Biophys. Acta* **190**, 323–336

BARBOSA, E., JOANNY, P. & CORRIOL, J. (1968) Uptake of some amino acids by rat brain slices: effect of various substrates. *Experientia* **24**, 1196–1197

BARBOSA, E., JOANNY, P. & CORRIOL, J. (1970) Accumulation active du tryptophane dans le cortex cérébral isolé du rat. *C.R. Séances Soc. Biol. Fil.* **164**, 345–350

CIVEN, M. & KNOX, W. E. (1960) The specificity of tryptophan analogues as inducers, substrates, inhibitors, and stabilizers of liver tryptophan pyrrolase. *J. Biol. Chem.* **235**, 1716–1718

CULLEY, W. J., SAUNDERS, R. N., MERTZ, E. T. & JOLLY, G. H. (1963) Effect of a tryptophan deficient diet on brain serotonin and plasma tryptophan level. *Proc. Soc. Exp. Biol. Med.* **113**, 645–648

CURZON, G. (1971) Effects of adrenal hormones and stress on brain serotonin. *Am. J. Clin. Nutr.* **24**, 830–834

DENCKLA, W. D. & DEWEY, H. K. (1967) The determination of tryptophan in plasma, liver and urine. *J. Lab. Clin. Med.* **69**, 160–169

EBER, O. & LEMBECK, F. (1958) Hydroxytryptamine content of the rat intestine at different tryptophan levels of nutrition. *Pfluegers Arch. Gesamte Physiol. Menschen Tiere* **265**, 563–566

FLECK, A., SHEPHERD, J. & MUNRO, H. N. (1965) Protein synthesis in rat liver: influence of amino acids in diet on microsomes and polysomes. *Science (Wash. D.C.)* **150**, 628–629

GÁL, E. M. & DREWES, P. A. (1962) Metabolism of 5-hydroxytryptamine (serotonin). II. Effect of tryptophan deficiency in rats. *Proc. Soc. Exp. Biol. Med.* **110**, 368–371

GARELIS, E. & SOURKES, T. L. (1973) The sites of origin in the central nervous system of monoamine metabolites measured in human cerebrospinal fluid. *J. Neurol. Neurosurg. Psychiatr.* **36**, 625–629

GRAHAME-SMITH, D. G. & PARFITT, A. G. (1970) Tryptophan transport across the synaptosomal membrane. *J. Neurochem.* **17**, 1339–1353

GREEN, A. R. & CURZON, G. (1968) Decrease of 5-hydroxytryptamine in the brain provoked by hydrocortisone and its prevention by allopurinol. *Nature (Lond.)* **220**, 1095–1097

KIELY, M. & SOURKES, T. L. (1972) Transport of L-tryptophan into slices of rat cerebral cortex. *J. Neurochem.* **19**, 2863–2872

KIM, J. H. & MILLER, L. L. (1969) The functional significance of changes in activity of the enzymes tryptophan pyrrolase and tyrosine transaminase, after induction in intact rats and in the isolated, perfused rat liver. *J. Biol. Chem.* **244**, 1410–1416

KNOX, W. E. (1951) Two mechanisms which increase *in vivo* the liver tryptophan peroxidase activity: specific enzyme adaptation and stimulation of the pituitary–adrenal system. *Br. J. Exp. Pathol.* **32**, 462–469

KNOX, W. E. (1966) The regulation of tryptophan pyrrolase activity by tryptophan. *Adv. Enzyme Regul.* **4**, 287–297

KORBITZ, B. C., PRICE, J. M. & BROWN, R. R. (1963) Quantitative studies on tryptophan metabolism in the pyridoxine-deficient rat. *J. Nutr.* **80**, 55–59

MADRAS, B. K. & SOURKES, T. L. (1965) Metabolism of alpha-methyltryptophan. *Biochem. Pharmacol.* **14**, 1499–1506

MADRAS, B. K. & SOURKES, T. L. (1968) Formation of respiratory $^{14}CO_2$ from variously labeled forms of tryptophan-^{14}C in intact and adrenalectomized rats. *Arch. Biochem. Biophys.* **125**, 829–836

MOIR, A. T. B., ASHCROFT, G. W., CRAWFORD, T. B. B., ECCLESTON, D. & GULDBERG, H. C. (1970) Cerebral metabolites in cerebrospinal fluid as a biochemical approach to the brain. *Brain* **93**, 357–368

MORAN, J. F. & SOURKES, T. L. (1963) Induction of tryptophan pyrrolase by alpha-methyl-tryptophan and its metabolic significance *in vivo. J. Biol. Chem.* **238**, 3006–3008

MUNRO, H. N (1968) Role of amino acid supply in regulating ribosome function. *Fed. Proc.* **27**, 1231–1237

ORAVEC, M. (1969) Mode of action of alpha-methyl-DL-tryptophan on hepatic glyconeogenesis, Ph. D. Thesis, McGill University

ORAVEC, M. & SOURKES, T. L. (1967) The influence of alpha-methyltryptophan and some tryptophan metabolites on hepatic glycogenesis. *Biochemistry* **6**, 2788–2794

ORAVEC, M. & SOURKES, T. L. (1969) Action of alpha-methyl-DL-tryptophan *in vivo* on catabolism of amino acids and their conversion to liver glycogen. *Can. J. Biochem.* **47**, 179–184

ORAVEC, M. & SOURKES, T. L. (1970) Inhibition of hepatic protein synthesis by alpha-methyl-DL-tryptophan *in vivo. Biochemistry* **9**, 4458–4464

OSBORNE, T. B. & MENDEL, L. B. (1914) Amino-acids in nutrition and growth. *J. Biol. Chem.* **17**, 325–349

POWANDA, M. C. & WANNEMACHER, R. W. (1970) Evidence for a linear correlation between the level of dietary tryptophan and hepatic NAD concentration and for a systematic variation in tissue NAD concentration in the mouse and the rat *J. Nutr.* **100**, 1471–1478

PRONCZUK, A. W., ROGERS, Q. R. & MUNRO, H. N. (1970) Liver polysome patterns of rats fed amino acid imbalanced diets. *J. Nutr.* **100**, 1249–1258

ROBERGE, A. G., MISSALA, K. & SOURKES, T. L. (1971) Les systèmes sérotoninergiques et le problème des faux transmetteurs. *Union Méd. Can.* **100**, 475–479

ROBERGE, A. G., MISSALA, K. & SOURKES, T. L. (1972) Alpha-methyltryptophan: effects on synthesis and degradation of serotonin in the brain. *Neuropharmacology* **11**, 197–209

SANKOFF, I. & SOURKES, T. L. (1962) The weight-depressing action of alpha-methyl-DL-tryptophan in the rat. *Can. J. Biochem. Physiol.* **40**, 739–747

SCHIMKE, R. T., SWEENEY, E. W. & BERLIN, S. M. (1965) The roles of synthesis and degradation in the control of rat liver tryptophan pyrrolase. *J. Biol. Chem.* **240**, 322–331

SIDRANSKY, H., BONGIORNO, M., SARMA, D. S. R. & VERNEY, E. (1967) The influence of tryptophan on hepatic polyribosomes and protein synthesis in fasted mice. *Biochem. Biophys. Res. Commun.* **27**, 242–248

SIDRANSKY, H., SARMA, D. S. R., BONGIORNO, M. & VERNEY, E. (1968) Effect of dietary tryptophan on hepatic polyribosomes and protein synthesis in fasted mice. *J. Biol. Chem.* **243**, 1123–1132

SOURKES, T. L. (1971a) Alpha-methyltryptophan (AMTP) and its actions on tryptophan metabolism. *Fed. Proc.* **30**, 897–903

SOURKES, T. L. (1971b) Effects of amino acid derivatives and drugs on the metabolism of tryptophan. *Am. J. Clin. Nutr.* **24**, 815–820

SOURKES, T. L. (1973) in *Treatment of Parkinson's Disease—The Role of Dopa Decarboxylase Inhibitors* (Yahr, M. D., ed.), pp. 13-36, Raven Press, New York

SOURKES, T. L. & TOWNSEND, E. (1955) Effects of alpha-methyl-DL-tryptophan on the oxidation of tryptophan. *Can. J. Biochem. Physiol.* **33**, 735–740

SOURKES, T. L., MISSALA, K. & ORAVEC, M. (1970a) Decrease of cerebral serotonin and 5-hydroxyindoleacetic acid caused by (–)-alpha-methyltryptophan. *J. Neurochem.* **17**, 111–115

SOURKES, T. L., MISSALA, K., ORAVEC, M. & ROBERGE, A. (1970b) A novel method of reducing the concentration of brain serotonin: use of alpha-methyltryptophan. *Laval Méd.* **41**, 897–898

WUNNER, W. H., BELL, J. & MUNRO, H. N. (1966) The effect of feeding with a tryptophan-free amino acid mixture on rat-liver polysomes and ribosomal ribonucleic acid. *Biochem. J.* **101**, 417–418

YUWILER, A. (1973) Conversion of D- and L-tryptophan to brain serotonin and 5-hydroxy-indoleacetic acid and to blood serotonin. *J. Neurochem.* **20**, 1099–1109

Discussion

Wurtman: A major problem in such studies is the lack of availability of starting materials. Apparently, L-α-methyltryptophan is still not commercially available.

I must repeat that the tryptophan oxygenase activity in the livers of untreated rats does not correlate with the amount of tryptophan in plasma or in the brain over 24 h (see pp. 15 and 16). Dr Munro has written extensively about the main determinant of the plasma concentrations of amino acids being not how much amino acid is in the body but the distribution of the amino acid between extracellular and intracellular compartments.

Since α-methyltryptophan produces such a long-lasting acceleration of the catabolism of tryptophan, probably by raising the activity of tryptophan oxygenase, is it likely that this compound is incorporated into the enzyme, by analogy with *p*-chlorophenylalanine (p. 346)?

Sourkes: There is no evidence for incorporation. We have studied only the metabolism so far.

Wurtman: What sort of mechanism would allow low molecular weight compounds to cause such prolonged elevations in enzymic activity?

Sourkes: We might unearth a clue by studying inhibitors of the decarboxylation of α-methyltryptophan. Such inhibition would block the synthesis of the possible false neurotransmitter 5-hydroxy-α-methyltryptamine (see p. 373).

Gessa: Several investigators have suggested an important correlation between the concentration of monoamines in the hypothalamus and the hypothalamic neurotransmitters which control the release of pituitary hormones. Are the concentrations of corticosteroids altered after α-methyltryptophan treatment?

Sourkes: I have no doubt that the output of corticosteroids is increased, at least in the acute phase. We have overcome the problem of corticosteroid release by using adrenalectomized rats; the effect of α-methyltryptophan is still evident in such animals but it does not last as long.

Fernstrom: The activity of tryptophan oxygenase is independent of normal variations in plasma concentrations of corticosteroids. Only large amounts have any effect.

Sourkes: Corticosterone is a weak steroid compared with cortisol, in this respect.

Fernstrom: You used L-[2-^{14}C]tryptophan to measure the oxygenase activity (see Table 1). Didn't some of this formamidase-labile radioactive atom enter the C_1 pool?

Sourkes: Certainly, it is very labile. A few years ago, it was claimed that folic acid is not involved in the ultimate metabolism of the formyl carbon atom. Possibly we did not deprive the animals sufficiently of folic acid to show an effect. Recent evidence (Case & Benevenga 1973) indicates that formate is metabolized in the one-carbon pool linked to folic acid in the rat liver.

At least our method of studying the metabolism of tryptophan *in vivo* seems to reflect many of the results that have been obtained by other methods of study, such as treating the animal with drugs or hormones and then directly measuring the activity of tryptophan oxygenase in the liver *in vitro*.

Curzon: What mechanism underlies the increase in the percentage of free tryptophan on giving the α-methyl compound? It could possibly displace tryptophan from albumin.

Sourkes: This free amino acid, present in excess, may be α-methyltryptophan itself. We do not know yet how long this compound remains in the plasma after its administration. We are about to determine that with the tedious but specific tryptophanase method (Sourkes *et al.* 1970).

If one uses a non-specific method (e.g. the glyoxylic acid method) for determining the amount of free tryptophan one could be including not only tryptophan but also α-methyltryptophan, tryptamine and α-methyltryptamine. We use the method of Denckla & Dewey (1967) which determines only the two indole-amino acids.

Wurtman: May I suggest a convention; instead of using the terms 'free' and 'bound' for what is in the plasma, we should use 'albumin-bound' or 'non-albumin-bound'.

Gessa: Results from our laboratory on the ultrafiltration dialysis with purified human albumin and prealbumin have shown that L-tryptophan binds to both albumin and prealbumin fractions with the same affinity for both.

Wurtman: One might expect that α-methyltryptophan would increase serum non-albumin-bound tryptophan by virtue of its ability to accelerate protein catabolism and, consequently, serum albumin would probably be depressed.

Sourkes: We have not checked that.

Gál: α-Methyltryptophan inhibited hydroxylation of both tryptophan and tyrosine *in vivo*. We agree with you, Dr Sourkes, that 5-hydroxy-α-methyltryptophan and its decarboxylated metabolite, the amine, are inhibitory *in vitro*. We found the K_m of L-α-methyltryptophan is 0.112 mM compared with 97 μM for tryptophan (Dr Kaufman finds 67 μM). (These values were obtained with 6-methyltetrahydropterin as cofactor.)

We have an inexplicable observation which we have verified many times: α-methyltryptophan seems to be a better substrate than L-tryptophan for septal tryptophan 5-hydroxylase.

Weiner: Does α-methyltryptophan inhibit tyrosine hydroxylase?

Gál: The evidence for that has come from our experiments *in vivo*.

Weiner: The depletion of noradrenaline could be due to displacement of the natural amine by a false transmitter rather than inhibition of synthesis.

References

CASE, G. L. & BENEVENGA, N. J. (1973) Metabolic interactions of methionine, *S*-methyl-cysteine and formate in rat liver preparations. *Fed. Proc.* **32**, 883 (Abstract)

DENCKLA, W. D. & DEWEY, H. K. (1967) The determination of tryptophan in plasma, liver and urine. *J. Lab. Clin. Med.* **69**, 160–169

SOURKES, T. L., MISSALA, K. & ORAVEC, M. (1970) Decrease of cerebral serotonin and 5-hydroxyindoleacetic acid caused by (–)-alpha-methyltryptophan. *J. Neurochem.* **17**, 111–115

Chairman's closing remarks

R. J. WURTMAN

Department of Nutrition and Food Science, Massachusetts Institute of Technology

It appears to be beyond controversy that the aromatic amino acids tryptophan, tyrosine and phenylalanine have special importance in the function of the brain. The synthesis of brain proteins requires all three compounds and may be limited by the availability of one of them, tryptophan, particularly after such treatment as the administration of α-methyltryptophan or of large doses of neutral amino acids which compete with tryptophan for brain uptake. The rate at which the brain produces the neurotransmitter 5-hydroxytryptamine is specifically controlled by its concentration of tryptophan while the syntheses of the neurotransmitters dopamine and noradrenaline are initiated by the intraneuronal hydroxylation of tyrosine and, perhaps, phenylalanine.

The complete characterization of the 'natural histories' of these three amino acids in brain will require a lot of additional information, much of which may not be obtainable by the experimental approaches currently available. For example, estimates of the following would be enormously useful but may be beyond our present reach:—

(1) The *concentrations* of these compounds within *neurons* and also within *specific populations of neurons* (e.g. the relatively few that actually convert tyrosine or tryptophan into their respective monoamines), and ultimately within *specific regions of those neurons* (e.g. the perikarya, which use the amino acids for protein synthesis, and the nerve terminals, which are probably the major locus of neurotransmitter synthesis). As we have seen, these concentrations are probably of major significance in determining the saturation, and thus the *in vivo* activities, of enzymes that control the utilization of the amino acids.

(2) The true *fluxes* of these amino acids into brain, into neurons as a group, into specific neurons, and finally into subcellular regions; the extents to which these compounds really are metabolized *in vivo* along pathways other than

hydroxylation and tRNA charging (e.g. transamination and decarboxylation without prior hydroxylation). Are there significant arteriovenous differences across the brain during any 24 h period (i.e. an interval that includes both fasting and postprandial states) for amino acids that are not converted irreversibly into monoamine neurotransmitters?

(3) The amino acid *patterns* within neuronal perikarya and terminals, and the extent to which these patterns, rather than the individual concentrations of particular amino acids or the charging of various transfer RNAs, control protein synthesis.

(4) The extent to which amino acids stored intraneuronally as proteins are *reused* (e.g. for the synthesis of monoamine neurotransmitters).

(5) Whether there exist *special mechanisms of uptake* for tryptophan into neurons that use this amino acid for the synthesis of 5-hydroxytryptamine, or for tyrosine or phenylalanine into catecholaminergic neurons. Professor Grahame-Smith's observations in this volume suggest that the uptake of amino acids into nerve terminals, like that into brain slices or intact whole brain, is mediated by a system that recognizes *families* of amino acids. If no special uptake system exists to allow the terminals of neurons that have a special need for tryptophan or tyrosine to concentrate these compounds without also concentrating leucine, isoleucine or valine, we might question the purpose of this arrangement.

If we focus our attention on the hydroxylations of tyrosine, tryptophan and phenylalanine within monoaminergic neurons, it seems clear that several questions formerly thought to have been resolved have once again been reduced to questions during this meeting:—

What factors really control tyrosine hydroxylase activity under physiological circumstances? After hearing the comments of Drs Kaufman, Carlsson and Weiner, one regards with dismay the large number of review papers that have appeared during the past few years enshrining the dogma of end-product inhibition. Several other control mechanisms discussed now bear intensive examination. Thus, it seems clear that increasing the nerve impulse flow through catecholaminergic neurons also increases tyrosine hydroxylase activity. However, it is most unclear how this is accomplished. I am reminded of my philosophy professor's description of a prototypic scientific 'explanation' from the late Middle Ages: 'Heat is hot because it has the calorific virtue'. To say that nerve impulse flow increases catecholamine synthesis is to describe a phenomenon, not to explain it.

Good evidence has also been presented for the existence of presynaptic receptors which would allow neurotransmitters released into synapses to feed back information to their cells of origin and suppress their own synthesis. In order to test this hypothesis cleanly, we really need an experimental preparation which is free from postsynaptic receptors. I am not aware that any such preparation now exists. Other factors described here which might affect tyrosine hydroxylase activity include ionic or calcium effects, or heparin-induced changes in the relation of the enzyme to intracellular membranes.

Another question which again becomes controversial is whether tyrosine or phenylalanine is the *physiological* precursor of brain catecholamines. It would be useful to quantify the fate of the tyrosine formed in brain by the hydroxylation of phenylalanine mediated by tyrosine hydroxylase. How much of this amino acid is subsequently converted into dopa? Is much catabolized by transamination?

If the end-product inhibition of tyrosine hydroxylase by the competition between catecholamines and the pteridine cofactor does not normally control the rate at which tyrosine is converted into dopa *in vivo*, should we not reconsider the possibility that the concentration of tyrosine (or perhaps phenylalanine) might be limiting? Again, we face the presently insurmountable problem of estimating the tyrosine concentration within a very small component of the brain, the nerve terminals of catecholaminergic neurons.

Several investigators initially chose to examine the possibility that tryptophan availability controlled the rate of tryptophan hydroxylation on the basis of what was apparently a false assumption, namely that the K_m of tryptophan hydroxylase for its amino acid substrate was unusually high, and vastly higher than that of, for example, tyrosine hydroxylase. We all observed good evidence for such control. Now we learn that, when the appropriate pteridine cofactor is used for enzyme assays, the K_m for tryptophan hydroxylase is not so unusually high after all. It seems that we have all been very fortunate to harbour this erroneous belief. In retrospect, perhaps we should not have been so impressed with the significance of an apparently high K_m value: the important thing is not that the enzyme is probably unsaturated *in vivo*—many enzymes are—rather, it is that brain tryptophan concentrations (and, consequently, the saturation of the enzyme) *do* change physiologically. The question of which plasma constituents determine brain tryptophan content under physiological conditions or in response to drugs or stress remains controversial. What most excites me is the evidence that all three major constituents of diet—carbohydrates, proteins and fat—affect plasma and brain tryptophan, and thus brain 5-hydroxytryptamine, in characteristic ways. This suggests a focal role for 5-hydroxytryptaminergic neurons in allowing the brain to sense the peripheral metabolism.

How are we to know whether a substrate-induced change in the cerebral concentration of 5-hydroxytryptamine (e.g. after carbohydrate consumption) makes any difference physiologically? When a neuron *contains* more 5-hydroxytryptamine, does it also *release* more and, if so, does it also produce a greater postsynaptic effect? Can brain functions thought by many to involve 5-hydroxytryptaminergic neurons (such as sleep, pituitary secretion, ovulation and pain thresholds) be affected by acute diet-induced changes in the cerebral concentration of the amine?

What is the role of aromatic amino acids in the control of brain protein synthesis? Are the transfer RNAs for tyrosine, phenylalanine and tryptophan normally saturated? If not, at what point does their desaturation retard protein synthesis, and does this relationship hold generally for all brain proteins? Might brain amino acid concentrations also affect protein synthesis by activating ribonucleases and thus disaggregating polyribosomes? Is this the mechanism by which circulating amino acids (in contrast to exogenous L-dopa and 5-hydroxytryptophan, which must first be decarboxylated) disaggregate brain polysomes? Is a transient suppression of brain protein synthesis likely to be important to the organism?

It seems that there are at least two forms in which exogenous aromatic amino acids can affect brain protein synthesis, that is, as the unchanged amino acids or as their decarboxylated products. Administered phenylalanine causes polysome disaggregation in the first way while dopa and 5-hydroxytryptophan are ineffective unless decarboxylated to their respective monoamines. The data now available suggest that these monoamines act on specific receptors to modify polysome aggregation and that these receptors may even be intracellular. How are we to explain this? Why should cells like glia—which presumably have never in their evolutionary history contained dopamine or 5-hydroxytryptamine —be endowed with specific intracellular receptors for these compounds which, when stimulated, affect something as important as protein synthesis? Are the effects of dopa and 5-hydroxytryptamine on brain polysomes simply toxicological or do they tell us something about how the brain normally functions?

A host of widely used drugs which happen to be synthetic aromatic amino acids affect brain amino acid concentrations and the syntheses of proteins and of monoamine neurotransmitters—that is, all three areas of interest to the members of this symposium. These drugs, which include such compounds as dopa, p-chlorophenylalanine, α-methyltryptophan, 5-hydroxytryptophan, α-methyltyrosine and α-methyldopa, have often been used on the assumption that they have only a single, specific effect, namely to increase or decrease neurotransmitter synthesis. Our discussions here show clearly that this view is no longer tenable.

I suggest that our present ignorance of the answer to many of these questions should not leave us despondent or pessimistic. Rather, it might be good to reflect on how fortunate we all are to be working in fields in which so much is known already and in which reasonably good paradigms and methods exist to help us learn more. I suspect that all of us can look forward to busy and interesting periods in our laboratories.

Index of contributors

Entries in **bold** *type indicate papers; other entries are contributions to discussions*

Indexes compiled by William Hill

Subject index

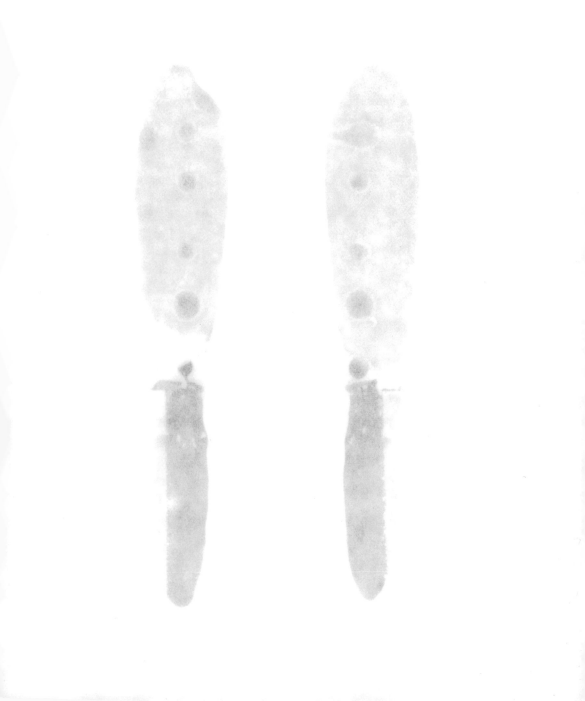